Handbook of Mathematical Calculations

is written for anyone who seeks a simplified form of the mathematics used in science. It serves as a brief and convenient reference for the mathematics encountered in scientific work. Mathematical calculations with practical examples are presented in a clear and concise manner.

HANDBOOK OF MATHEMATICAL CALCULATIONS will be of special interest to precollege high school students and undergraduate students in the general sciences. It is also designed as a refresher and an aid to students, technicians, and researchers in the fields of premedicine, health science, elementary chemistry and physics, with solved problems from biomedical, environmental, and earth sciences.

The book's four sections are uniquely designed to aid scientists on all levels. The Mathematical Review contains much of the basic information essential to the understanding and working of mathematical problems. Mathematical Applications will be helpful to those working in a broad range of scientific investigations that require mathematics in the compilation and interpretation of data—teaching, research, and preparation of reports. The "language of statistics" used in scientific work is briefly but very clearly explained in the section on Statistics. The Appendix provides a good reference on a wide variety of standard tables, terms, formulas, conversion factors, and other reference material necessary for quantification of data.

KAREN ASSAF is a research associate, School of Public Health, University of Texas, Houston. She was formerly an Instructor of Earth Science, Miami-Dade Junior College.

SÁID A. ASSAF is A [illegible]
Codirector, [illegible]
University [illegible]
He was fo[illegible]
Howard Hu[illegible]
sistant Prof[illegible]
at the Unive[illegible]
Both Karen and Sáid Assaf are graduates of Iowa State University.

Handbook of Mathematical Calculations

for science students and researchers

KAREN ASSAF
SÁID A. ASSAF

with an introduction by
KEITH M. HUSSEY

Handbook of Mathematical Calculations

for science students and researchers

THE IOWA STATE UNIVERSITY PRESS
Ames 1974

About the Authors

Karen Assaf received her undergraduate and graduate degrees from Iowa State University. Her M.S. degree was in Earth Science with minors in Statistics and Education. Having taught Earth Science both at Iowa State and Miami-Dade Junior College, her exposure to large numbers of students who shied away from mathematics in science motivated her to write this book. She is currently a research associate, School of Public Health, University of Texas at Houston.

Sáid A. Assaf holds a Ph.D. degree in Biochemistry from Iowa State University and is currently Associate Professor of Medicine and Codirector of the Hematology Research Laboratories at the University of Texas Medical School, Houston, Texas. Dr. Assaf has been honored as a visiting investigator of the Howard Hughes Medical Institute in Miami where he also held the position of Assistant Professor in the Department of Medicine and the Department of Biochemistry at the University of Miami School of Medicine. He has published widely in biochemical journals and has lectured on his work in various universities and national as well as international biochemical and medical symposia.

Library of Congress Cataloging in Publication Data

Assaf, Karen, 1941–
Handbook of mathematical calculations.

Bibliography: p.
1. Mathematics—Handbooks, manuals, etc.
I. Assaf, Sáid A., joint author. II. Title.
QA40.A85 510'.2'02 73-16451
ISBN 0-8138-1135-X

Composed and printed in the U.S.A. by
Science Press

First edition, 1974

TO OUR CHILDREN:

Dena, Rema, Nabila, and Jenin

Contents

Preface

The chief aim of the authors has been to present material on the basic concepts of mathematics in the fundamental and applied sciences in a manner that is clear and simple, yet concise. Simplification of the mathematics employed in the fundamental and applied sciences was done with three points in mind:

(1) To reach the student of science who shies away from some courses because he believes he learns things better in words than equations.

(2) To reach the working technicians, nurses, physicians, and scientists who have forgotten the basic principles involved in calculations and/or the mathematical manipulation of information.

(3) To provide a convenient outline with examples and practical problems for instructors involved in teaching mathematics in fundamental and applied science courses or in training programs.

Introduction

One might well ask why, with all the excellent books on and about mathematics, the authors felt constrained to write still another. Even a quick review of the table of contents will show that this book is significantly different from the vast majority of those mathematics books published to date.

The reason for the significant difference is that the book was written for a much broader spectrum of readers than probably any other mathematics book you have ever seen. It was written for just about anybody who has a need to read or to use mathematics in his daily activities but who has either not had the mathematics background or has not used it often enough to remember how to use it. These people are legion and are present in all walks of life. In this book they are told what the many symbols mean and how to use mathematics in the solution of scientific problems.

The book developed, through a series of changes, from a very narrowly directed version aimed at the needs of secondary school science teachers into the final almost encyclopedic version that should be a welcome addition to the working library of all science teachers, researchers, and other professional scientists.

The organization of the book is very practical. The first part, Mathematical Review, contains much of what was in the original version. Therefore, it is to that part that most secondary school teachers will be referring with greatest frequency for information and assistance. The review contains much of the basic information essential to understanding and working of mathematical problems. It will serve as a very practical reference for the many occasional users of mathematics who do not remember the methods.

The second part, Mathematical Applications, might well be compared with that part of any educational pursuit in which one learns how to use the essential tools. Certainly mathematics are part of the language of science in any area of scientific investigation that requires measurement. In Mathematical Applications the authors have endeavored to cover the mathematics necessary to teaching, research, preparation of reports, and writing of theses and dissertations. Those who have a fair background in mathematics will find this treatment to be of great assistance in a broad range of scientific investigations that require mathematics in the compilation of data and their interpretation. For those who have to read scientific reports, this part of the book will provide great assistance through its descriptions of laws and procedures, and definitions of terms.

In the third part of the book, the authors show the use of statistics in scientific investigations. Although this practice is not new, it is one with which many teachers have little if any proficiency. For many years, the use of statistics was confined almost exclusively to sophisticated investigations. Today it is commonly used in secondary school science courses. Therefore, access to some of the basic concepts of statistics is essential to the science teachers who have never had formal instruction in that discipline. The authors have done an excellent job of presenting those concepts in understandable language.

The last part of the book, Appendix, is a "gold mine" of information on a wide variety of more specialized techniques, terms, measurements, tables, conversions, procedures, etc., including some that are just too lengthy to be put in the body of the book. Some are just essential to a particular area of investigation, while others are of quite common usage. Together they serve to make up a section of this excellent book that is of equal importance with the others.

In summary, the book is uniquely designed to serve scientists at all levels from secondary school students to postgraduate research workers, and to attain the goals as set forth by the authors in the preface.

The authors are deeply indebted to all those kind people who encouraged them to undertake the task and those who read and offered valuable criticism of the work in its developing stages. Particular indebtedness is acknowledged to the Literary Executor of the late Sir Ronald A. Fisher, F.R.S., to Dr. Frank Yates, F.R.S., and to Oliver Boyd, Edinburgh, for permission to use tables, as credited, from their book *Statistical Tables for Biological, Agricultural and Medical Research;* to Dr. Allen L. Edwards for permission to use a table, as credited in the Appendix, from his book *Experimental Design in Psychological Research*, published by Holt, Rinehart and Winston; to Dr. S. M. Selby, editor, and the publishers of the 48th edition of the *Handbook of Chemistry and Physics* for permission to use the tables, as credited in the Appendix; to Dr. Paul G. Stocher, editor, and Merck & Company for permission to use the Tables from *The Merck Index*, 8th edition, as credited in the Appendix; and to Houghton Mifflin, publisher of *Modern Algebra and Trigonometry, Structure and Method*,

Book 2, by Dolciani, Berman and Wooten, as credited in the Appendix. The kindness of these people has saved the authors countless hours of tedious work.

Keith M. Hussey
Head, Department of Earth Science
Iowa State University

PART 1

Mathematical Review

1.1 MEASUREMENT

Any measurement must always be made in standard, well-defined units. A *unit* is the value, or quantity, or size of a weight or measure by which values of other weights and measures are fixed. The value of a unit is arbitrarily determined and set forth in a law. For example, three feet equal one yard. A *standard* is the actual physical reproduction of a unit. The yard is a unit of length but the standard yard is a bronze bar with fine lines engraved exactly thirty-six inches apart on gold studs set in the bar. All yards must be measured according to this standard.

Thus the two essential elements of measurement are (1) the adoption of a unit of measure and (2) the operation of counting the number of times the unit is applied to the given magnitude.

In making a measurement, the size of the unit should be chosen with reference to the size of the magnitude to be measured. For instance, the dimensions of a sheet of writing paper are ordinarily given in inches or centimeters, the dimensions of buildings are given in feet or meters, and distances between cities are expressed in miles or kilometers. To measure the distance to the nearest fixed star, the mile is too small a unit. Instead, the light year, the distance which light travels in a year moving at a rate of 186,000 miles per second, is used by astronomers.

The direct measurement of distances and heights is often inconvenient or impossible. This is true, for example, of the height of a mountain or the width of a lake. Such problems as finding the distance across a river stimulated the interest of the ancients in developing methods of indirect measurement. These investigations led slowly to the scientific procedures which are the foundation of surveying and astronomy. The very word geometry means earth measurement. By a combination of the methods of geometry and trigonometry it has become possible to measure continents and the earth. Indirect measurement has enabled man to reach out into the universe and to determine the distance, the size, and the motions of the heavenly bodies.

In measurement, lengths are measured by rulers, tapes, or yardsticks.

Volume is sometimes measured by standard containers. The physical art of measuring involves some problems. There is a built-in error of measurement in that there is no way to measure exactly. Lengths are compared and the preciseness of measurement is limited by the ability to perceive small differences. Precision instruments for measurement are designed to show small differences that cruder instruments cannot show. However, there is some degree of error in most instruments.

1.2 NUMBERS

Decimal Numerals

The first widely recognized system which introduced special symbols to represent number units was the *Arabic system.* This system has gained universal use because it was *based on the realization that a comparatively few number symbols could be conveniently combined to represent any number, no matter how large, by utilizing the zero.*

The Arabic system employs ten figures or characters (1, 2, 3, 4, 5, 6, 7, 8, 9, and 0) which are referred to as digits. For ease in reading larger numbers, it is customary to consider a large number as consisting of groups of three digits, starting from the right. In all numbers such a group of three is called a *period.* The period at the right is called the units period (see below); the next period to the left is the thousands period; the third, the millions period; the fourth, the billions period. In addition to the above-mentioned periods, there are those which are rarely used because there is little need to represent numerically the vast quantities for which they stand. They are, in order, the trillions period, the quadrillions, quintillions, sextillions, septillions, and octillions.

The first figure of each period is called the units place; the second figure, the tens place; the third figure the hundreds place. For example,

billion	million	thousand	units
660	408	250	070

To read a number, read the last figure to the left in the period of greatest magnitude with its place name, and then add the name of that period. Then do the same for each succeeding period. The numbers written above would read "six hundred sixty billion, four hundred eight million, two hundred fifty thousand, seventy."

Roman Numerals

The *Roman* numeral system differs from the Arabic system in that letters and not figures are used as the symbols of enumeration. Today, the Roman system of numeration has only a few uses. In science, the

Roman numerals most commonly used are those between one and ten. They are sometimes used in books for indicating sequence in lists and outlines. Occasionally they are used in numbering tables, charts, chapters, etc. (e.g., Chapter X, Table II), and sometimes they are accompanied by letters (Table Ia, Table Ib).

Roman Numerals

I	II	III	IV	V	VI	VII	VIII	IX	X	L	C	D	M
1	2	3	4	5	6	7	8	9	10	50	100	500	1000

The Romans wrote their numerals from left to right, from the largest to the smallest. The numerals were then added up to get the number represented. If a numeral was out of this order (a smaller before a larger), it was subtracted from the following numeral instead of being added to it. The following are the only subtractions permitted:

IV = 4 XL = 40 CD = 400
IX = 9 XC = 90 CM = 900

Although this system is basically additive, the principle of multiplication is used in writing large numbers. For example, a bar over a number multiplies the number by one thousand (e.g., $\overline{\text{CLXXV}}\text{XLIII} = 175{,}043$).

Binary System

An unusual and useful system of numbers is the *binary system.* The *number two is the base*, and therefore is represented by the symbol 10. In this system the fundamental numbers are 1, 10. *The basic additional fact in this system is 1 + 1 = 10.*

Decimal:	1	2	3	4	5	6
Binary:	1	10	11	100	101	110

Decimal:	7	8	9	10	11	12	
Binary:	111	1000	1001	1010	1011	1100	. . .

Binary notation is used in many electronic computers. Sometimes a computer is arranged to convert numbers from one notation to another, so that it may use binary notation internally but communicate with human beings in decimal language.

To write the binary equivalent of a decimal number, the number must be written in terms of powers of 2 so that in translation it will appear in terms of powers of 10 in the binary system. This can be done most easily by first dividing by the highest power of 2 contained in the given number, and then dividing the remainder by the highest power of 2 contained in the remainder. Continue this division of remainders until a remainder less than 2 is obtained.

Powers of 2

$2^1 = 2$, $2^2 = 4$, $2^3 = 8$, $2^4 = 16$, $2^5 = 32$,
$2^6 = 64$, $2^7 = 128$, $2^8 = 256$, $2^9 = 512$, etc.

Convert the decimal number 175 into the binary system.

$$\begin{array}{r} 1 \\ 128\overline{)175} \\ \underline{128} \\ 47 \end{array} \quad \begin{array}{r} 1 \\ 32\overline{)47} \\ \underline{32} \\ 15 \end{array} \quad \begin{array}{r} 1 \\ 8\overline{)15} \\ \underline{8} \\ 7 \end{array} \quad \begin{array}{r} 1 \\ 4\overline{)7} \\ \underline{4} \\ 3 \end{array} \quad \begin{array}{r} 1 \\ 2\overline{)3} \\ \underline{2} \\ 1 \end{array}$$

Thus $175 = 1 \times 2^7 + 1 \times 2^5 + 1 \times 2^3 + 1 \times 2^2 + 1 \times 2 + 1 \times 1$ which in the binary system becomes $1 \times 10^7 + 1 \times 10^5 + 1 \times 10^3 + 1 \times 10^2 + 1 \times 10 + 1 \times 1 = 10101111$.

To convert a binary number to decimal form, read the number noting the place values in decimal form and then translate into binary form.

For example, $1100 = 1 \times 10^3 + 1 \times 10^2 \longrightarrow 1 \times 2^3 + 1 \times 2^2 = 8 + 4 = 12$

Convert the binary number 10101 into the decimal system:

$$10101 = 1 \times 10^4 + 1 \times 10^2 + 1 \times 1 \longrightarrow 1 \times 2^4 + 1 \times 2^2 + 1 \times 1$$
$$16 + 4 + 1 = 21$$

The above process of converting from the decimal system to the binary system can be used to convert any number from one base to another; i.e., decimal to base 5 (quinary) or base 8 and vice versa.

Examples and Problems

State the following numbers in Roman numerals.

a. 999 CMXCIX
b. 1972 MCMLXXII
c. 1827 MDCCCXXVII

State the following decimal numbers in the binary system.

a. 14 1110
b. 25 11001
c. 128 10000000
d. 80 1010000
e. 19 10011
f. 64 1000000

State the following binary numbers in the decimal system.

a. 1100011 99
b. 100001 33
c. 101011111 351

d. 1111 15
e. 110010 50
f. 1010 10

1.3 METRIC SYSTEM

The metric system is based on the meter which is not an arbitrary unit but theoretically one ten-millionth of the distance from the equator to the North Pole. The modern standard for the meter is equal to 1,650,763.73 wavelengths of the orange-red light of excited krypton of the mass number 86. As standardized by the United States Bureau of Standards, a meter measures 39.37 inches.

One thing that makes the metric system simpler than the English system (i.e., inches, feet, pounds, etc.) is that it is a decimal system. Another advantage of the metric system is the direct relationship of the units of length, mass, and capacity to one another.

In the metric system of measurements, the unit for length is the meter, the unit for capacity is the liter, and the unit for weight or mass is the gram. If one knows that

a meter is 39.37 inches long,
a liter is 0.946 of a quart, and
a gram is 0.035 ounces,

one can tell right away the value of any weight or measurement in the metric system in comparison with the English system. Unlike the English system in which every weight and measure has a separate name that does not give any clue to its value, the names of the units in the metric system tell one exactly what they are.

There are six commonly used prefixes that can be attached to any of the units and each of these prefixes means a number.

Latin prefixes:
- milli means 1/1000 (one-thousandth) of the unit, i.e., 10^{-3}*
- centi means 1/100 (one-hundredth) of the unit, i.e., 10^{-2}
- deci means 1/10 (one-tenth) of the unit, i.e., 10^{-1}

Greek prefixes:
- deka means 10 (ten) times the unit, i.e., 10^{1}
- hecto means 100 (one hundred) times the unit, i.e., 10^{2}
- kilo means 1000 (one thousand) times the unit, i.e., 10^{3}

Any of these prefixes when attached to meter, liter, or gram tells exactly what the value of the measurement is. For example:

---

*10^{-3} is ten to the power minus 3. For explanation see Section 1.6 on Exponents.

10 millimeters = 1 centimeter
10 centimeters = 1 decimeter
10 decimeters = 1 meter
10 meters = 1 dekameter
10 dekameters = 1 hectometer
10 hectometers = 1 kilometer

The above table may also be written as

1 meter = 1000 millimeters
= 100 centimeters
= 10 decimeters
= 0.1 dekameter
= 0.01 hectometer
= 0.001 kilometer

Thus all units of length, volume, and mass in the metric system are multiples of or subdivisions of the meter, liter, or gram. To change a unit to a larger or smaller unit, one divides or multiplies by ten or powers of ten, which is why the metric system is so simple and is called the decimal system.

Measurements of area are made by multiplying length times width just as is done in the English system. In measuring volume, the cubic meter is used. For small volumes of liquids there is the milliliter. A milliliter of water weighs approximately one gram. The measurement "cc" which is used by doctors and chemists is the abbreviation for "cubic centimeter" and is used as a substitute for a milliliter.

The most commonly used multiples and subdivisions of the metric system and their abbreviations are: for length, the kilometer (km), meter (m), centimeter (cm), and millimeter (mm); for capacity, the liter (l) and the milliliter (ml); for area, the square meter (m^2), square centimeter (cm^2), and square millimeter (mm^2); for volume, the cubic meter (m^3), cubic centimeter (cm^3), and cubic millimeter (mm^3). The most commonly used weights are the metric ton (1000 kilograms), the kilogram (kg), the gram (g), and the milligram (mg).

Sometimes when measurements are for scientific use, it is convenient to multiply or divide the basic units by more than a thousand, so some new prefixes have been introduced:

tera means 1,000,000,000,000 (one trillion) times the unit, i.e., 10^{12}
giga means 1,000,000,000 (one billion) times the unit, i.e., 10^9
mega means 1,000,000 (one million) times the unit, i.e., 10^6
micro means 1/1,000,000 (one-millionth) of the unit, i.e., 10^{-6}
nano means 1/1,000,000,000 (one-billionth) of the unit, i.e., 10^{-9}

pico means 1/1,000,000,000,000 (one-trillionth) of the unit, i.e., 10^{-12}
fento means 1/1,000,000,000,000,000 (one over 10^{15}) of the unit, i.e., 10^{-15}

So many scientific things are measured in micrometers (one-millionth of a meter) now that scientists have made up a shorter word, micron (μ). *Nanometer* (nm)* [formerly millimicron (mμ)] is a word used for a measurement even smaller than a micron. A nanometer is a thousandth of a millionth of a meter. This may seem too small to measure anything, but there are waves of light even shorter than a millimicron.

The metric system of measurement was carefully designed by scientists to facilitate computation. However, one of the things to guard against (unless only the ability to translate is one's goal) is that of learning the metric system entirely in a verbal manner. The metric system must be learned directly and not as a matter of translation.

To learn the metric system effectively, one should be familiar with meter sticks, metric tape, a liter block, and several liter measures varying in capacity from the milliliter to the liter. (See Fig. 1.1.) Considerable use of these materials should serve to give that direct contact with the system which will make it a useful part of one's experience. The United States government is considering joining most countries in the world in adopting the metric system which is the system invariably used in higher education.

To show the superiority of the metric system over the English system, convert a number such as 86 centimeters to inches, then to feet, then to yards.

86 cm = 0.86 meter
8.6 decimeters
86 cm = 33.8582 inches
2.49 feet
0.94 yards

To reinforce the fact that the metric system is simpler than the English system, Table 1.1 and Table 1.2 are presented to illustrate that *all that is needed in the metric system is to understand what the unit means* and not to memorize how much; i.e., in the case of the meter, a milli of it is the millimeter, which is 1/1000 of a meter, and a kilo of it is the kilometer, which is 1000 meters. In the English system no such simple relationship can be derived from the names of the units. For example, the yard does

*The term nanometer (10^{-9} m) is now preferred to millimicron. Therefore, wavelengths are now given in nm rather than mμ.

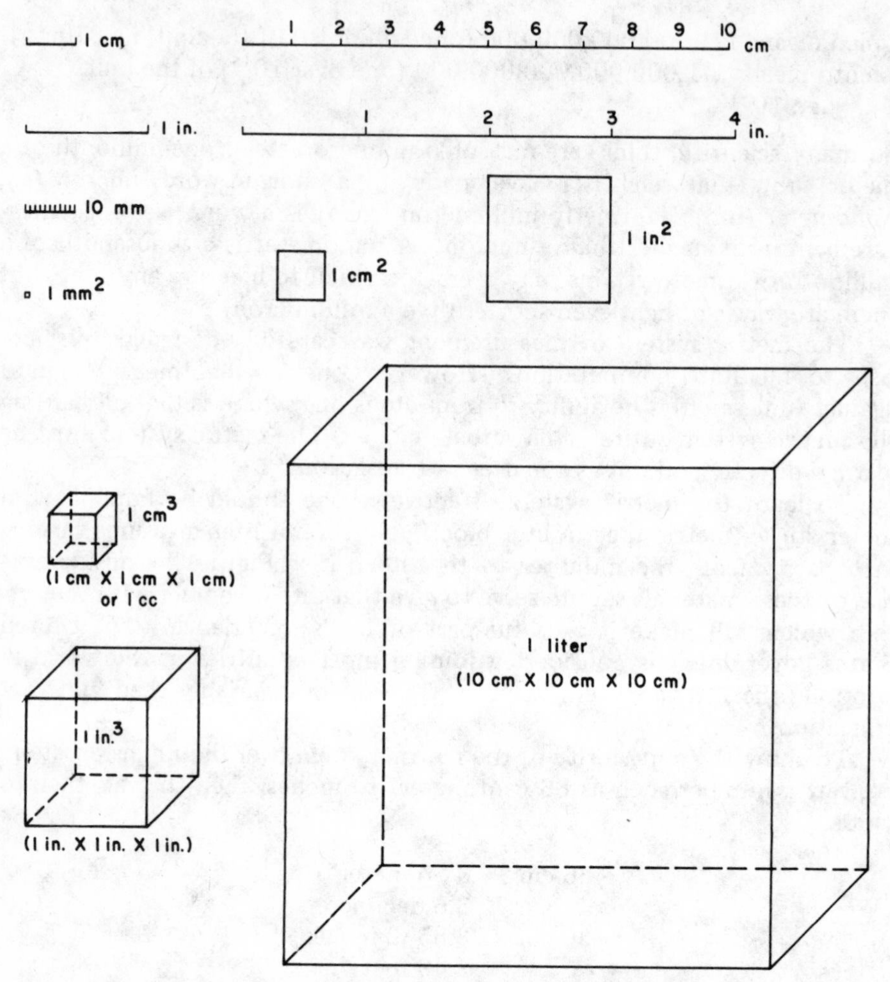

Fig. 1.1. Comparison of English and metric units. (Reduced by 33%.)

not tell how many feet there are in a yard, nor does the word inch tell its relation to the foot.

Conversion from units of the English system to the metric system and vice versa seems difficult. However, if one understands the decimal system and how to read tables, conversion between the two systems will require only the process of multiplication or division. Tables can be set up showing the relationships in the two systems between the units of length, area, volume, and mass. Table 1.3 is presented as an example for the relationship of units of length in the two systems. (See Appendix for more complete conversion tables.)

Table 1.1. Illustrative Table Showing the Relationship of the Units of Length, Area, Volume, and Weight in the Metric System

Measurement	*Units*		
Length	one meter	= 100 centimeters	= 1000 millimeters
Area	one meter^2	= 100 cm^2	= 1000 mm^2
Volume	one meter^3	= 100 cm^3	= 1000 mm^3
Weight	one kilogram	= 1000 grams	= 1,000,000 milligrams

Table 1.2. Illustrative Table Showing the Relationship of the Units of Length, Area, Volume, and Weight in the English System

Measurement	*Units*		
Length	one yard	= 3 feet	= 36 inches
Area	one yard^2	= 9 square feet	= 1296 square inches
Volume	one yard^3	= 27 cubic feet	= 57,656 cubic inches
Weight	one ton	= 2000 pounds	= 32,000 ounces

Note: Notice the clumsiness of the values of the English system compared to that in Table 1.1 for the metric system.

Table 1.3. Relationship between Units of Length in the English and Metric System

Unit	*in.*	*ft*	*yd*	*mile*	*cm*	*meter*	*km*
in.	1	0.083	0.027	. . .	2.54	0.0254	. . .
ft	12	1	0.333	0.00189	30.48	0.3048	. . .
yd	36	3	1	0.000568	91.44	0.9144	0.00091
mile	63,360	5,280	1760	1	. . .	1609.35	1.609
cm	0.3937	0.0328	0.01093	. . .	1	0.01	. . .
meter	39.370	3.28083	1.09361	0.00621	100	1	0.001
km	39,370	3280.83	1093.61	0.62137	100,000	1000	1

Examples and Problems

Using Table 1.4, make the following conversions within the metric system:

99 kilometers to meters
move decimal to right 3 places,
thus 99 km = 99,000 m

2793 milliliters to liters
move decimal to left 3 places,
thus 2793 ml = 2.793 l

6335 grams to milligrams
move decimal to right 3 places,
thus 6335 g = 6,335,000 mg

400 kilograms to grams
move decimal to right 3 places,
thus 400 kg = 400,000 g

2.50 milliliters = _?_ liters
= 0.0025 l

0.096 centimeter = _?_ millimeter
= 0.96 mm

Table 1.4. Conversion of Units within the Metric System

Move decimal of \ *To obtain*	*Kilo*	*Hecto*	*Deka*	*Meter, liter, or gram*	*Deci*	*Centi*	*Milli*
Kilo	...	→1	→2	→3	→4	→5	→6
Hecto	←1	...	→1	→2	→3	→4	→5
Deka	←2	←1	...	→1	→2	→3	→4
Meter, liter, or gram	←3	←2	←1	...	→1	→2	→3
Deci	←4	←3	←2	←1	...	→1	→2
Centi	←5	←4	←3	←2	←1	...	→1
Milli	←6	←5	←4	←3	←2	←1	...

To convert from one unit to another in the metric system, simply move the decimal point to the right (→) or to the left (←) the number of places that separate the units to be converted.

← To divide, move the decimal point to the left

(This process converts small units into larger ones, for example, meter to kilometer.)

Kilo Hecto Deka Meter, liter, or gram . Deci Centi Milli

Decimal point

To multiply, move the decimal point to the right →

(This process converts large units into smaller units, for example, meter to millimeter.)

0.096 centimeter = __?__ meter
= 0.00096 m

5 yards = __?__ meters
= 4.57 m

1.0 English ton = __?__ pounds
= 2000 lb

1 pound = __?__ grams
= 453.59 g

1 pound = __?__ kilograms
= 0.454 kg

1 metric ton = __?__ kilograms
= 1000 kg

1 English ton = __?__ kilograms
= 908 kg

What is the volume in liters of a rectangular box 2.0 m long, 55 cm wide, and 30 mm high?

2 m = 200 cm
55 cm = 55 cm
30 mm = 3.0 cm
$200 \times 55 \times 3.0$ = 33,000 cu cm or cc
Since 1000 cc equal 1 liter, the volume of the box is 33 liters.

What is the volume of a box in liters whose dimensions are 4.00 yd by 20 in. by 0.500 m?

4.00 yd = 144 in. = 366 cm
20 in. = 50.8 cm
0.500 m = 50.0 cm
$366 \times 50.8 \times 50.0 = 743{,}740$ cu cm or cc
= 743.74 liters
$= 7.44 \times 10^2$ liters

An airplane is moving at a rate of 300 mi/hr. What is this in centimeters (cm) per minute (min)? (Given 5280 ft/mi, 12 in./ft, and 2.54 cm/in.).

Since mathematically this / (per slash sign) means divided by or over, then

$$\frac{300 \text{ mi}}{1 \text{ hr}} \times \frac{5280 \text{ ft}}{1 \text{ mi}} \times \frac{12 \text{ in.}}{1 \text{ ft}} \times \frac{2.54 \text{ cm}}{1 \text{ in.}} \times \frac{1 \text{ hr}}{60 \text{ min}} = 80{,}520 \text{ cm/min.}$$

An oil tanker was unloading in Houston 2000 barrels of oil from Kuwait. Each barrel was labeled: contents 500 kg. How many pounds is this? How many metric tons? How many English tons? (Given 454 g/lb and 1000 g/kg.)

2000 barrels $\times$ 500 kg = 1,000,000 kg which is 1000 metric tons.

To convert to pounds and tons in the English system, first convert to grams.

Thus 1,000,000 kg = 1,000,000 $\times$ 1000 g = 1,000,000,000 g.

Since 1 lb is equal to 454 g, this value in pounds is

$$\frac{1{,}000{,}000{,}000}{454} = 2{,}202{,}000 \text{ lb}$$

and since each 2000 lb = 1 English ton

$$\frac{2{,}202{,}000}{2000} = 1101 \text{ English tons.}$$

An exchange college student said that the speed limit in his hometown, Beirut, Lebanon, is 40 km/hr while it is 25 mi/hr in Ames, Iowa. Which town has a higher speed limit?

1 km = 5/8 mi

Therefore the speed limit in Beirut in miles/hour is $40 \times 5/8 = 25$ mi/hr

and since 1 mile = 8/5 mi

then the speed limit in Ames in km/hr is

$25 \times 8/5 = 40$ km/hr

and hence the speed limit in both towns is equal.

What is the area of a square piece of land in units of centimeters and in kilometers if one of its sides is 450 yd? (Given 3 ft/yd, 12 in./ft, and 2.54 cm/in.).

$$450 \text{ yd} \times \frac{3 \text{ ft}}{1 \text{ yd}} \times \frac{12 \text{ in.}}{1 \text{ ft}} \times \frac{2.54 \text{ cm}}{1 \text{ in.}} = 41{,}148 \text{ cm (one side)}$$

The area is 41,148 cm × 41,148 cm = 213,159,904 cm^2
= 2131.6 km^2.

An area of a desk is 200 $inch^2$ (i.e., 200 square inches). What is it in square centimeters? (Given 1 in. = 2.54 cm).

Note: squaring one side of an equality requires squaring the other side

Thus $(1 \text{ in.})^2 = (2.54 \text{ cm})^2$

$(1 \text{ in.})^2 = (2.54)^2 \times \text{cm}^2$

$(1 \text{ in.})^2 = 6.45 \text{ cm}^2$

$(200 \text{ in.}^2) = 6.45 \times 200 \text{ cm}^2$

$= 1290 \text{ cm}^2$.

A similar method is applied in the interconversion of units of volume. Simply change the square sign $(\ )^2$ to cube $(\ )^3$ as illustrated in the following example.

What is the volume of a rectangular-shaped container in cubic centimeters if its height is 10 ft, its long side is 72 in., and its short side is 48 in.?

The sides in feet are 10 ft × 6 ft × 4 ft = 240 ft^3

1 ft = 30.48 cm and 1 ft^3 = $(30.48 \text{ cm})^3$ = 28,316.8 cu cm

Therefore 240 ft^3 = 28,316.8 cu cm × 240 ft = 6,796,032.0 cu cm
= 6.796 cu m.

A baby weighs 7 lb 8 oz at birth. How many kilograms does it weigh?

1 lb ≅ 454 g

454 g × 7.5 lb = 3405 g

Since each 1000 g equals 1 kg, the baby weighs 3.4 kg.

The density of water in the English system is 62.4 lb/cu ft. What is it in g/cu cm (i.e., g/ml)?

1 lb ≅ 454 g

1 ft = 30.48 cm; 1 ft cubed = $(30.48 \text{ cm})^3$

$$\frac{62.4 \text{ lb} \times 454 \text{ g}}{(30.48)^3} = \frac{28{,}329.6 \text{ g}}{28{,}316.85 \text{ cu cm}} = \sim 1.0 \text{ g/cu cm or } 1.0 \text{ g/ml.}$$

Some balances weigh accurately up to 1 mg. What is this in ounces?

1 g = 1000 mg

1 oz = 28.35 g = 28,350 mg

Therefore 1 mg = $\frac{1}{28{,}350}$ of an ounce, or 2.835×10^{-4} of an ounce.

Using an analytical microscope, a hematologist determined the diameter of a blood cell and found it to be 8 microns (μ). What is this value in centimeters and in inches?

1 micrometer = one-millionth of a meter = $\frac{1}{1{,}000{,}000}$ m which is the same as $\frac{1}{10^6}$ m.

$8\ \mu = 8$ micrometers $= 8 \times \frac{1}{10^6}$ m

1 cm = one-hundredth of a meter $= \frac{1}{100}$ m which is the same as $\frac{1}{10^2}$ m.

Therefore the equivalent of 8 μ in centimeters is equal to one-hundredth of

$$8 \times \frac{1}{10^6} = \frac{1}{10^2} \times 8 \times \frac{1}{10^6} = \frac{8}{10^8} = 8 \times 10^{-8}$$

which is 0.08 of a millionth of a centimeter.

Looking in the text, we find that each centimeter = 0.3937 in.

Thus the answer in inches is $0.3937 \times 8 \times 10^{-8} = 3.1496 \times 10^{-8}$ in. which is approximately one-third of a hundred-millionth of an inch. One can now see why small values of a measurement are more meaningful using small units of the metric system such as the micron in this example.

A physicist measured a wavelength of electromagnetic radiation and found it to be 2×10^{-10} m. What is this value in millimeters, microns, and Angstroms (A)?

1 m = 1000 mm

2×10^{-10} m $= 1000 \times 2 \times 10^{-10} = 2 \times 10^{-7}$ mm

1 m = one million $\mu = 1 \times 10^6\ \mu$

2×10^{-10} m $= 1 \times 10^6 \times 2 \times 10^{-10} = 2 \times 10^{-4}\ \mu$

1 m = 10 billion A $= 10 \times 10^9$ A $= 1 \times 10^{10}$ A

2×10^{-10} m $= 1 \times 10^{10} \times 2 \times 10^{-10} = 2$ A.

Eight picograms of folic acid and B_{12} were found in a patient. What does this value mean in terms of a gram, a milligram, and a microgram (μg)?

A picogram = one-trillionth of a gram which is $\frac{1}{10^{12}}$ of a gram, therefore 8 pico-

grams $= 8 \times \frac{1}{10^{12}} = 8 \times 10^{-12}$ g.

Remembering this answer and remembering that a gram = 1000 mg, then 8 picograms $= 1000 \times 8 \times 10^{-12} = 8 \times 10^{-9}$ mg.

Since a gram = one million μg $= 10^6\ \mu$g, 8 picograms $= 10^6 \times 8 \times 10^{-12} = 8 \times 10^{-6}\ \mu$g.

1.4 FRACTIONS

Common Fractions

Parts of whole numbers or integers are termed fractions. For example, the value 1/3 refers to a quantity which is one of three equal parts of a given unit. The two numbers in the fraction are called the terms of the fraction. The number over the line is called the *numerator*, and the one below the line is called the *denominator*. The denominator shows into how many equal parts the unit has been divided and the numerator shows how many of these parts have been taken.

Computations which involve fractions are the same as those with integers. Fractions may be added, subtracted, multiplied, and divided. However, to add or subtract fractions, they must have the same denominators.

Addition: Add the numerators and place the total over a common denominator. Be careful to add only like items. Usually it is most desirable to find the lowest common denominator which is the smallest number divisible by all given denominators. Then both numerator and denominator of each fraction is multiplied by the number of times its denominator is contained in the common denominator.

$$\frac{a}{b} + \frac{c}{d} = \frac{ad + bc}{bd} \qquad \frac{1}{2} + \frac{3}{4} = \frac{4 + 6}{8} = \frac{10}{8} = \frac{5}{4}$$

Examples: $$\frac{1}{a} + \frac{4}{b} + \frac{1}{4ab} = \frac{4b + 16a + 1}{4ab}$$

$$\frac{1}{a} + \frac{1}{2} + \frac{x}{b} + \frac{3x}{ab} = \frac{2b + ab + 2ax + 6x}{2ab}$$

Subtraction: Obey the same rules as addition except for the sign. Subtract the numerator of the smaller fraction from the numerator of the larger fraction and place answer over common denominator.

Multiplication: Multiply the numerators of all the fractions to find the numerator of the product, then multiply the denominators of all the fractions to find the denominator of the product. Reduce to lowest terms.

$$\frac{a}{b} \times \frac{c}{d} = \frac{ac}{bd} \qquad \frac{1}{2} \times \frac{1}{4} = \frac{1}{8} \qquad \frac{z}{3a} \times \frac{4}{3z} \times \frac{9a}{2x} = \frac{36\,az}{18\,azx} = \frac{2}{x}$$

Division: Dividing a number by a fraction is the same as multiplying by its reciprocal.* When a fraction is to be divided by a whole number, first interpret the whole number as a fraction having one for its denominator, invert to get its reciprocal and multiply. Thus in all division of fractions, invert the divisor and multiply.

---

*By definition, the reciprocal of a number is 1 divided by the number. In general, when "a" is a fraction = to $\frac{1}{4}$, its reciprocal is 1/a is = $\frac{1}{\frac{1}{4}}$, and it has the same value as the fraction inverted (i.e., 1/1 × 4/1 = 4/1). The method of handling division when fractions are involved is called the reciprocal method and it points out the reciprocal relation, or inverse relation between multiplication and division.

$$\frac{\frac{a}{b}}{c} = \frac{a}{1} \times \frac{c}{b} = \frac{ac}{b}$$

$$\frac{\frac{b}{c}}{a} = \frac{b}{c} \times \frac{1}{a} = \frac{b}{ca}$$

$$\frac{\frac{1}{2}}{\frac{1}{4}} = \frac{1}{2} \times \frac{4}{1} = \frac{4}{2} = 2$$

Multiprocess of multiplication, addition, and/or subtraction:

Multiplication and division should precede addition or subtraction unless the process is interrupted by parentheses ( ).

Example:

$$\left(\frac{1}{4}+\frac{1}{6}+\frac{1}{3}\right)\left(\frac{4}{8}+\frac{3}{15}\right)-\frac{1}{4}+\frac{1}{2}\times\frac{1}{3}-\left(\frac{1}{8}-\frac{1}{6}\right)\frac{7}{5}+\frac{1}{2} =$$

$$\left(\frac{6}{24}+\frac{4}{24}+\frac{8}{24}\right)\left(\frac{60}{120}+\frac{24}{120}\right)-\frac{1}{4}+\frac{1}{6}-\left(\frac{3}{24}-\frac{4}{24}\right)\frac{7}{5}+\frac{1}{2}=$$

$$\left(\frac{18}{24}\right)\left(\frac{84}{120}\right)-\frac{3}{12}+\frac{2}{12}-\left(-\frac{1}{24}\right)\frac{7}{5}+\frac{1}{2}=$$

$$\left(\frac{3}{4}\right)\left(\frac{7}{10}\right)-\frac{1}{12}-\left(-\frac{7}{120}\right)+\frac{1}{2}=$$

$$\left(\frac{21}{40}\right)-\frac{1}{12}+\frac{7}{120}+\frac{1}{2}=$$

$$\frac{63}{120}-\frac{10}{120}+\frac{7}{120}+\frac{60}{120}=\frac{120}{120}=1$$

The value of a fraction is altered if

1. a constant (K) is added to both the numerator and the denominator

$$\frac{a}{b} \neq \frac{a+K}{b+K}$$

2. both numerator and denominator are raised to the same power or reduced to the same root.

$$\frac{a}{b} \neq \frac{a^2}{b^2} \quad \text{and} \quad \frac{a}{b} \neq \frac{\sqrt{a}}{\sqrt{b}}$$

The value of a fraction is retained if both its numerator and denominator are multiplied or divided by the same term.

$$\frac{a}{b} = \frac{Ka}{Kb} \quad \text{but this does not equal } K\,\frac{a}{b}$$

$$\text{which is } = \frac{Ka}{b}$$

$$\frac{a}{b} = \frac{a/K}{b/K}$$

Lastly, $0/x$ equals zero, provided that x does not itself equal 0; and $x/0$ is undefined.

Decimal Fractions

A *decimal fraction* is a fraction whose denominator is 10 or a multiple of 10. The distinguishing feature of a decimal is the omission of the denominator in the written form. A decimal is written with a period, called the decimal point, placed before the numerator, as 0.3, 0.06, 0.232. The number of places to the right of the decimal point determines the denominator as shown in the following table:

1 place = tenths (0.1) or (10^{-1})
2 places = hundredths (0.01) or (10^{-2})
3 places = thousandths (0.001) or (10^{-3})
4 places = ten-thousandths (0.0001) or (10^{-4})
5 places = hundred-thousandths (0.00001) or (10^{-5})
6 places = millionths (0.000001) or (10^{-6})
and so on.

Thus

$$0.3 = \frac{3}{10}, \quad 0.06 = \frac{6}{100}, \quad \text{and} \quad 0.232 = \frac{232}{1000}.$$

Addition and subtraction of decimals are performed in the same manner as addition and subtraction of integers. However, it is necessary to stress that the decimal points of the fractions be placed under each other so that tenths are placed under tenths, hundredths under hundredths, and so on. The fractions are then added or subtracted and the decimal point placed in the answer beneath the other decimal points.

Example:

```
  11.40
 635.7335
 373.9041
---------
1021.0376
```

Multiplication of decimal fractions is performed in the same manner as multiplication of common fractions; i.e., the numerators are multiplied to find a new denominator. The number of decimal places needed in the product is the total of the number of decimal places* in the multiplicand and in the multiplier.

Example:

```
  0.06        635.7335
   0.3           11.40
 0.018       254293400
               6357335
             6357335
             7247.361900
```

To divide by a decimal, it is necessary to convert the decimal in the divisor to an integer and then proceed as in the division of integers. This is done by moving the decimal point in the divisor and in the dividend as many places to the right as there are places in the divisor.

Example:

```
          13 00.                 87.01
  0.05.)65.00.         5.15.)448.13.30
                             412 0
                              36 13
                              36 05
                                 8 30
                                 5 15
                                 3 15 remainder
```

Because of the place value in the decimal notation, moving the decimal point one place to the right multiplies a number by 10, two places to the right multiplies it by 100, etc.; moving the point one place to the left divides a number by 10, two places to the left divides by 100, etc.

A decimal fraction may be changed to a common fraction by writing the numerator over the denominator and (if desired) reducing to lowest terms:

$$0.125 = \frac{125}{1000} = \frac{1}{8}$$

A common fraction may be changed to a decimal fraction by dividing the numerator by the denominator.

$$\frac{3}{8} = 8\overline{)3.000}^{\,0.375} = 0.375$$

*The number of places to the right of the decimal point.

Examples and Problems

Compute the following sums and products:

a. $\left(-\frac{3}{8}\right)+\frac{2}{5}=-\frac{15}{40}+\frac{16}{40}=\frac{1}{40}$

b. $(-2.4)+(-3.86)=-6.26$

c. $2\frac{1}{2}+(-3)+\left(-\frac{1}{8}\right)=\frac{5}{2}+\left(-\frac{3}{1}\right)+\left(-\frac{1}{8}\right)=\frac{20}{8}+\frac{-24}{8}+\left(-\frac{1}{8}\right)$

$=\frac{20}{8}-\frac{25}{8}=-\frac{5}{8}$

d. $\frac{2}{3}\left(-\frac{6}{7}\right)=-\frac{12}{21}=-\frac{4}{7}$

e. $\left(-3\frac{1}{2}\right)\left(-1\frac{2}{3}\right)=\left(-\frac{7}{2}\right)\left(-\frac{5}{3}\right)=\frac{35}{6}$

f. $(-2)\left(-\frac{3}{4}\right)\left(-\frac{7}{15}\right)=\left(-\frac{2}{1}\right)\left(-\frac{3}{4}\right)\left(-\frac{7}{15}\right)=\frac{-42}{60}=-\frac{7}{10}$

g. $\frac{16}{4}\div\frac{2}{4}=\frac{16}{4}\times\frac{4}{2}=\frac{64}{8}=8$

h. $2(5+9)=2\times 5+2\times 9=10+18=28$
or $2(5+9)=2\times 14=28$

Change the following decimals to fractions:

a. $0.3=\frac{3}{10}$

b. $0.008=\frac{8}{1000}=\frac{1}{125}$

c. $0.36=\frac{36}{100}=\frac{9}{25}$

d. $0.376=\frac{376}{1000}=\frac{47}{125}$

e. $0.358=\frac{358}{1000}=\frac{179}{500}$

f. $0.88=\frac{88}{100}=\frac{22}{25}$

g. $0.52=\frac{52}{100}=\frac{13}{25}$

h. $0.43=\frac{43}{100}$

Change the following fractions to decimals:

a. $\frac{4}{5}=0.80$

b. $\frac{5}{20}=0.25$

c. $\frac{19}{25}=0.76$

d. $\frac{1}{8}=0.125$

e. $\frac{5}{8}=0.625$

f. $\frac{8}{10}=0.8$

g. $\frac{1}{3}=0.33\frac{1}{3}$ (mixed decimal)

h. $\frac{4}{9}=0.44\frac{4}{9}$

i. $\frac{17}{21}=0.8095$

1.5 SIGNIFICANT FIGURES

When one counts objects accurately, every figure in the numeral expressing the total number of objects must be taken at its face value. Such figures may be said to be absolute. But, when a measurement is recorded, the last figure to the right must be taken to be an approximation, an admission that the limit of necessary accuracy has been reached and that any further figures to the right would be nonsignificant—i.e., meaningless or needless. Significant figures then are consecutive figures that express the value of a concrete number accurately enough for a given purpose. The accuracy varies with the number of significant figures, which are all absolute in value except the last—and this is properly called uncertain.

Any of the digits in a valid concrete number must be regarded as significant. Whether zero is significant, however, depends upon its position or upon known facts about a given number. The interpretation of zero may be summarized as follows:

1. Any zero between digits is significant.
2. Initial zeros to the left of the first digit are never significant; they are included to merely show the location of the decimal point and thus give place value to the digits that follow.
3. One or more final zeros to the right of the decimal point may be taken to be significant.
4. One or more final zeros in a whole number—i.e., immediately to the left of the decimal point—sometimes merely serve to give place value to the digits to the left, but they should be considered significant unless shown by the data to be nonsignificant.

Computation Rules:

Rule 1. When recording a measurement, retain as many figures in a result as will give only one uncertain figure.

Rule 2. When rejecting superfluous figures in the result of a calculation, add one to the last figure retained if the following figure is 5 or more.

Rule 3. In adding or subtracting a number of quantities, extend the significant figures in each term and in the sum or difference only to the point corresponding to that uncertain figure farthest to the left relative to the decimal point.

Example: Sum the following three terms, assuming the last figure in each to be uncertain: 0.0121, 25.64, 1.05782.

Solution:
```
 0.01
25.64
 1.06
```

Explanation: The second term (25.64) has its first uncertain figure (the 4) in the hundredths place. Thus it is useless to extend the digits of the other terms beyond the hundredths place even though they are given to the ten-thousandths place in the first term and to the hundred-thousandths place in the third term.

Rule 4. In multiplication or division, the percentage precision of the product or quotient cannot be greater than the percentage precision of the least precise factor entering into the computation. Thus, in computations involving multiplication or division, or both, retain as many significant figures in each factor and in the numerical result as are contained in the factor having the larger percentage of deviation.

Example: Multiply the following three terms, assuming the last figure in each to be uncertain: 0.0121, 25.64, 1.05782.

Solution: $0.0121 \times 25.6 \times 1.06 = 0.328$

Explanation: The answer is 0.328 utilizing three significant figures in the factors multiplied and the product.

0.0121 = three significant figures because the zero is merely giving place value to the digits that follow
25.6 = three significant figures
1.06 = three significant figures because the zero is between digits

If the first term is assumed to have a possible variation of 1 in the last place, it has an actual deviation of 1 unit in every 121 units, and its percentage deviation would be $1/121 \times 100 = 0.8$. Similarly, the possible percentage deviation of the second term would be $1/2564 \times 100 = 0.04$, and that of the third term would be $1/105{,}782 \times 100 = 0.0009$. The first term, having the largest percentage of deviation, therefore governs the number of significant figures which may be properly retained in the product, for the product cannot have a precision greater than 0.8%.

Rule 5. In carrying out the operations of multiplication or division by use of logarithms, retain as many figures in the mantissa of the logarithm of each factor as are properly contained in the factors themselves under Rule 4. Thus, in solving the example given under Rule 4, the logarithms of the factors are expressed as follows:

$$\begin{aligned}
\log\ 0.0121 &= 8.083 - 10 \text{ (i.e., negative log)} \\
\log 25.64 &= 1.409 \\
\log\ 1.05782 &= 0.024 \\
\hline
&\ 9.516 - 10 \\
\text{antilog } 9.516 - 10 &= 0.328
\end{aligned}$$

As stated above, zeros appearing in front of a number are not considered significant figures, while those appearing after a number are considered as significant figures.

0.062	= 2 significant figures
450	= 3 significant figures
0.0820	= 3 significant figures
0.001341	= 4 significant figures
2.006	= 4 significant figures
56000.0	= 6 significant figures

Thus, for example, the number 100 can actually be stated to contain 1, 2, or 3 significant figures.

1×10^2	= 1 significant figure
1.0×10^2	= 2 significant figures
100	= 3 significant figures

The statement of the population of Kuwait as being 615,000, for instance, does not obviously indicate the exact number of people. The measurement is only accurate to the thousands and it should be properly written as 6.15×10^5.

Many scientists will express their findings in powers of 10 so that the number of significant figures is apparent. The total number of digits stated indicates the number of significant figures in the answer. For example:

Resulting Answer	Expressed as Powers of 10	Number of Significant Figures
34,000	3.4×10^4	2
	3.40×10^4	3
	3.400×10^4	4
0.00097	9.7×10^{-4}	2
	9.70×10^{-4}	3
	9.700×10^{-4}	4

Add 27.93, 15.00, and 635.7335.

$$\begin{array}{r} 27.93 \\ 15.00 \\ \underline{635.73} \\ 678.66 \end{array}$$

Add 60.40, 0.063, and 666.6224.

$$\begin{array}{r} 60.40 \\ 0.06 \\ \underline{666.62} \\ 727.08 \end{array}$$

Subtract 1.550 from 3.3152.

$$\begin{array}{r} 3.315 \\ \underline{-1.550} \\ 1.965 \end{array}$$

Multiply $0.25 \times 10.11 \times 1.09$.

Use only two significant figures because 0.25 has the largest percentage of deviation (4.0%).

$0.25 \times 10.0 \times 1.1 = 2.75$ answer = 2.8

Divide 0.00751 by 0.12.

Use two significant figures because 0.12 has the largest percentage of deviation (8.3%).

```
         0.625
0.12 |0.0075        answer = 0.63
        72
        --
         30
         24
         --
          60
          60
          --
```

Examples and Problems

What is the average of three successive weighings of a compound?

```
  27.21 g
  28.07 g
  28.56 g
3 |83.8400
  27.9467    Correct answer is 27.95 g, rounded to 4 significant figures.
```

Sum the following.

```
  9.72
  6.2
722.003
-------
737.923    Correct answer is 737.9 because 6.2 is significant to only tenths.
```

Multiply $300 \times 4.3 = 12{,}900$.

Since the minimum significant figures are 2 (4.3), the product should be expressed as 13,000.

1.6 EXPONENTS

An *exponent* is a small raised digit to the right of a number indicating the number of times that number is to be multiplied by itself to form a product.

For example:

$$2^3 = 2 \times 2 \times 2$$

$$7^4 = 7 \times 7 \times 7 \times 7$$

The power of a number is named by an exponent.

In dealing with very large or very small numerical values, scientists often use what is called *scientific notation* or *powers of ten.* Thus, if the distance from the earth to the sun is given as 92,000,000, it may be recorded as $92 \times 1{,}000{,}000$ or 92×10^6 miles. Astronomical data may involve billions, trillions, or even larger values. For example, the weight of the earth in pounds is about 13,000,000,000,000,000,000,000,000 or 13×10^{24} pounds, or a light year is about 6,000,000,000,000 miles or 6×10^{12} miles.

This form of writing such large numbers using the laws of exponents and the base number of ten puts the significant figures before the reader more prominently and it is read more easily than the ordinary long form.

The simple powers of ten are 10, 100, 1,000, 10,000 and so on. The alternate way of writing 10 is:

10 is 10^1 (ten-to-the-first-power) and then

100 is 10×10, or 10^2 (ten-to-the-second-power, or ten-squared)

1,000 is $10 \times 10 \times 10$, or 10^3 (ten-to-the-third-power, or ten-cubed)

70,000 is $7 \times 10 \times 10 \times 10 \times 10$, or 7×10^4 (seven-times-ten-to-the-fourth)

635,000 is $635 \times 10 \times 10 \times 10$, or 635×10^3, or 6.35×10^5

Remember that 10^0 equals one.

It is obvious that the exponents of these numbers are the number of zeros after the first number. For numbers smaller than one, the exponents are negative numbers.

$\frac{1}{10}$ (0.1) is written as 10^{-1} (ten-to-the-minus-one)

$\frac{1}{100}$ (0.01) is written as 10^{-2} (ten-to-the-minus-two)

$\frac{6}{1000}$ (0.006) is written as 6×10^{-3} (six-times-ten-to-the-minus-three)

Some general rules:

1. $x^a \times x^b = x^{(a+b)}$ Note: exponents are added
2. $x^a \times x^{-b} = x^{(a-b)}$ Note: exponents are subtracted
3. $x^{-a} \times x^b = x^{(b-a)}$
4. $x^{-a} \times x^{-b} = x^{(-a+-b)}$ or $x^{-(a+b)}$
5. $x^{-a} \times x^{-a} = x^{-2a}$
6. $(xy)^a = x^a y^a$
7. $(x^a)^b = x^{ab}$ Note: exponents are multiplied
8. $\frac{x^a}{x^b} = x^{(a-b)}$ Note: exponents are subtracted

9. $\left(\frac{x}{y}\right)^a = \frac{x^a}{y^a}$

10. $\frac{1}{x^a} = x^{-a}$ Note: exponent in the denominator gains minus sign when brought up to numerator

11. $x^{1/a} = \sqrt[a]{x}$

12. $x^{a/b} = \sqrt[b]{x^a}$

13. $x^0 = 1$

14. $x^a \times y^b = x^a \times y^b$ Note: combining exponents not possible as in $2^3 \times 3^4$, see example below

Examples and Problems

$10^2 \times 10^4 = 10^6 = 1{,}000{,}000$

$2^3 \times 3^4 = 8 \times 81 = 648$

$\frac{1}{10^4} = 10^{-4} = 0.0001$

$\frac{10^6}{10^4} = 10^6 + 10^{-4} = 10^2 = 100$

$x^{1/2} = \sqrt{x}$ i.e., the square root of x

$x^{4/9} = \sqrt[9]{x^4}$ i.e., the 9th root of x to the 4th power

Numbers written in scientific notation.

$181.6 = 1.816 \times 10^2$

$0.0031416 = 3.1416 \times 10^{-3}$

$186{,}200 = 1.862 \times 10^5$

$0.57118 = 5.7118 \times 10^{-1}$

$9500 = 9.5 \times 10^3$

Add $6 \times 10^{-3} + 4.5 \times 10^{-1} + 2.3 \times 10^{-2}$.

Put terms in the form of most positive exponent (−1 in this case) and then add.

$0.06 \times 10^{-1} + 4.5 \times 10^{-1} + 0.23 \times 10^{-1}$

Placing the common factor 10^{-1} outside,

$10^{-1} \times (0.06 + 4.5 + 0.23) = 4.79 \times 10^{-1} = 0.479.$

- - - -

$4 \times 10^2 + 9 \times 10^{-2} + 6 \times 10^{-1}$

$4 \times 10^2 + 0.0009 \times 10^2 + 0.006 \times 10^2$

$10^2 \times (4 + 0.0009 + 0.006) = 4.0069 \times 10^2 = 400.69$

Subtract $7.843 \times 10^{-2} - 9.6 \times 10^{-4}$.

Since 10^{-2} is more positive (or less negative) than 10^{-4}, we place the latter in a 10^{-2} form.

$7.843 \times 10^{-2} - 0.096 \times 10^{-2} =$

$10^{-2} \times (7.843 - 0.096) = 7.747 \times 10^{-2} = 0.07747$

Multiply $5 \times 10^{-2} \times 7.2 \times 10^{3} \times 1 \times 10^{5}$.

First add the exponents of 10.

$-2 + 3 + 5 = 6$; thus, we have 10^{6}.

Then multiply $5 \times 7.2 \times 1 = 36$.

The answer is 36×10^{6} or 3.6×10^{7} which is 36,000,000.

Multiply $40{,}000 \times 2{,}500 \times 20 \times 0.02$.

First, convert to exponents

$40{,}000 = 4 \times 10^{4}$

$2{,}500 = 25 \times 10^{2}$ or 2.5×10^{3}

$20 = 2 \times 10^{1}$

$0.02 = 2 \times 10^{-2}$

Then, add the exponents of 10

$4 + 3 + 1 - 2 = 6$

Multiply $4 \times 2.5 \times 2.0 \times 2.0 = 40$

The answer is 40×10^{6} or 4.0×10^{7} which is 40,000,000.

- - - -

$$\frac{(1.5 \times 10^{-12})(4 \times 10^{3})}{2 \times 10^{-23} \times 1.5 \times 10^{19}}$$

Add exponents of 10 in both numerator and denominator and multiply.

$$\frac{(1.5 \times 4) \times 10^{-9}}{(2 \times 1.5) \times 10^{-4}} = \frac{6.0 \times 10^{-9}}{3.0 \times 10^{-4}} = 2 \times 10^{-5}$$

Note: When 10^{-4} is subtracted from 10^{-9}, its sign is changed and thus the answer is 10^{-5}.

- - - -

$\sqrt{1600} = \sqrt{16} \times \sqrt{10^2} = \sqrt{16} \times \sqrt{100} = 4 \times 10 = 40$

$\sqrt{0.25} = \sqrt{25} \times \sqrt{10^{-2}} = \sqrt{25} \times \sqrt{0.01} = 5 \times 0.1 = 0.5$

- - - -

$(7 \times 10^{5}) \times (8 \times 10^{4}) = (7 \times 8) \times (10^{5} \times 10^{4}) = 56 \times 10^{9}$

$(2.0 \times 10^{3}) \times (4.1 \times 10^{-2}) = (2.0 \times 4.1) \times (10^{3} \times 10^{-2}) = 8.2 \times 10^{1}$

$(7.3 \times 10^{10}) \times (3.0 \times 10^{9}) = (7.3 \times 3.0) \times (10^{10} \times 10^{9})$

$= 21.9 \times 10^{19} = 2.19 \times 10^{20}$

- - - -

$$\frac{10^6}{10^4} = 10^{6-4} = 10^2 \qquad \text{because} \qquad \frac{10^6}{10^4} = \frac{1{,}000{,}000}{10{,}000} = 100 = 10^2$$

$$\frac{16.0 \times 10^5}{4.0 \times 10^2} = \frac{16}{4} \times 10^{5-2} = 4 \times 10^3$$

$$\frac{21.0 \times 10^2}{7.0 \times 10^{-7}} = \frac{21}{7} \times 10^{2-(-7)} = 3 \times 10^9$$

- - - -

$$5.0 \times 10^4 + 7 \times 10^3 = 5.0 \times 10^4 + 0.7 \times 10^4 = 5.7 \times 10^4$$
$$(5 \times 10^2)^3 = 5^3 \times (10^2)^3 = 125 \times 10^6 = 1.25 \times 10^8$$
$$(7 \times 10^6)^2 = 7^2 \times (10^6)^2 = 49 \times 10^{12} = 4.9 \times 10^{13}$$

- - - -

$$\frac{10{,}000 \times 4{,}600 \times 0.006}{200 \times 0.03} = \frac{10^4 \times 4.6 \times 10^3 \times 6 \times 10^{-3}}{2 \times 10^2 \times 3 \times 10^{-2}}$$
$$= 10^4 \times 4.6 = 46{,}000$$

- - - -

$$\frac{24{,}000\,(0.0002 + 5 \times 10^{-6})}{0.080} = \frac{24 \times 10^3\,(200 \times 10^{-6} + 5 \times 10^{-6})}{8 \times 10^{-2}}$$
$$= \frac{24 \times 10^3\,(205 \times 10^{-6})}{8 \times 10^{-2}} = \frac{4920 \times 10^{-3}}{8 \times 10^{-2}}$$
$$= 615 \times 10^{-1} = 61.5$$

1.7 EXTRACTING THE ROOT OF A NUMBER

The mathematical operation which is the reverse of raising a number to a power is *extracting the root of a number.* The *root* of a number is any of the equal factors which when multiplied together produce that number. If a number, 36 for example, has two equal roots, the process of finding these roots is called extracting the square root. If a number, such as 125, has three equal roots, the process of finding these roots is called extracting the cube root, etc.

The radical sign, $\sqrt{}$, is used to indicate that the root of a number is to be extracted. The number whose root is to be extracted is placed under the sign in this manner, $\sqrt{64}$. This reads, "the square root of 64." If other than the square root is to be extracted, the number of that root, called the index, is placed on the radical sign in this manner, $\sqrt[3]{125}$. This is read, "the cube root of 125."

Square Root

In extracting the square root of a number, we try to find a factor which when multiplied by itself will yield the number. The various steps in this process can best be learned and understood by seeing how a square root of an integer is extracted, as in the two following examples:

Find the square root of 4489.

Step 1. Place the number under the radical and separate the number into periods of two, beginning at the right.

$$\sqrt{44,89}$$

Step 2. Find the square root of the perfect square which is less than and nearest to the first period at the left, 44. This must be done by inspection. In this case, the square root of the nearest perfect square, 36, is 6. Write the 6 over the first period.

$$\begin{array}{l} \;\;6 \\ \sqrt{44,89} \end{array}$$

Step 3. Square the 6 and write the square under the first period. Subtract and the remainder is 8. (If the number in the first period is a perfect square there will be no remainder.)

$$\begin{array}{l} \;\;6 \\ \sqrt{44,89} \\ \;\;\underline{36} \\ \;\;\;\;8 \end{array}$$

Step 4. Bring down the next period, 89, and place it after the remainder, 8. The new dividend is now 889.

$$\begin{array}{l} \;\;6 \\ \sqrt{44,89} \\ \;\;\underline{36} \\ \;\;\;\;8\;89 \end{array}$$

Step 5. Multiply the 6 by 2 and write the answer, 12, as a partial divisor of the dividend, 889. Be sure to leave room to insert the rest of the divisor.

$$\begin{array}{r|l} & \;\;6 \\ & \sqrt{44,89} \\ & \;\;36 \\ \hline 12 & \;\;8\;89 \end{array}$$

Step 6. It is now necessary to complete the partial divisor by inserting a number after 12. The complete divisor is to be divided into the dividend, 889. The quotient must be a number which when multiplied by the complete divisor will be equal to or less than the dividend, 889. Here the number needed to complete the divisor is 7. Write this number after the partial divisor, 12, making the complete divisor 127. Divide 127 into 889; the quotient is 7. Write the 7 in the answer over the second period, 89. Multiply 7 by 127; the product is 889. Subtracting, there is no remainder. The square root of 4489 is 67. Check: $67 \times 67 = 4489$.

$$\begin{array}{r|r} & 6\;\;7 \\ & \sqrt{44,89} \\ & 36\;\;\;\;\; \\ \hline 127 & 8\,89 \\ & \underline{8\,89} \end{array}$$

Find the square root of 2,985,984.

Solution:

$$\begin{array}{r|r} & 1\;\;7\;\;2\;\;8 \\ & \sqrt{2,98,59,84} \\ & 1\;\;\;\;\;\;\;\;\;\;\;\;\;\; \\ \hline 27 & 1\,98\;\;\;\;\;\;\;\;\; \\ & 1\,89\;\;\;\;\;\;\;\;\; \\ \hline 342 & 9\,59\;\;\;\;\; \\ & 6\,84\;\;\;\;\; \\ \hline 3448 & 2\,75\,84 \\ & \underline{2\,75\,84} \end{array}$$

The square root of a decimal is found in much the same manner as the square root of a whole number, with one exception: the periods are formed by grouping pairs of digits to the right of the decimal point.

$$\begin{array}{r} .\;6\;\;3\;\;5 \\ \sqrt{.40,32,25} \end{array}$$

The square root of a number which consists of an integer and a decimal is found by pointing off pairs of digits to both the left and right of the decimal point.

$$\begin{array}{r} 1\;\;1.\;4\;\;0 \\ \sqrt{1,29.96,00} \end{array}$$

To find the square root of a fraction, find the square root of the numerator and the square root of the denominator.

$$\sqrt{\frac{121}{1600}} = \frac{\sqrt{121}}{\sqrt{1600}} = \frac{11}{40}$$

Very often the number whose square root is to be extracted is not a perfect square. The complete root of such a number can never be found. With such a number one finds the square root to so many decimal places. As many pairs of zeros are added to the number as decimal places are called for.

$$\begin{array}{c} 2\ \ 1.\ 8\ \ 1 \\ \sqrt{4{,}76.00{,}00} \end{array}$$

When taking a root of an exponential number, the root of the number is extracted and the power of the number is divided by the specified root.

$$\sqrt[2]{16 \times 10^{12}} = \sqrt[2]{16 \times 10^{12/2}} = 4 \times 10^{6}$$

$$\sqrt[3]{729 \times 10^{27}} = \sqrt[3]{729 \times 10^{27/3}} = 9 \times 10^{9}$$

Examples and Problems

```
        6  0. 4  0
      √36 48.16 00
        36
1204 | 0 48 16
         48 16
         -----
         - - - -
```

```
      2  7. 9  3
    √7 80.08 49
      4
47   | 3 80
       3 29
549  |  51 08
        49 41
5583 |  1 67 49
        1 67 49
        -------
        - - - -
```

```
        2  5. 2  1  3
      √6 35.73 35 00   etc.
        4
45   | 2 35
       2 25
502  |  10 73
```

```
         2  6  0  4.
       √6 78 08 16.
         4
26   | 2 78
       2 76
5204 |     2 08 16
           2 08 16
           -------
           - - - -
```

```
            3  3. 1  2  5
         √10 97.26 56 25
            9
63     | 1 97
         1 89
661    |    8 26
            6 61
6622   |  1 65 56
          1 32 44
66245  |    33 12 25
            33 12 25
            --------
            - - - -
          10 04
5041   |    69 35
            50 41
50423  |  18 94 00
          15 12 69
50426  |   3 81 31 00
```

1.8 PROGRESSION

Arithmetic Progression

If numbers are so arranged in a series that each number differs from the number preceding it by a like amount, the series of numbers is called an arithmetic progression. The numbers in the series or progression may be successively larger or successively smaller. The numbers in the progression are called *terms.* The first and last terms are called the *extremes.* The sum of all the terms is called the *sum of the series.* The differences between the terms is called the *common difference.*

To find either extreme, deduct one from the total number of terms. Then, multiply the common difference by this total number of terms minus one to find the total of common differences. Add the total of common differences and the first term to find the last term; to find the first term, subtract the total of common differences from the last term.

Example: The first term in an arithmetic progression is 15, the common difference is 3, the number of terms is 25. Find the last term.

Solution:
$25 - 1 = 24$ Number of missing terms
$24 \times 3 = 72$ Total of common differences
$15 + 72 = 87$ Last term

The *common difference* of a progression may be found by dividing the difference between the extremes by the number of terms, less one.

Example: The first term of an arithmetical progression is 11, the last term is 291, and the number of terms is 41. Find the common difference.

Solution:
$291 - 11 = 280$ Total of common differences
$280/(41 - 1) = 7$ Common difference

The *number of terms* in a progression is found by dividing the difference between the extremes by the common difference and then adding one to the answer.

Example: The first term of a progression is 9, the last term is 450, and the common difference is 9. Find the number of terms.

Solution:
$450 - 9 = 441$ Difference between first and last terms
$441/9 = 49$ Number of terms less one
$49 + 1 = 50$ Number of terms

The *sum of the terms* in a progression is found by multiplying the sum of the first and last terms by half the number of terms.

Example: The first term of a progression is 52, the last term is 12, and the number of terms is 8. Find the sum of the terms.

Solution:	$52 + 12 = 64$	Sum of the first and last terms
	$8/2 = 4$	One-half the number of terms
	$64 \times 4 = 256$	Sum of the series

Geometric Progression

If numbers in a series are so arranged that each number will be either larger or smaller than the one preceding by the same ratio, the series of numbers is called a geometric progression. As in arithmetic progression, the numbers in a geometric progression are called *terms;* the first and last terms are the *extremes.* The *common ratio* is the quotient of any terms divided by the term preceding it. An *increasing geometric progression* is one in which the common ratio is greater than 1; therefore, each term will be greater than the preceding one. A *decreasing geometric progression* is one in which the common ratio is less than 1; therefore, each term will be smaller than the preceding term. The sum of all the terms is called the *sum of the series.*

The *last term* of a geometric progression is the product of the first term multiplied by the common ratio, raised to a power which is one less than the number of terms.

Example: The first term of a geometrical progression is 2, the common ratio is 2, and the number of terms is 8. Find the last term.

Solution: $2 \times 2^7 = 2 \times 128 = 256$

The *first term* is found by dividing the last term by the common ratio raised to a power one less than the number of terms.

Example: The last term of a geometrical progression is 1024, the common ratio is 2, and the number of terms is 10. Find the first term.

Solution: $1024/2^9 = 1024/512 = 2$

To find the *common ratio*, divide the last term by the first term. The answer is a number of which the root must be found. The power of the root is determined by subtracting one from the number of terms.

Example: The first term of a geometrical progression is 6, the last term is 162, and the number of terms is 4. Find the common ratio.

Solution: $162/6 = 27$
$4 - 1 = 3$
$\sqrt[3]{27} = 3$, Common ratio

To find the *sum of the terms* in a geometric progression, when the common ratio is less than one, the first term is multiplied by one minus the ratio raised to the power of the number of terms. This is divided by 1 - ratio.

$$\text{Sum} = \frac{\text{first term } (1 - \text{ratio}^{\text{number of terms}})}{(1 - \text{ratio})}$$

When the common ratio is more than one, the

$$\text{Sum} = \frac{\text{first term } (\text{ratio}^{\text{number of terms}} - 1)}{(\text{ratio} - 1)}$$

When the common ratio is one, the

$$\text{Sum} = (\text{number of terms})\ (\text{first term})$$

Example: The common ratio equals −3. The first term equals 3, and the last and fourth term is −8.

Solution:

$$\text{Sum} = \frac{3\,[1 - (-3^4)]}{1 - (-3)} = \frac{3\,[1 - (+81)]}{4}$$

$$= \frac{3\,(-80)}{4} = \frac{-240}{4} = -60$$

Example: The common ratio equals 3. The first term equals 4, and the fourth and last term is 108.

Solution:

$$\text{Sum} = \frac{4\,(3^4 - 1)}{3 - 1} = \frac{4\,(81 - 1)}{2} = \frac{4\,(80)}{2} = \frac{320}{2} = 160$$

Example: The common ratio equals 1. The first term equals 4, and the last and fifth term is 4.

Solution: $\text{Sum} = 5\,(4) = 20$

Examples and Problems

Determine if the progression is arithmetic or geometric.

5, 11, 17, 23, 29 arithmetic, common difference = 6

$-\frac{3}{2}, -\frac{2}{3}, \frac{1}{6}$ arithmetic, common difference = $\frac{5}{6}$

4, 12, 36, 108 geometric, common ratio = 3

3, −9, 27, −81 geometric, common ratio = −3

$5x^3, 10x^2, 20x, 40$ geometric, common ratio = $\frac{2}{x}$

The 25th term of the progression 1, 3, 9 is

$1 \times 3^{24} = 3^{24}$ or 282,429,536,481

The 25th term of the progression 3, 9, 27 is

$3 \times 3^{24} = 3 \times 282{,}429{,}536{,}481$ or 847,288,609,443.

The sum of the first 22 terms of the progression 12, 14, 16 is:

Find last term: $21 \times 2 = 42$
$42 + 12 = 54$

Find sum: $54 + 12 = 66$
$11 \times 66 = 726$

Find the tenth term and the sum of the first ten terms of the following:

a. An arithmetic progression where 3 is the first term and the common difference is −5.

Tenth term: $10 - 1 = 9$
$9 \times -5 = -45$
$3 + -45 = -42$ is 10th term

Sum of terms: $-42 + 3 = -39$
$10/2 = 5$
$-39 \times 5 = -195$

b. A geometric progression where the first term is 1 and 3 is the multiplier.

Tenth term: $1 \times 3^9 = 19683$

Sum of terms: $\frac{1\,(3^{10} - 1)}{3 - 1} = \frac{1\,(59049 - 1)}{2} = \frac{59048}{2} = 29524$

1.9 RATIO, PROPORTION, AND PERCENT

A *ratio* is the relationship between two numbers or quantities which can be compared because they are expressed in like units. To make such a comparison, one quantity is divided by the other and the quotient is the ratio. The ratio of a to b can be written $a{:}b$ or a/b. The quantities that are compared are called the terms of the ratio. The first term is called the antecedent and the second term is called the consequent. Such common expressions as "four out of five" and "every third person" indicate the fractions 4/5 and 1/3, respectively. Also, the Greek letter π is used as a symbol to represent the ratio of the length of the circumference of a circle to its diameter $\left(\text{i.e., } \frac{22}{7} \cong 3.143 \cong \pi\right)$.

To further explain the concept of ratio, consider the set of numbers 10, 20, 30, 40 which has an important property in common with the set 2, 4, 6, 8.

$$
\begin{aligned}
10 &\text{ is } 2 \times 5 \\
20 &\text{ is } 4 \times 5 \\
30 &\text{ is } 6 \times 5 \\
40 &\text{ is } 8 \times 5
\end{aligned}
$$

The numbers of the first set are all five times larger than the corresponding numbers of the second set. (See Figs. 1.2 and 1.3.) This is what is

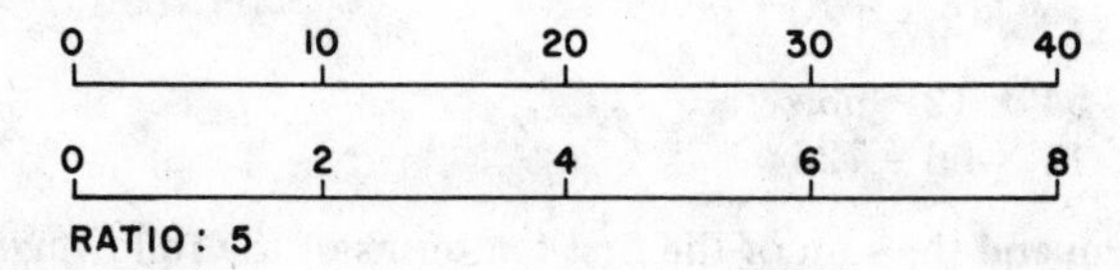

Fig. 1.2. Proportional sets and constant ratio. If two sets of numbers are proportional, they can be arranged on two different number lines so that all the members of one set correspond to members of the other set. The ratio (number in the first set) : (number in the second set) is constant.

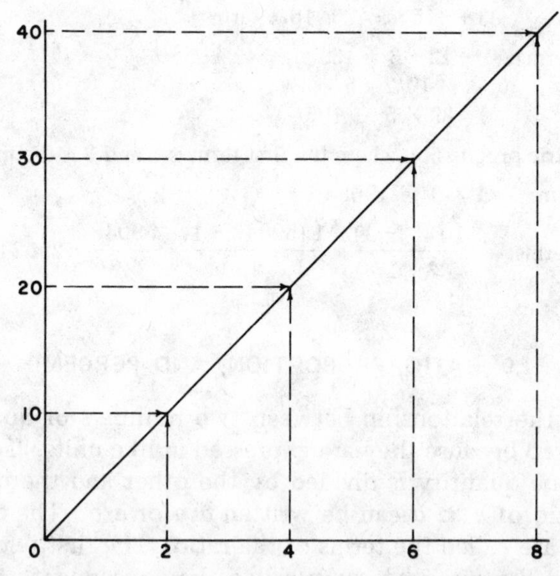

Fig. 1.3. Proportional sets plotted on regular graph paper. The regular kind of graph paper is like two number lines at right angles to each other. Proportional sets give a straight line.

meant by ratio: one number in one set is compared to a corresponding number in another set. If the ratio of all the corresponding pairs in the two sets are equal, the sets are said to be proportional.

A *proportion* is a statement asserting that one quantity has the same relation to another quantity as a third quantity has to a fourth; in short, a statement of the equality of two ratios. For example, 1.2 is to 6 as 2 is to 10 or $\frac{1}{5}$. If two sets of numbers are proportional, the larger one set is, the larger is the other. An example is a car traveling at a steady rate of speed (40 mph). Provided the car stays at this steady rate, the distance traveled will be proportional to the time taken.

Distance in miles	40	80	120	160
Total time in hours	1	2	3	4

Once it is known that two sets are proportional, it is easy to find an unknown term in either of them.

To find a missing proportional, one must first write the proportion using the fraction bar form and insert an x for the missing term. Then solve as a simple equation. In a proportion, the two outer numbers are called the *extremes*, and the two inner numbers are called the *means*. In a proportion such as 1:2 = 4:8, the product of the two extremes ($1 \times 8 = 8$) is equal to the product of the two means ($2 \times 4 = 8$). From this is derived the basic rule of proportion: *the product of the extremes equals the product of the means.* That is, the product of the numerator of the one and the denominator of the other always equals the product of the denominator of one and the numerator of the other, i.e., cross products are equal. For example, 3 is to 9 as 12 is to what? Using the above stated procedure, $\frac{3}{9} = \frac{12}{x}$, $3x = 108$, $x = \frac{108}{3} = 36$; therefore, $\frac{3}{9} = \frac{12}{36}$.

It is also true in a proportion that the reciprocals of the proportion are equal. Since $\frac{2}{4} = \frac{4}{8}$, then $\frac{4}{2} = \frac{8}{4}$. Also, the numerator of one fraction equals the product of its denominator and the other fraction. If $\frac{a}{b} = \frac{c}{d}$, then $a = \frac{bc}{d}$ and $c = \frac{ad}{b}$.

Example: If $\frac{6}{15} = \frac{2}{5}$

$$\text{then } 6 = 15 \times \frac{2}{5} \left(\text{i.e., } \frac{15 \times 2}{5}\right) = 6$$

$$\text{and } 2 = 5 \times \frac{6}{15} \left(\text{i.e., } \frac{5 \times 6}{15}\right) = 2$$

And, the denominator of one fraction equals the quotient of its numerator divided by the other fraction. Again, if $\frac{a}{b} = \frac{c}{d}$, then $b = \frac{ad}{c}$ and $d = \frac{bc}{a}$.

Example: If $\frac{6}{15} = \frac{2}{5}$

$$\text{then } 15 = 6 \div \frac{2}{5} \left(\text{i.e., } \frac{6 \times 5}{2}\right) = 15$$

$$\text{and } \quad 5 = 2 \div \frac{6}{15} \left(\text{i.e., } \frac{2 \times 15}{6}\right) = 5$$

Percentage is a convenient ratio that expresses a number of parts in every 100 of the same kind. It is the proportionate part of a quantity expressed as an equivalent part of one hundred. In the concept of percentage, 100 is taken as the unit and quantities are looked upon as parts of 100. For example, $\frac{1}{2} = \frac{50}{100}$ = or 50 percent. In percentage statements, the 1/100 fraction is replaced by %, the percent sign. To write a percent as a fraction, divide the percent number by 100 and then reduce the expression to its lowest terms. For example, $12\% = \frac{12}{100} = \frac{3}{25}$, and $25\% = \frac{25}{100} = \frac{1}{4}$. In the same context, to convert any fraction to % simply multiply the fraction by 100, the answer obtained after completion of the multiplication is followed by the percent sign. For example, $\frac{1}{4} = \frac{1}{4} \times 100 = \frac{100}{4} = 25\%$.

Another form of percent is the *parts per million* (ppm). To convert percent to ppm simply multiply the percent by 10,000. For example, if the percent of sodium fluoride in water is 0.0002% then there are $0.0002 \times 10{,}000 = 2$ ppm of this inorganic substance in the water. In analogy, *parts per billion* (ppb) is obtained by multiplying percent by 10,000,000. Thus it is obvious that the conversion of ppm to percent requires a division of the ppm value by 10,000 and in the case of ppb conversion to percent, one divides by 10,000,000.

Example: Finding percent from a proportion.

If there are 36 white mice in a population sample of 90, what percent of the mice are white?

$$\frac{36}{90} = \frac{x}{100}$$

$$90x = 3600$$

$$x = \frac{3600}{90} = 40\% \text{ of the mice are white.}$$

Finding ppm from percent.

The percent of lead in the soil crust of Miami International Airport is 0.5%. What is this value in ppm?

$0.5 \times 10{,}000 = 5000$ ppm

A good example of proportion is encountered in chemical formulas. A chemical formula is an expression that indicates the proportions of elements which are present in a chemical compound. The formula tells what elements are present in the compound and in what ratio the atoms of these elements are present. The formula H_2SO_4 (sulfuric acid) indicates that two atoms of hydrogen, one of sulfur, and four of oxygen are present in each molecule of the acid.

Interpolation is also based on the concept of proportion, assuming that the differences between tabular values (the values down the side of a table) are proportional to the differences between the tabulated values (the value looked up on the table). If x_1 and x_2 are the consecutive tabular values of a variable, and y_1 and y_2 are the corresponding tabulated values, the x is any value of the variable between x_1 and x_2 and y is the corresponding tabulated value of x. The relationship of these values is as follows:

$$\frac{x - x_1}{x_2 - x_1} = \frac{y - y_1}{y_2 - y_1}$$

Examples and Problems

For example: In the sample section of a sine table, find the sine of 38° 34′ where the sine values listed for 38° 30′ and 38° 40′ are 0.6225 and 0.6248, respectively.

Arrange the table this way

$$10' \left\{ \begin{array}{l} 4' \left\{ \begin{array}{l} 38°30' \\ 38°34' \end{array} \right. \\ \quad\; 38°40' \end{array} \right. \qquad \left. \begin{array}{c} \text{Sine} \\ 0.6225 \\ y \\ 0.6248 \end{array} \right\} 23$$

Angle	Sine
38° 30′	0.6225
38° 34′	y
38° 40′	0.6248

$x_1 = 38°30'$ $\quad y_1 = 0.6225$

$x_2 = 38°40'$ $\quad y_2 = 0.6248$

$x = 38°34'$ $\quad y = ?$

Solution: $\dfrac{38°34' - 38°30'}{38°40' - 38°30'} = \dfrac{y - 0.6225}{0.6248 - 0.6225}$

$\dfrac{4}{10} = \dfrac{y - 0.6225}{0.0023}$

$10\,(y - 0.6225) = 0.0092$
$10y - 6.225 = 0.0092$
$10y = 0.0092 + 6.225$
$10y = 6.2342$
$y = 0.62342$

Therefore, sine $38°34' = 0.6234$

- - - -

$x = 16{:}x \qquad \dfrac{x}{1} = \dfrac{16}{x} \qquad x^2 = 16 \qquad x = 4$

- - - -

$\dfrac{22x^2}{16} = \dfrac{4x}{2} \qquad 44x^2 = 64x \qquad \dfrac{x^2}{x} = \dfrac{64}{44} \qquad x = \dfrac{16}{11}$

- - - -

$\dfrac{15u}{11} = \dfrac{45}{55} \qquad 825u = 495 \qquad u = \dfrac{495}{825} = \dfrac{99}{165} = \dfrac{3}{5}$

- - - -

The ratio of 200 immature white cells in a population of 10,000 cells is

$\dfrac{200}{10{,}000} = \dfrac{1}{50}$

The percent of immature white cells as described above is

$10{,}000 = 100\%$
$200 = ?$

$\dfrac{10{,}000}{100\%} = \dfrac{200}{x}$

$10{,}000\,x = 20{,}000 \qquad x = 2\%$ or $\dfrac{200}{10{,}000} = \dfrac{x}{100\%} \qquad \dfrac{200 \times 100\%}{10{,}000} = 2\%$

Twenty-five percent of the students are above average. The ratio of the students above average is

$\dfrac{25\%}{100\%} = \dfrac{1}{4}$, i.e., 1 out of 4

Compare the size of 10% of 90 with 90% of 10.

10% of 90 = 9
90% of 10 = 9

- - - -

50 = ?% of 275   7% of ? = 168

$x\% = \dfrac{50}{275} \times \dfrac{100}{1} = 18.2\% \qquad x = \dfrac{168}{.07} = 2400$

- - - -

8% of 255 = ?
$0.08 \times 255 = 20.40$

28 out of 60 is ?%
28 = ?% of 60

$$x\% = \frac{28}{60} \times \frac{100}{1} = 46.67\%$$

A blood smear shows 1 platelet for every 10 red cells. The red cell count was 4,500,000, i.e., 4.5×10^6 cells. What is the platelet count?

10 red cells for every 1 platelet

4.5×10^6 red cells for every ? platelets

$$\frac{10}{1} = \frac{4.5 \times 10^6}{x} \qquad 10x = 4.5 \times 10^6 \qquad x = \frac{4.5 \times 10^6}{10}$$

$x = 4.5 \times 10^5$ or 450,000 platelets

Write the following as percents:

a. $\frac{1}{2} = 0.5 = 50\%$

b. $0.625 = \frac{625}{1000} = \frac{62.5}{100} = 62.5\%$

c. $\frac{20}{200} = \frac{10}{100} = 10\%$

d. $\frac{7\frac{1}{2}}{100} = 7\frac{1}{2}\%$

e. $\frac{2}{3} = 66\frac{2}{3}\%$

f. $\frac{80}{1000} = \frac{8}{100} = 8\%$

g. $1 = \frac{1}{1} = \frac{100}{100} = 100\%$

h. $\frac{13}{20} = 65\%$

A patient's hemoglobin was found to be 6 g. Given that 14.5 g of hemoglobin = 100%, what is the percent of hemoglobin in this patient's blood?

14.5 = 100%
6 = ?%

$$\frac{14.5}{100\%} = \frac{6}{x} \qquad 14.5x = 600 \qquad x = \frac{600}{14.5} = 41.4\%$$

If a medical technologist found 15 reticulocyte (r) cells per 1000 red blood cells, what is the percent of reticulocytes?

15 r cells per 1000
? r cells per 100

$$\frac{15}{1000} = \frac{?}{100} \qquad 1000x = 1500 \qquad x = 1.5\%$$

If 15.5 g of hemoglobin in 100 ml of blood is considered normal and a patient had 13.95 g of hemoglobin per 100 ml, what percent of the normal does this represent?

15.5 = 100%
13.95 = ?%

$$\frac{15.5}{100\%} = \frac{13.95}{?\%} \qquad 15.5x = 1395 \qquad x = \frac{1395}{15.5} = 90\% \text{ of normal}$$

1.10 LOGARITHMS

A *logarithm* is the power to which some base number must be raised to express a given number. Tables of logarithms are thus compilations of these exponents.

Although any number could be made the base for a set of logarithms, three numbers are most often used:

1. 10—logarithms to the base 10 are most frequently used and are called *common logarithms.* These logarithms are usually symbolized as log, e.g., log 2 = 0.3010.

2. 2—logarithms to the base 2 ($\log_2$) have an application to processes wherein some property increases by doubling.

3. e (2.718 . . . the natural base)—logarithms to the base e ($\log_e$ or ln) are also known as *Napierian logarithms. Natural logarithms* describe processes that increase or decrease in a naturally exponential manner.

Logarithms to the base e are directly proportional to logarithms to the base 10, but are larger by a factor of about 2.303.

$$\ln n = 2.303\ (\log_{10} n)$$
$$\log_{10} n = 0.4343\ (\ln n)$$

Thus when changing from one base to another, either multiply by 0.4343 or by 2.303.

The most commonly used base for logarithms is the base 10. Every positive number can be expressed either exactly or nearly exactly as a power of 10. When a number is expressed in such a manner, the corresponding exponent is called the common logarithm of the number or the logarithm of the number to the base 10. Thus, since $10^2 = 100$ and $10^{1.4914} = 31$, the logarithm of 100 is 2 and the logarithm of 31 is 1.4914.

$$
\begin{array}{llll}
10^3 = 1000 & \log 1000 & = & 3 \\
10^2 = 100 & \log 100 & = & 2 \\
10^1 = 10 & \log 10 & = & 1 \\
10^0 = 1 & \log 1 & = & 0 \\
10^{-1} = 0.1 & \log 0.1 & = & -1 \\
10^{-2} = 0.01 & \log 0.01 & = & -2 \\
10^{-3} = 0.0001 & \log 0.001 & = & -3
\end{array}
$$

If a number is not an exact power of 10, its common logarithm can be expressed only approximately. Thus the logarithm of a number between 100 and 1000 is 2 plus a continuing decimal fraction.

The integral part of the logarithm is called the *characteristic;* the decimal part is called the *mantissa.* Only the mantissa of the logarithm is given in a table of logarithms; the characteristic is found by means of the following two rules:

1. The characteristic of the logarithm of a number greater than 1 is one less than the number of places to the left of the decimal point. The characteristic can be determined by counting the number of places the decimal point must be shifted so that one nonzero digit will be to the left of it. For example, the characteristic of log 7335.5 is 3; the characteristic of log 7.3355 is 0; and the characteristic of log 1000.24 is also 3.

2. The characteristic of the logarithm of a decimal number between 0 and 1 is the negative of a number having a numerical value one more than the number of zeros between the decimal point and the first significant figure. Thus the characteristic of log 0.0073355 is - 3; the characteristic of 0.6357335 is - 1.

The mantissa of a logarithm is always positive; the characteristic may be either positive or negative. For example, log 23.60 = 1.3729 and log 0.07335 = - 2 + 0.8653. The last logarithm is better written $\bar{2}.8653$ (using the superscript-bar system for the negative characteristic) with the understanding that only the 2 is negative. Another common method for expressing this logarithm is 8.8653 - 10. This form is obtained by adding 10 to and then subtracting 10 from the logarithm. The latter presentation is more practical.

To determine the *antilogarithm* of a given common logarithm, the number corresponding to its mantissa is read off from the Table of Antilogarithms. The actual antilogarithm is then obtained by inserting a decimal point into this number in a position defined by the characteristic of the logarithm. Obviously, the rules that governed the assignment of the characteristic to the logarithm in the first place operate in reverse when the characteristic is used to locate the decimal point in the antilogarithm, so that

1. If the characteristic is positive, the number of figures that must

precede the decimal point in the antilogarithm is one greater than the characteristic.

2. If the characteristic is negative, then between the decimal point and the number from the Table of Antilogarithms there must be inserted a number of zeros which is one less than the characteristic.

Example: Find the antilog of 4.5145.

Find the mantissa of the given logarithm in the tables. The corresponding number is 327. Since the characteristic of the given logarithm is 4, it is known that 5 digits of the required number precede the decimal point. Therefore, antilog 4.5145 = 32700.

Since the antilog of 0.699 in the Table of Antilogarithms is 5, the
antilog 0.6990 = 5
antilog 1.6990 = 50
antilog $\bar{1}.6990 = 0.5$
antilog $\bar{2}.6990 = 0.05$

Logarithmic notation greatly simplifies and speeds many common arithmetic operations. The quantities in the computation are expressed in logarithmic form and the desired operations are performed (see below); the result is then converted from logarithms to numerical form (i.e., antilogarithms).

Numerical Operation	Corresponding Manipulation with Logarithms
Multiplication	Add logarithms $\log ab = \log a + \log b$
Division	Subtract logarithms $\log \frac{a}{b} = \log a - \log b$
Raise to a power	Multiply logarithm by desired power $\log a^n = n \log a$
Extract a root	Divide logarithm by root $\log \sqrt[n]{a} = \frac{\log a}{n}$

Example: Evaluate $x = \dfrac{38.26 \times 0.1020 \times 0.0864 \times 100}{1.675}$

The several multiplications are accomplished in a single step by addition of the logarithms.

$$\begin{aligned}
\log\ 38.26 &= 1.5828 \\
\log\ 0.1020 &= 9.0086 - 10 \\
\log\ 0.0864 &= 8.9365 - 10 \\
\log 100. &= 2.0000 \\
\hline
&\ \ 21.5279 - 20
\end{aligned}$$

After eliminating 20 from the characteristic and remainder, subtract log 1.675.

$$\begin{array}{lr} & 1.5279 \\ \log 1.675 & -0.2240 \\ \hline \log x & 1.3039 \end{array}$$

Thus $x = 20.14$

There are two types of logarithmic graph paper: (1) *semilog paper* with a vertical logarithmic and a horizontal arithmetic scale and (2) *double-log* or *log-log paper* with both scales logarithmic. Papers of both types are available with 1, 2, 3 or even more cycles on a sheet, each cycle representing a 10-fold range of numbers (i.e., a single $\log_{10}$ unit). The space occupied by a single cycle is fixed, so that the freedom of choice with a logarithmic scale is limited to deciding where the decimal point will be placed. For example, if the "1" at the beginning of the lowest log cycle is to be 0.01, the numeral "1" at the beginning of the next cycle must be 0.1, etc. The numbers plotted are not logarithms; they are the actual values of whatever quantity is to be plotted logarithmically, written for convenience on a scale of logarithms. The same values could be plotted directly on regular arithmetic coordinate paper with a free choice of scale if the numerical values of the numbers are looked up in the Table of Logarithms.

To determine the slope from a logarithmic plot on Figure 1.4B, read off two values of y and divide the difference between these numbers by the difference in the corresponding x values. For example,

$$\frac{1.15 - 0.69}{15 - 5} = \frac{0.46}{10} = 0.046$$

In Figure 1.4C, however, one must read two values of y, look up their log values, and divide the difference between the log values by the corresponding x values (not the log of x because the x axis is not logarithmic). Thus

$$\frac{1.36 - 0.9}{20 - 10} = \frac{0.46}{10} = 0.046$$

To determine the slope in the log-log graph of Figure 1.5, one must divide the difference in the log of two y values by the difference in the log of the two corresponding x values. Thus from this graph,

$$\begin{array}{ll} x \log 2 = 0.301 & \log\ 3 = 0.477 \\ y \log 8 = 0.903 & \log 27 = 1.431 \end{array}$$

$$\frac{1.431 - 0.903}{0.477 - .301} = \frac{0.528}{0.176} = 3$$

Note: Also see logarithmic plots under Section 2.4 on graphs and Section 2.5 on Slope.

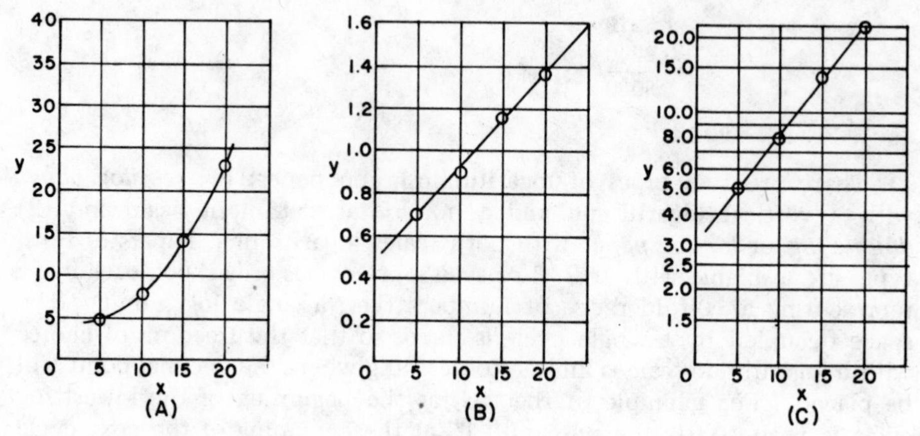

Fig. 1.4. Hypothetical illustration plotting y and $\log y$ as a function of x.

x	5	10	15	20
y	5	8	14	23
$\log y$	0.699	0.90	1.15	1.36

(A) y plotted against x on regular graph paper
(B) $\log y$ plotted against x on regular graph paper
(C) y plotted against x on semilogarithmic paper
In graph A, the experimental measurements are plotted directly.
In graph B, y is looked up in the Tables of Logarithms.
In graph C, y is plotted directly on the logarithmic scale of the ordinates.

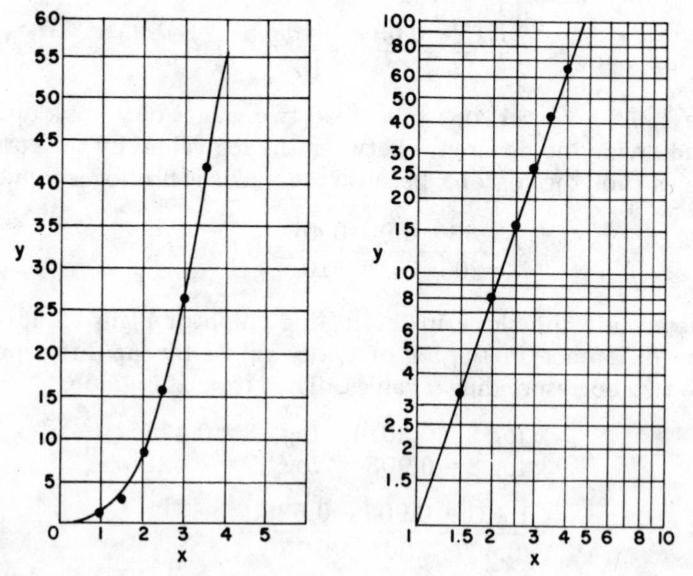

Fig. 1.5. Hypothetical illustration plotting $y = x^3$ on regular and log-log graph paper.

Data:							
x	1	1.5	2	2.5	3	3.5	4
y	1	3.38	8	16.6	27	42.9	64

Examples and Problems

log 10,000 = 4
log 329 = 2.5172
log 59 = 1.7709
log 0.000051 = $\bar{4}.2924$
log 1.6×10^{-6} = $\bar{5}.7959$

log 45 = 1.6532
log 450 = 2.6532
log 4500 = 3.6532
log 4.5 = 0.6532

log 20 = log 2.0×10^{1}
= 0.3010 + (+1)
= 1.3010

log 0.002 = log 2.0×10^{-3}
= log 2.0
= 0.3010 + (−3)
= −2.6990
or = $\bar{3}.3010$
or = 7.3010 − 10

log 0.0093 = log 9.3×10^{-3}
= log 9.3 = 0.9685
= 0.9685 + (−3)
= −2.0315
or = $\bar{3}.9685$
or = 7.9685 − 10

log 0.45 = $\bar{1}.6532$ or 9.6532 − 10
log 0.045 = $\bar{2}.6532$ or 8.6532 − 10
log 0.0045 = $\bar{3}.6532$ or 7.6532 − 10

Antilog of

1.6294 = 42.6
2.6294 = 426.0
3.6294 = 4260.0
0.6294 = 4.26
$\bar{1}.6294$ or 9.6294 − 10 = 0.426
$\bar{2}.6294$ or 8.6294 − 10 = 0.0426
$\bar{3}.6294$ or 7.6294 − 10 = 0.00426

Antilog of

8.5371 − 10 = 0.03444
7.5371 − 10 = 0.003444
6.5371 − 10 = 0.0003444
5.5371 − 10 = 0.00003444

Multiply 43×0.0016

log 43 = 1.6365
+log 0.0016 = +$\bar{3}.2041$
−2.8406

antilog = 0.06928 = 0.069

Multiply 589×0.00056

log 589 = 2.7701
+log 0.00056 = +$\bar{4}.7482$
−1.5183

antilog = 3.298×10^{-1} = 0.33

Multiply 5.37×7.91

log 5.37 = 0.7300
+log 7.91 = +0.8982
1.6283

antilog = 42.48 = 42.5

Multiply 0.09271×0.000517

log 0.09271 = 8.9671 − 10
+log 0.000517 = +6.7135 − 10
15.6806 − 20

or

5.6806 − 10

antilog = 0.0000479

Divide 840 by 0.002

$$\begin{array}{rl} \log 840 &= 2.9243 \\ -\log 0.002 &= -\bar{3}.3010 \\ \hline & 5.6233 \end{array}$$

antilog = 420,100

Divide 0.0061 by 329

$$\begin{array}{rl} \log 0.0061 &= \bar{3}.7853 \\ -\log 329 &= -2.5172 \\ \hline & -5.2681 \end{array}$$

antilog = 1.854×10^{-5}

Divide 23.68 by 3.37

$$\begin{array}{rl} \log 23.68 &= 1.3744 \\ -\log \ 3.37 &= -0.5276 \\ \hline & 0.8468 \end{array}$$

antilog = 7.028

Divide 0.0719 by 0.000531

$$\begin{array}{rl} \log 0.719 &= \bar{2}.8567 \\ -\log 0.000531 &= \bar{4}.7251 \\ \hline & 2.1316 \end{array}$$

antilog = 135.4 = 135

$\log (7.15)^3 = 3 \times \log 7.15$

$$\begin{array}{rl} \log 7.15 &= 0.8543 \\ & \times \quad 3 \\ \hline & 2.5629 \end{array}$$

antilog = 366

$$\sqrt[3]{15.06} = \frac{\log 15.06}{3} = \frac{1.1778}{3} = 0.3926$$

antilog = 2.470

1.11 SLIDE RULE

The slide rule consists of three parts: (1) the upper and lower bars (stator), (2) the slide, and (3) the indicator (cursor) on which there is a hairline which is used as an aid to adjust the slide and read scales. The scales on the slide and bars are so arranged that they work together to solve problems.

On the left end of the slide rule, each scale is named by a letter (A, B, C, D, L, S, T). The slide rule may have many different scales for specialized purposes, but for multiplying and dividing any numbers only two of these scales are really necessary. These are usually labeled C and D. The D scale is on the main body of the rule (stator), while the C scale is the lower scale on the sliding strip. These two scales are exactly alike.

On an ordinary ruler scale (inches or centimeters), the units are equally spaced. But the units on the C and D scales of a slide rule are not equally spaced. The distance between 1 and 2, for example, on these scales is longer than the distance between 5 and 6 because the scales are logarithmic. The numbers on the scales are unequally spaced so that their logarithms (to the base 10) are equally spaced.

Adding the logarithms of any numbers has the same result as multiplying the numbers themselves. Subtracting one logarithm from another produces the same result as the division of the numbers (see logarithm section). The procedure in multiplying and dividing when using log tables is to first look up the logarithms of the numbers, add or subtract as required, and then look up the antilogarithm of the result. On a slide rule, however, this procedure has already been allowed for by the special arrangement of the numbers on the C and D scales.

The position of the decimal point is not given on the slide rule—just the right figures of the answer are given. The decision of where to place the decimal point can be worked out by common sense or by the following method.

When multiplying, add the number of figures in front of the decimal point of each number. If the end of the sliding scale is to the left of the rule, this sum is the number of figures in front of the decimal point in the answer. However, if the sliding scale is to the right, subtract one.

When dividing, subtract the number of figures in the divisor from the number in the dividend (i.e., subtract the number of figures on the bottom from the number of figures on the top). This gives the number of figures before or after the decimal point if the sliding scale remains to the left. If the scale is to the right, add one.

Multiplying:

Slide to the left

$0.08 \times 162.0 = 12.96$

Number of figures = −1 plus +3 = +2
(Note: When the number is a decimal fraction, count the number of zeros before the first figure.)

Slide to the right

$15 \times 0.5 = 7.5$

Number of figures = 2 + 0 = 2 − 1 (because slide to right) = 1

Dividing:

Slide to the left

$$\frac{0.00072}{4.102} = \sim 0.00018$$

Number of figures = −3 minus +1 = −4
Place first figure in answer in fourth place after decimal.

Slide to the right

$$\frac{924.0}{3.2} = 288.75$$

Number of figures = 3 - 1 = 2 + 1 (because slide to right) = 3

Using a slide rule is a fast as well as convenient method for computation once one has become familiar with the scales and the required operations. It should be remembered that ordinary slide rules are nowhere as accurate as 4-figure log tables. Accuracy on the slide rule depends simply on how accurate the user is in judging a fraction of a scale division. Thus the longer the slide rule the more accurate the results because all the scale divisions are larger.

Using The Slide Rule

To multiply

Rule: Set the index of the C scale (the number 1 mark) over one of the factors (i.e., numbers of the problem) on the D scale. Move the hairline over the other factor on the C scale. The answer is read directly below on the D scale.

To multiply 2.7 by 7.21, align the 1 mark of the C scale against the 2.7 mark on the D scale. Move the line marker to 7.21 on the C scale. This lines up with about 19.5 on the lower D scale, which is the answer. (Multiplied by hand the answer is 19.467.)

To divide

Rule: Set the divisor on the C scale opposite the dividend on the D scale. The result is read on the D scale under the index (the 1 mark) of the C scale.

To divide 9.2 by 0.15, line up 9.2 on the D scale with 0.15 on the C scale. Move the line marker to line up with the one mark on the C scale. It is opposite 6.13 on the D scale. The real answer obviously is larger, i.e., 61.3.

Continued Products $(a \times b \times c)$

Rule: Set hairline over a on D scale.
Move index of C under hairline.
Move hairline over b on C scale.
Move index of C under hairline.
Move hairline over c on C scale.
Move index of C under hairline.
Read result under hairline on D scale.
Note: Move hairline and index alternately until all numbers have been multiplied.

$2 \times 3 \times 4$

Set hairline on 2 on D scale.

Move left-hand index on C scale under hairline.
Move hairline over 3 on C scale.
Move right-hand index on C scale under hairline.
Move hairline over 4 on C scale.
Move left-hand index of C scale under hairline.
Answer is under hairline on D scale, i.e., 24.

Combined Multiplication and Division $\frac{a \times b \times c}{d \times e \times f}$

Rule: Set a on D scale and d on C scale.
Move hairline to b on C scale.
Move slide to set e on C scale under hairline.
Move hairline to c on C scale.
Move slide to set f on C scale under hairline.
Read result under index of C on D scale if there are the same number of factors in numerator and denominator.

$$\frac{25 \times 7}{7 \times 5} = \frac{175}{35} = 5$$

Read result under hairline on D scale if there is one more factor in numerator than the denominator.

$$\frac{6 \times 7 \times 2}{5 \times 4} = \frac{84}{20} = 4.2$$

Read the result under the C index on the D scale if there is one more factor in the denominator than the numerator.

$$\frac{17 \times 4}{2 \times 3 \times 4} = \frac{68}{24} = 2.83$$

Proportion $\frac{x}{y} = \frac{a}{b}$

Rule: Set x on C scale opposite y on D scale.
Under a on C scale, read b on D scale.

Reciprocal $\frac{1}{a}$

Rule: Set a on C scale, read reciprocal directly above on C1 scale (which is an inverted scale in that it increases from right to left).

Square Root $\sqrt{x}$

Rule: Set x on A scale, read square root on D scale

or

Set x on B scale, read square root on C scale (on reverse side of slide).

Cube Root $\sqrt[3]{x}$

Set x on K, read cube root on D.

1.12 ALGEBRAIC EQUATIONS

In a linear equation of the first degree the exponent of the unknown x is 1.

$$ax + b = 0$$

The equation $ax + b = 0$ can be solved by first subtracting b from both sides, making $ax = -b$, and then dividing both sides by a, so that x is found to be equal to $\frac{-b}{a}$.

To solve a linear equation (1) multiply both sides by the lowest common denominator if the equation involves fractions, (2) perform indicated operations and collect like terms, (3) add and subtract from both sides such values that the resulting equation will have the unknown on one side and the constant term on the other, and (4) divide by the coefficient of the unknown to obtain the solution.

Example: Solve $3(2a - 1) - 2(a - 6) = a - 10$

(1) no fractions involved

(2) $6a - 3 - 2a + 12 = a - 10$
$4a + 9 = a - 10$

(3) $4a - a = -10 - 9$
$3a = -19$

(equation fits the linear form, i.e., $3a + 19 = 0$)

(4) $a = -\frac{19}{3}$ or $-6\frac{1}{3}$

Example: Solve $\frac{3}{4}(3t - 1) + 2 = \frac{1}{2}(7t - 5)$

(1) $4 \times \frac{3}{4}(3t - 1) + 4 \times 2 = 4 \times \frac{1}{2}(7t - 5)$

$3(3t - 1) + 8 = 2(7t - 5)$

(2) $9t - 3 + 8 = 14t - 10$
$9t + 5 = 14t - 10$

(3) $9t - 14t = -10 - 5$
$-5t = -15$ (linear form $-5t + 15 = 0$)

(4) $t = \frac{-15}{-5} = 3$

The general equation of the second degree is $ax^2 + bx + c = 0$. The unknown x appears in two terms, one with an exponent of 2 and one with an exponent of 1. An equation in which the highest power of the unknown is 2 is also called a *quadratic equation.* It should be noted that the equation may contain both the first and second powers, as well as constant terms, or it may contain the second power alone.

When like terms are collected and combined in a complete quadratic equation, there will, in general, be three terms: one containing the square and one the first power of the unknown and one which does not contain the unknown. If a represents the coefficient of the square term, b that of the first power, and c the third term, any complete quadratic equation can be written in the form

$$ax^2 + bx + c = 0*$$

as stated above where x represents any unknown and the coefficients a, b, and c may be positive, negative, numerical or literal, monomial or polynomial.† In any given equation of the above form, the signs and values of a, b, and c will be known. When these are substituted in the following formula and the indicated calculations carried out, the results will be the solutions of the given equation.

$$x = \frac{-b \pm \sqrt{b^2 - 4ac}}{2a}$$

Quadratic equations can also be solved by factoring and by a method called completing the square. This method consists of transposing all terms to the left side of the equation, dividing through by the coefficient of the square term, then adding to the constant (and to the right side) a number sufficient to make the left side a perfect trinomial square (e.g., to complete the square in $2x^2 + 8x + 2 = 0$, divide both sides by 2, obtaining $x^2 + 4x + 1 = 0$. Now add 3 to both sides, obtaining $x^2 + 4x + 4 = (x + 2)^2 = 3$. The general quadratic formula for the roots of all quadratic equations in terms of a, b, and c is obtained by the methods of completing

*The constant a cannot equal zero, because if $a = 0$ the equation is linear.

†Monomial in x = a single term of the form cx^n where c is a constant and n is zero or a positive number.

Polynomial in x = a function which is the sum of a finite number of the monomial terms.

the square. The formula is used because the method of completing the square is usually not convenient.

Example: Solve $x^2 - 7x + 12 = 0$

$a = 1 \quad b = -7 \quad c = 12$

Substituting these values in the formula

$$x = \frac{-(-7) \pm \sqrt{(-7)^2 - (4 \times 1 \times 12)}}{2 \times 1}$$

$$= \frac{7 \pm \sqrt{49 - 48}}{2}$$

$$= \frac{7 \pm 1}{2}$$

Using the plus sign $x = \frac{8}{2} = 4$

Using the minus sign $x = \frac{6}{2} = 3$

In using the quadratic formula, the equation should be cleared of all fractions, all terms must be transposed to the left side, and like terms must be combined. If the coefficients contain a common factor, the equation should be divided by this factor.

Equations of higher degrees are named for the highest power that occurs in the equation: (Graphic examples of first-, second-, and third-degree equations are shown in Fig. 1.6.)

cubic or third-degree equation

$$x^3 + ax^2 + bx + c = 0$$

quartic or fourth-degree or bi-quadratic equation

$$x^4 + ax^3 + bx^2 + cx + d = 0$$

Equations of the second, third, and fourth degree can be solved by algebraic solutions. The solutions for the cubic and quartic are found by formula as in the case of the quadratic but the formulas are not as simple as the quadratic formula and are beyond the scope of this book.

Examples and Problems

Solve for x:

$\frac{x + 19}{5} = 3 + \frac{x}{4}$ This equals $\frac{x + 19}{5} = \frac{12 + x}{4}$.

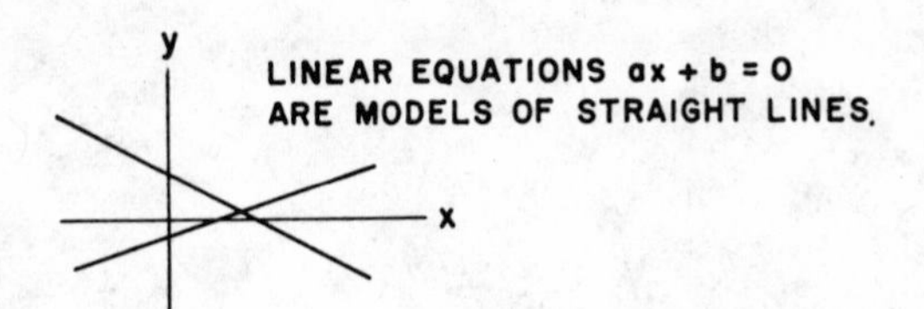

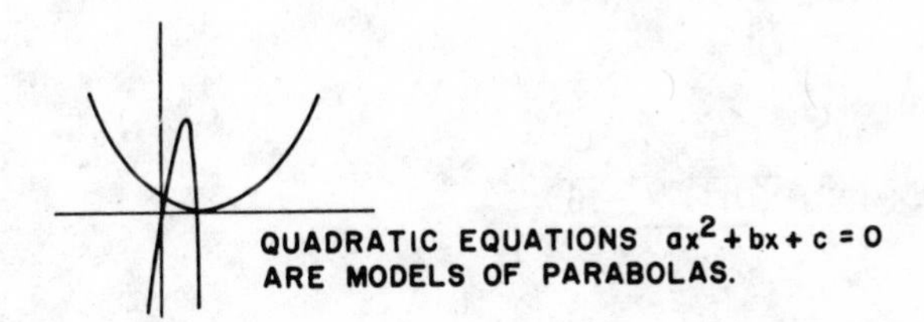

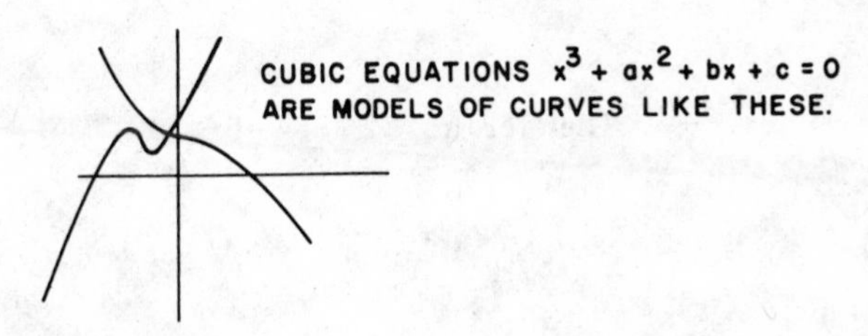

Fig. 1.6. Graphic examples of first-, second-, and third-degree algebraic equations.

- - - -

$$5x + 60 = 4x + 76$$
$$5x - 4x = 76 - 60$$
$$x = 16$$

- - - -

$$\frac{4 - 5x}{6} - \frac{1 - 2x}{3} = \frac{13}{42}$$
$$\frac{4 - 5x - 2 + 4x}{6} = \frac{13}{42}$$
$$42\,(4 - 5x - 2 + 4x) = 6(13)$$
$$168 - 210x - 84 + 168x = 78$$
$$-42x = 78 + 84 - 168$$
$$-42x = -6$$
$$x = \frac{-6}{-42} = \frac{6}{42} = \frac{1}{7}$$

$2 = y + (x - 1)z$

$$2 = y + xz - z$$
$$2 + z - y = xz$$
$$\frac{2 + z - y}{z} = x$$

$x^2 - 3x - 10 = 0$

$$x = \frac{-(-3) \pm \sqrt{(-3)^2 - 4(1 \times -10)}}{2 \times 1}$$

$$= \frac{+3 \pm \sqrt{9 - (-40)}}{2 \times 1}$$

$$= \frac{+3 \pm \sqrt{49}}{2}$$

$$= \frac{+3 \pm 7}{2} = \frac{-4}{2} \text{ or } \frac{10}{2} \qquad \text{Therefore } x = -2 \text{ or } +5$$

$16x^2 - 24x + 9 = 0$

$$x = \frac{-(-24) \pm \sqrt{(-24)^2 - 4(16 \times 9)}}{2 \times 16}$$

$$= \frac{+24 \pm \sqrt{576 - 576}}{32}$$

$$= \frac{+24 \pm 0}{32} = \frac{3}{4}$$

Therefore, $x = \frac{3}{4}$

1.13 SOLVING SIMULTANEOUS EQUATIONS

Simultaneous equations are two or more equations that must be solved together so that it will be known what values of the unknowns will solve all the equations at the same time. In solving two equations with two unknowns, one must combine the equations into one which contains one unknown.

Rules for solving two equations:

1. Clear both equations of fractions and transpose so that all terms containing unknowns are on the left side and all other terms are on the right.

2. If neither unknown has the same coefficient in both equations, multiply one or both equations so that the coefficient of one unknown is the same in both equations.

3. Add or subtract the equations so that one of the unknowns is eliminated.

4. Solve the resultant equation for the unknown which it contains.

5. Place this answer in one of the original equations and solve this equation for the other unknown.

Examples and Problems

Twice a certain number plus five times a certain number is equal to 19. Twice the first number and three times the second number is equal to 13. What are the unknown numbers?

x = 1st number y = 2nd number

Therefore $2x + 5y = 19$

$2x + 3y = 13$

Subtract 2nd equation from 1st

$2y = 6$

$y = 3$

Substitute $y = 3$ in 1st equation

$2x + 5(3) = 19$

$2x + 15 = 19$

$2x = 4$

$x = 2$

Test answers in 2nd equation

$2(2) + 3(3) = 13$

$4 + 9 = 13$

Solve $x - 2y = 13$ equation 1

$4x + 3y = 8$ equation 2

Multiply 1st equation by 3, 2nd equation by 2

$3x - 6y = 39$

$8x + 6y = 16$

Add to eliminate the unknown y

$$11x = 55$$

$$x = \frac{55}{11} = 5$$

Therefore $x = 5$. Substitute this value in equation 1, and solve for y.

$$5 - 2y = 13$$
$$-2y = 8$$
$$y = -4$$

Test answers in equation 2.

$$4(5) + 3(-4) = 8$$
$$20 + \quad -12 = 8$$

In the case of three equations with three unknowns:

$$x + 4y + 3z = 1 \quad \ldots\ldots. 1$$
$$2x + 5y + 4z = 4 \quad \ldots\ldots. 2$$
$$x - 3y - 2z = 5 \quad \ldots\ldots. 3$$

Divide each term of each equation by the coefficient of x.

$$x + 4y + 3z = 1 \quad \ldots\ldots. 1$$

$$x + \frac{5y}{2} + 2z = 2 \quad \ldots\ldots. 2$$

$$x - 3y - 2z = 5 \quad \ldots\ldots. 3$$

Subtract (or add depending on signs) equation number 3 from equations 1 and 2.

$$7y + 5z = -4 \quad \ldots\ldots. 1$$

$$\frac{11}{2}y + 4z = -3 \quad \ldots\ldots. 2$$ Note: x has now been eliminated.

Divide each term of each equation by the coefficient of y.

$$y + \frac{5}{7}z = -\frac{4}{7} \quad \ldots\ldots. 1$$

$$y + \frac{8}{11}z = -\frac{6}{11} \quad \ldots\ldots. 2$$

Subtract (or add depending on signs) equation 2 from equation 1.

$$\left(\frac{5}{7} - \frac{8}{11}\right)z = \left(-\frac{4}{7}\right) + \frac{6}{11}$$ Note: y has now been eliminated.

Solve for z.

$$\left(\frac{55 - 56}{77}\right)z = \frac{-44 + 42}{77}; \quad \frac{-1}{77}z = \frac{-2}{77}; \quad -z = -2; \quad z = 2$$

Substitute this value into either of the equations with two unknowns.

$$7y + 5z = -4$$
$$7y + 5(2) = -4$$
$$7y = -14$$
$$y = -2$$

Substitute the values for y and z into one of the original equations.

$$x + 4y + 3z = 1$$
$$x + 4(-2) + 3(2) = 1$$
$$x - 8 + 6 = 1$$
$$x - 2 = 1$$
$$x = 3$$

Check the solutions for x, y, and z by substituting all three in one of the original equations that wasn't used in the previous step.

$$x - 3y - 2z = 5$$
$$(3) - 3(-2) - 2(2) = 5$$
$$3 + 6 - 4 = 5$$
$$9 - 4 = 5$$
$$5 = 5$$

The above procedure can be used for N equations and N unknowns.

Substitution Method

The object of this method is to first solve for one of the unknowns in terms of the other unknown in the simplest equation given. This answer is then substituted into the other equation to obtain the value. For example:

$$6x - 4y = 10$$
$$-x + y = 2$$

Solve for y in terms of x in the second equation.

$$y = 2 + x$$

Now substitute this value for y in the first equation and obtain the value of x.

$$6x - 4(2 + x) = 10$$
$$6x - 8 - 4x = 10$$
$$2x = 18$$
$$x = 9$$

Now substitute the value for x in the second equation to obtain the value for y.

$$-9 + y = 2$$
$$y = 11$$

The solutions of the two equations are thus $x = 9$, $y = 11$.

Examples and Problems

Solve the following pair of equations for x and y:

$$\begin{aligned} 3x - 4y &= 13 \qquad \text{answer: } x = 11 \\ 5x - 7y &= 20 \qquad \qquad \quad\; y = 5 \end{aligned}$$

$$\begin{aligned} x - y &= 6 \qquad \text{answer: } x = 4 \\ x + 2y &= 0 \qquad \qquad \quad\; y = -2 \end{aligned}$$

$$\begin{aligned} 2x + 9y &= 11 \qquad \text{answer: } x = 4 \\ 5x - 6y &= 18 \qquad \qquad \quad\; y = \frac{1}{3} \end{aligned}$$

Some examples by method of addition or subtraction:

$$\begin{aligned} x + y &= 9 \\ x - y &= 1 \end{aligned}$$

$$\begin{aligned} 2x &= 10 \\ x &= 5 \end{aligned}$$

$$\begin{aligned} 5 + y &= 9 \\ y &= 9 - 5 \\ y &= 4 \end{aligned}$$

Therefore $x = 5$, $y = 4$.

$$\begin{aligned} 3x + 2y &= 17 \\ x + 4y &= 19 \end{aligned}$$

$$\begin{aligned} 6x + 4y &= 34 \\ x + 4y &= 19 \end{aligned}$$

$$\begin{aligned} 5x &= 15 \\ x &= 3 \end{aligned}$$

$$\begin{aligned} 3 + 4y &= 19 \\ 4y &= 16 \\ y &= 4 \end{aligned}$$

Therefore $x = 3$, $y = 4$.

$$2\frac{1}{4}x + 3\frac{3}{4}y = 39$$

$$6\frac{3}{4}x - 2\frac{1}{4}y = 9$$

$$\left.\begin{aligned} \frac{9x}{4} + \frac{15y}{4} &= 39 \\ \frac{27x}{4} - \frac{9y}{4} &= 9 \end{aligned}\right\} \text{Multiply by 4}$$

$$9x + 15y = 156$$
$$27x - 9y = 36$$
$$27x + 45y = 468$$
$$27x - 9y = 36$$
$$54y = 432$$
$$y = 8$$

$$9x + 15(8) = 156$$
$$9x + 120 = 156$$
$$9x = 36$$
$$x = 4$$

Therefore $x = 4$, $y = 8$.

Some examples eliminating by substitution:

$$x + 2y = 16$$
$$2x + 3y = 26$$

$$x = 16 - 2y$$

$$2(16 - 2y) + 3y = 26$$
$$32 - 4y + 3y = 26$$
$$-y = -32 + 26$$
$$-y = -6$$
$$y = 6$$

$$x + 2(6) = 16$$
$$x + 12 = 16$$
$$x = 16 - 12 = 4$$

Therefore $x = 4$, $y = 6$.

$$3x + 2y = 29$$
$$2x + 4y = 38$$

$$3x = 29 - 2y$$

$$x = \frac{29 - 2y}{3}$$

$$2\left(\frac{29 - 2y}{3}\right) + 4y = 38$$

$$\frac{58 - 4y}{3} + 4y = 38$$

$$58 - 4y + 12y = 114$$
$$8y = 56$$
$$y = 7$$

$$3x + 2(7) = 29$$
$$3x + 14 = 29$$

$$3x = 15$$
$$x = 5$$

Therefore $x = 5$, $y = 7$.

$$5x - 2y = 28$$
$$13x - 3y = 108$$

$$5x = 28 + 2y$$

$$x = \frac{28 + 2y}{5}$$

$$13\left(\frac{28 + 2y}{5}\right) - 3y = 108$$

$$\frac{364 + 26y}{5} - 3y = 108$$

$$364 + 26y - 15y = 540$$
$$11y = 176$$
$$y = 16$$

$$13x - 3(16) = 108$$
$$13x - 48 = 108$$
$$13x = 156$$
$$x = 12$$

Therefore $x = 12$, $y = 16$.

1.14 TRIGONOMETRY

Trigonometry was first developed as a tool for astronomers. The name trigonometry which means *triangle measurement* indicates the first use made of this subject.

Trigonometry is the study of certain ratios called trigonometric functions. The applications of these functions serve the purpose of solving problems involving triangles as well as developing notions needed in other branches of mathematics, physics, mechanics, and engineering.

There are six trigonometric functions of the angle θ upon which the whole subject of trigonometry is based. The definitions are based on the right triangle and are valid only for acute angles between 0° and 90°. When the angle θ is placed in standard position* (see Fig. 1.7) at the center of a unit circle on rectangular coordinates,† the trigonometric functions of θ are defined as follows:

---

*Standard position is actually an unusual and special situation as illustrated in Figure 1.7.

†In measuring the angle from one line to another, it is essential to define which direction of rotation shall be considered as positive. The positive direction of rotation is from the positive x axis to the positive y axis with the vertex of the angle at the origin, so that the angle from the positive x axis to the positive y axis is +90° and not −90°.

$$\text{sine } \theta \text{ (written } \sin\theta) = \frac{\text{ordinate}}{\text{hypotenuse}} \text{ or } \frac{\text{side opposite } \theta}{\text{hypotenuse}}$$

$$\text{cosine } \theta \text{ (written } \cos\theta) = \frac{\text{abscissa}}{\text{hypotenuse}} \text{ or } \frac{\text{side adjacent } \theta}{\text{hypotenuse}}$$

$$\text{tangent } \theta \text{ (written } \tan\theta) = \frac{\text{ordinate}}{\text{abscissa}} \text{ or } \frac{\text{side opposite } \theta}{\text{side adjacent } \theta}$$

$$\text{cotangent } \theta \text{ (written } \cot\theta) = \frac{\text{abscissa}}{\text{ordinate}} \text{ or } \frac{\text{side adjacent } \theta}{\text{side opposite } \theta}$$

$$\text{secant } \theta \text{ (written } \sec\theta) = \frac{\text{hypotenuse}}{\text{abscissa}} \text{ or } \frac{\text{hypotenuse}}{\text{side adjacent } \theta}$$

$$\text{cosecant } \theta \text{ (written } \csc\theta) = \frac{\text{hypotenuse}}{\text{ordinate}} \text{ or } \frac{\text{hypotenuse}}{\text{side opposite } \theta}$$

The ratios depend solely upon the size of the angle θ for their values and are therefore functions of θ. These six trigonometric functions are not independent but are connected by numerous relations. The following are the *reciprocal functions* of sine and cosecant, cosine and secant, and tangent and cotangent.

$$\sin\theta = \frac{1}{\csc\theta}, \quad \text{or } \csc\theta = \frac{1}{\sin\theta}, \quad \text{or } \sin\theta \csc\theta = 1$$

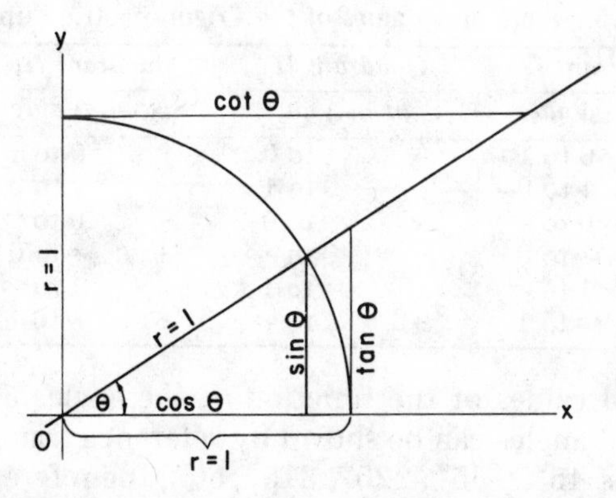

Fig. 1.7. Representation of trigonometric functions on the unit circle (circle with radius = 1). Note that the functions listed below have a denominator equal to 1 because the radius of the unit circle is equal to 1

$$\sin = \frac{\text{side opp. } \theta}{\text{hypotenuse}} \quad \cos = \frac{\text{side adj.}}{\text{hypotenuse}} \quad \tan = \frac{\text{side opp.}}{\text{side adj.}} \quad \cot = \frac{\text{side adj.}}{\text{side opp.}}$$

$$\cos\theta = \frac{1}{\sec\theta}, \quad \text{or } \sec\theta = \frac{1}{\cos\theta}, \quad \text{or } \cos\theta\ \sec\theta = 1$$

$$\tan\theta = \frac{1}{\cot\theta}, \quad \text{or } \cot\theta = \frac{1}{\tan\theta}, \quad \text{or } \tan\theta\ \cot\theta = 1$$

The signs of the trigonometric functions of an angle in any quadrant depend only upon the signs of the abscissa and ordinate of any point on its terminal side. (See Fig. 1.8 and Table 1.5.)

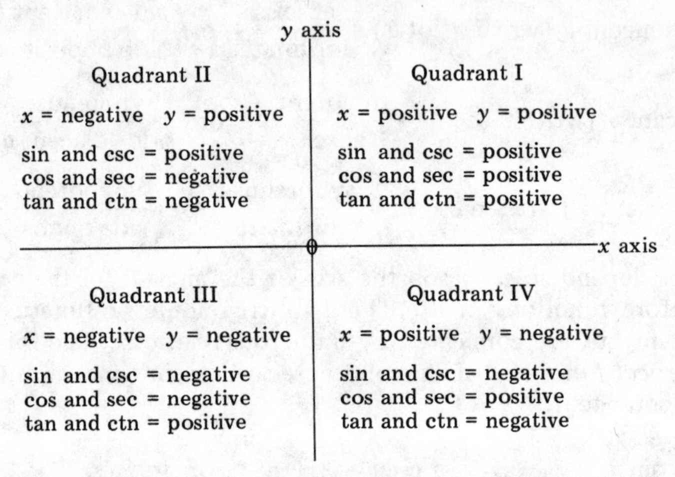

Fig. 1.8. Signs of the functions in the quadrants.

Table 1.5. Signs and Values of the Trigonometric Functions

Function	*Quadrant I*		*Quadrant II*		*Quadrant III*		*Quadrant IV*	
	Sign	*Value*	*Sign*	*Value*	*Sign*	*Value*	*Sign*	*Value*
sin	+	0 to 1	+	1 to 0	−	0 to 1	−	1 to 0
cos	+	1 to 0	−	0 to 1	−	1 to 0	+	0 to 1
tan	+	0 to ∞	−	∞ to 0	+	0 to ∞	−	∞ to 0
cot	+	∞ to 0	−	0 to ∞	+	∞ to 0	−	0 to ∞
sec	+	1 to ∞	−	∞ to 1	−	1 to ∞	+	∞ to 1
csc	+	∞ to 1	+	1 to ∞	−	∞ to 1	−	1 to ∞

The numerical values of the function of the angles 30°, 45°, and 60° and certain related angles can be shown by reference triangles.

For the angles 45°, 135°, 225°, 315°, etc., the reference triangle *OBC* (Fig. 1.9) is an isosceles (equal sides) right-angled triangle. So, if *OB* and *BC*, the equal sides of the triangle, have the numerical values of their lengths made equal to one, the hypotenuse *OC* will be the length $\sqrt{2}$.

For the angles 60°, 120°, 240°, 300°, etc., the reference triangle *ODE* (Fig. 1.10) is a 30°-60° right triangle with the vertex of the 60° angle at

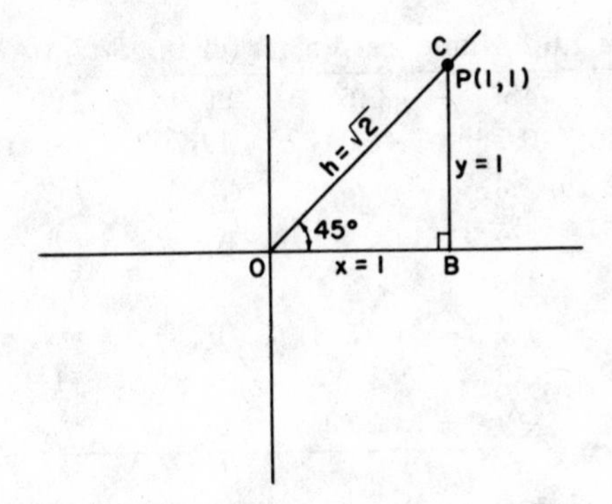

Fig. 1.9. Reference Triangle *OBC*.

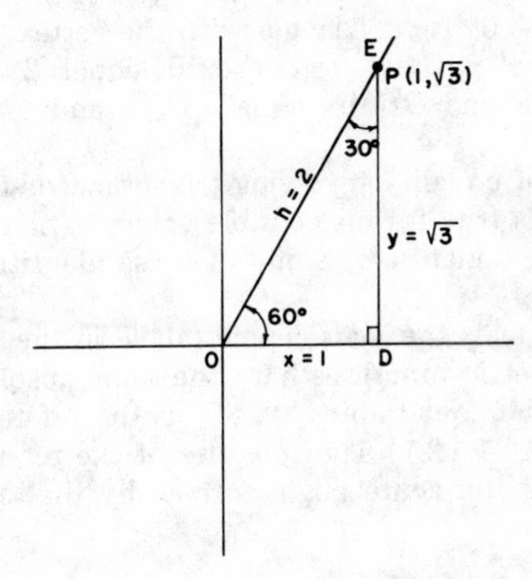

Fig. 1.10. Reference Triangle *ODE*.

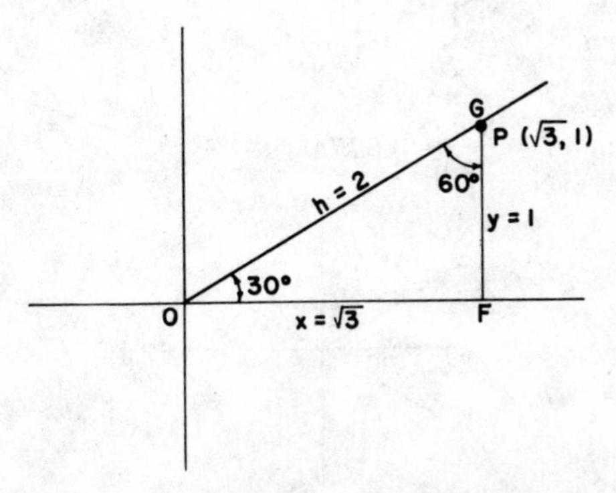

Fig. 1.11. Reference Triangle *OFG*.

Table 1.6. Numerical Values for Important Angles

Degrees	0°	30°	45°	60°	90°	180°	270°	360°
Radians	0	0.5236	0.7854	1.0472	1.5708	3.1416	4.7124	6.2832
sin	0	1/2	$\sqrt{2}/2$	$\sqrt{3}/2$	1	0	-1	0
cos	1	$\sqrt{3}/2$	$\sqrt{2}/2$	1/2	0	-1	0	1
tan	0	$\sqrt{3}/3$	1	$\sqrt{3}$	∞	0	∞	0
cot	∞	$\sqrt{3}$	1	$\sqrt{3}/3$	0	∞	0	∞
sec	1	$2\sqrt{3}/3$	$\sqrt{2}$	2	∞	-1	∞	1
csc	∞	2	$\sqrt{2}$	$2\sqrt{3}/3$	1	∞	-1	∞

the origin. If the point P on the hypotenuse side equals 2, the numerical value of the lengths OD and OE are equal to 1 and $\sqrt{3}$.

For the angles 30°, 150°, 210°, 330°, etc., the reference triangle OFG (Fig. 1.11) is a 30°-60° right triangle with the vertex of the 30° angle at the origin. If point P on the hypotenuse side equals 2 the numerical value of the length of OF and OG are equal to $\sqrt{3}$ and 1, respectively. (See Table 1.6.)

An expression of equality involving trigonometric functions of one or more angles which is true for all possible values of the unknown angles is called a trigonometric identity. Some of these identities are listed in the Appendix.

A reference angle is the acute angle (angle in the first quadrant) for which the trigonometric functions have the same absolute values as for a given angle in another quadrant, e.g., 30° is the reference angle of 150° and 210°. (See Fig. 1.12.) The measure of the reference angle can be found by measuring the acute angle formed by the terminal side of the angle 0 and the x-axis.

Example: Find the sin of 210° and the tan of 210°. (See Table 1.7.)

$$\text{reference angle} = 210° - 180° = 30°$$

$$\sin 210° = -\sin 30° = -\frac{1}{2} = -0.5$$

$$\tan 210° = \tan 30° = \frac{\sqrt{3}}{3} = 0.5774$$

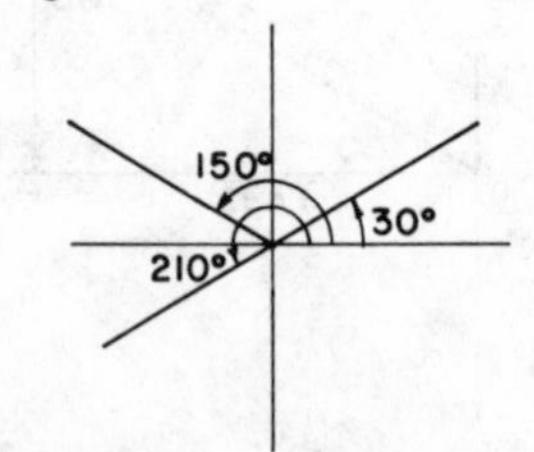

Fig. 1.12. Reference angles.

Table 1.7. Value of Trigonometric Functions in the Four Quadrants

Value of	When θ Terminates in Quadrant I	II	III	IV
$\sin\theta$	$\sin\theta'$	$\sin\theta'$	$-\sin\theta'$	$-\sin\theta'$
$\cos\theta$	$\cos\theta'$	$-\cos\theta'$	$-\cos\theta'$	$\cos\theta'$
$\tan\theta$	$\tan\theta'$	$-\tan\theta'$	$\tan\theta'$	$-\tan\theta'$
$\cot\theta$	$\cot\theta'$	$-\cot\theta'$	$\cot\theta'$	$-\cot\theta'$
$\sec\theta$	$\sec\theta'$	$-\sec\theta'$	$-\sec\theta'$	$\sec\theta'$
$\csc\theta$	$\csc\theta'$	$\csc\theta'$	$-\csc\theta'$	$-\csc\theta'$

θ = angle in standard position
θ' = reference angle (an acute first quadrant angle)
(Also see Fig. 1.8.)

Law of Cosines

In a triangle the square of the length of any side equals the sum of the squares of the lengths of the other two sides minus twice the product of these two sides and the cosine of their included angle. (See Fig. 1.13.)

$$c^2 = a^2 + b^2 - 2ab\cos C \qquad \cos C = \frac{a^2 + b^2 - c^2}{2ab}$$

$$b^2 = a^2 + c^2 - 2ac\cos B \qquad \cos B = \frac{a^2 + c^2 - b^2}{2ab}$$

$$a^2 = b^2 + c^2 - 2bc\cos A \qquad \cos A = \frac{b^2 + c^2 - a^2}{2ab}$$

Law of Sines

In any triangle the sides are proportional to the sines of the opposite angles. (See Fig. 1.13.)

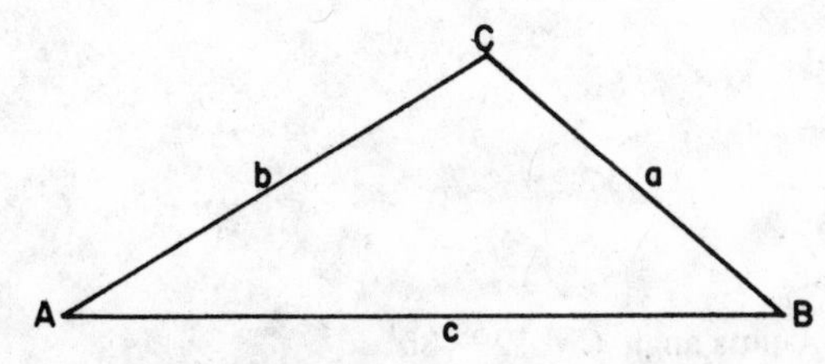

Fig. 1.13. Reference triangle. Sides = a, b, and c. Angles = A, B, and C.

$$\frac{a}{\sin A} = \frac{b}{\sin B} = \frac{c}{\sin C}$$

$$a = \frac{c \sin A}{\sin C} \quad \text{or} \quad \frac{b \sin A}{\sin B}$$

$$b = \frac{c \sin B}{\sin C} \quad \text{or} \quad \frac{a \sin B}{\sin A}$$

$$c = \frac{b \sin C}{\sin B} \quad \text{or} \quad \frac{a \sin C}{\sin A}$$

Area of a Triangle

The area of a triangle equals one-half the product of the lengths of two sides times the sine of their included angle.

$$A = \frac{1}{2}\, ab \sin C$$

$$A = \frac{1}{2}\, ac \sin B$$

$$A = \frac{1}{2}\, bc \sin A$$

The area of a triangle can also be found if one side and two angles are known.

$$A = \frac{1}{2}\, a^2\, \frac{\sin B \sin C}{\sin A}$$

$$A = \frac{1}{2}\, c^2\, \frac{\sin A \sin B}{\sin C}$$

$$A = \frac{1}{2}\, b^2\, \frac{\sin A \sin C}{\sin B}$$

Example: Find sides a and c and angle B in an oblique triangle, given side $b = 35$, angle $A = 98°\ 40'$, and angle $C = 30°\ 55'$.

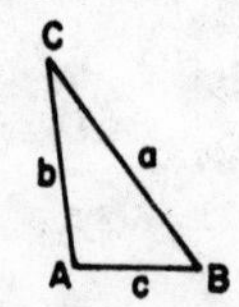

Calculations:

Since angle A plus angle $C = 129°\ 35'$
angle $B = 180° - 129°\ 35' = 50°\ 25'$

$$\text{Side } a = \frac{b \sin A}{\sin B} \qquad \sin A = 0.98858 \qquad \sin B = 0.77070$$

$$a = \frac{35\ (0.98858)}{0.7707} = \frac{34.5903}{0.7707} = 44.894$$

$$\text{Side } c = \frac{b \sin C}{\sin B} \qquad \sin C = 0.51379 \qquad \sin B = 0.77070$$

$$c = \frac{35\ (0.5179)}{0.7707} = \frac{17.98265}{0.7707} = 23.333$$

Therefore $B = 50^\circ\ 25'$
$a = 44.894$
$c = 23.333$

1.15 CALCULUS

Calculus comes from a word meaning simply *to calculate*, but in mathematics it includes only a specialized type of calculation. Calculus deals with the ways things change in relation to one another. *Differential calculus* involves finding the rate at which a variable quantity is changing; integral calculus involves finding a function when its rate of change is given.

The best way of showing how things change is to present them on a graph. The value of y for any value of x can be found simply by looking at the graph. The slope can also be found. On a straight line, the slope is the same anywhere along it. When the graph is a curved line, the slope varies from point to point along it. But, at any point, the slope can be found.

In calculus, a curved line is imagined to be made up of an infinitely large number of infinitely small straight lines joined end to end. The slope of any point then becomes the slope of the infinitely small straight line at that point. Over the small portion of the graph covered by the straight line, both x and y change by tiny amounts. The symbols used for the small changes in the values of x and y are δ or Δ meaning delta-x and delta-y (Δy). Delta-x (Δx) is the horizontal change and delta-y is the vertical change. (It should be noted that delta-x does not mean delta multiplied by x but should be regarded as a single symbol meaning the increment in x.)

The slope is delta-y divided by delta-x, i.e., $\frac{\Delta y}{\Delta x}$. When Δx and Δy become so small that they are almost equal to zero, they are called dx and dy. The ratio of these two quantities may not be small, however, may have a finite value. The gradient of the sloping graph is $\frac{dy}{dx}$; $\frac{dy}{dx}$ is the rate of change of y with respect to x and it is called the *derivative*. The process of finding the derivative $\frac{dy}{dx}$* is called *differentiation* with respect to x and is the fundamental operation of differential calculus. There is a simple rule for differentiating mathematical expressions like $y = x^3$, $y = 2x^2$, and $y = 5x^3$. The rule is multiply the number in front of the x by the power of x, then reduce the power of x by one.

$$\frac{dy}{dx} = nx^{n-1} \quad \text{if} \quad y = x^n$$

or

$$\frac{dy}{dx} = cnx^{n-1} \quad \text{if} \quad y = cx^n$$

where c is a constant and
n is any integer.

Thus the derivatives of the mathematical expressions stated previously would be $3x^2$, $4x$, and $15x^2$, respectively. The result is written as a ratio $\frac{dy}{dx}$ which should be looked upon as a single symbol. (See Fig. 1.14.) Any two varying quantities can be represented by y and x, for example, speed and time, or pressure and temperature. The derivative of the elevation of a roadway is equal to the slope or incline of the road. Differentiating

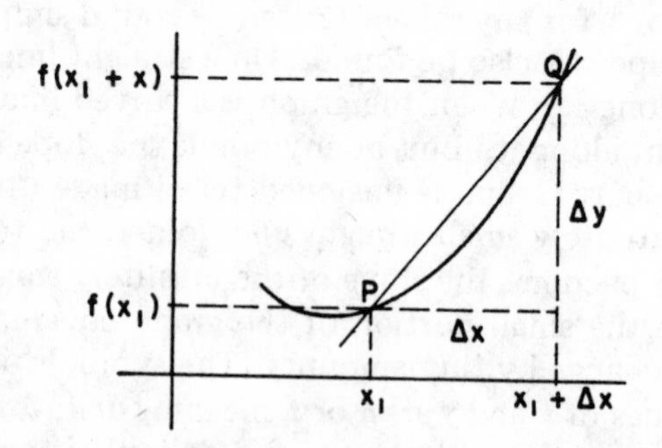

Fig. 1.14. Derivative of a function. The ratio $\frac{\Delta y}{\Delta x}$ is the slope of the line PQ.

*In addition to $\frac{dy}{dx}$ (read dy over dx), various notations are used to denote the derivative of a function f of x. The ones most commonly used are $f'(x)$ (read f prime of x) which is equivalent to y' (read y prime), and $D_x y$ (read dx of y). All these symbols have the same meaning: the instantaneous rate of change of the function f as x changes.

shows how quickly one quantity changes when the other quantity changes. Differentiating y with respect to x means finding how quickly y changes when x changes. The differentiation rule can be proven by drawing graphs of y and x and finding the slope which is the same as the derivative $\frac{dy}{dx}\left(\text{i.e.,} \frac{y_2 - y_1}{x_2 - x_1}\right)$.

Some rules for derivatives where c is a constant and u and v are functions of x:

$$\frac{dc}{dx} = 0$$

$$\frac{d\,(cu)}{dx} = c\,\frac{du}{dx}$$

$$\frac{d\,(u + v)}{dx} = \frac{du}{dx} + \frac{dv}{dx}$$

$$\frac{d\,(uv)}{dx} = u\,\frac{dv}{dx} + v\,\frac{du}{dx}$$

$$\frac{du^n}{dx} = nu^{n-1}\,\frac{du}{dx}$$

$$\frac{dcx^n}{dx} = cnx^{n-1}$$

$$\frac{d\left(\frac{u}{v}\right)}{dx} = \frac{v\,\frac{du}{dx} - u\,\frac{dv}{dx}}{v^2}$$

Examples of some easy derivatives:

when

$y = c$ $\qquad \frac{dy}{dx} = 0$

$y = x^2$ $\qquad \frac{dy}{dx} = 2x$

$y = \frac{1}{x}$ $\qquad \frac{dy}{dx} = \frac{-1}{x^2}$

$y = \sin x$ $\qquad \frac{dy}{dx} = \cos x$

$$y = \cos x \qquad \frac{dy}{dx} = -\sin x$$

$$y = \ln x \qquad \frac{dy}{dx} = \frac{1}{x}$$

$$y = \tan x \qquad \frac{dy}{dx} = \sec^2 x$$

$$y = \log_a x \qquad \frac{dy}{dx} = \frac{1}{x} \log_a e$$

Examples and Problems

Find $\frac{dy}{dx}$ if $y = x^3 + 7x^2 - 5x + 4$

(Find the derivatives of the separate terms and add the results.)

Thus $$\frac{dy}{dx} = \frac{d(x^3)}{dx} + \frac{d(7x^2)}{dx} + \frac{d(-5x)}{dx} + \frac{d(4)}{dx}$$

$$= 3x^2 + 14x - 5x^0 + 0$$
$$= 3x^2 + 14x - 5$$

What is the slope of the curve $y = x^3 - 6x + 2$ where it crosses the y axis?

The slope at point (x, y) is equal to $\frac{dy}{dx} = 3x^2 - 6$.

The curve crosses the y axis when $x = 0$.
Thus the slope at that point is $m = (3x^2 - 6)_{x=0} = -6$

A body moves in a straight line according to the law of motion. If $s = t^3 - 4t^2 - 3t$, find its acceleration at each instant when the velocity is zero.

Note: If $s = f(t)$ gives the position of a moving body at time t, then the first derivative ds/dt gives the velocity and the second derivative d^2s/dt^2 gives the acceleration of the body at time t.

$$\text{velocity} = v = 3t^2 - 8t - 3 = \frac{ds}{dt}$$

$$\text{acceleration} = a = 6t - 8 = \frac{dv}{dt}$$

The velocity equals zero when $3t^2 - 8t - 3 = (3t + 1)(t - 3) = 0$

that is, when $t = -\frac{1}{3}$ or $t = +3$.

The corresponding values of acceleration are

when $t = -\frac{1}{3}$, $a = -10$

when $t = +3$, $a = 10$

A *differential* (dx or dy) is one part of the derivative $\frac{dy}{dx}$ taken separately. The relation between the differential and the derivative can be approximated as follows:

$$dy = \frac{dy}{dx} \times dx$$

According to this equation, the differential of y is equal to the derivative of y with respect to x times the differential of x. The symbol $\frac{dy}{dx}$ has the significance of any fraction such as $\frac{1}{2}$. It can also be written as the identical equation $\frac{dy}{dx} = \frac{dy}{dx}$, and can be transposed by the ordinary rules of algebra to $dy = \frac{dy}{dx} \times dx$. Thus the differentials dy and dx and the derivative $\frac{dy}{dx}$ can be treated in calculus as if they were ordinary algebraic quantities and can be subjected to ordinary mathematical operations.

$$d\,(u + v) = du + dv$$
$$d\,(cu) = c\,du$$

A *differential equation* is one in which an unknown function occurs along with its derivatives of various orders. It is an equation for finding the required unknown function (i.e., a law expressing the dependence of certain variables on others). If the unknown function depends on a single variable, the differential equation is called an *ordinary differential equation*. If the unknown function depends on several variables and the equation contains derivatives with respect to some or all of these variables, the differential equation is called a *partial differential equation*.

Every differential equation has in general not one but infinitely many solutions; every differential equation defines a whole class of functions that satisfy it. The basic problem of the theory of differential equations is to investigate the functions that satisfy the differential equation to enable one to form a sufficiently broad understanding of the properties of all the functions satisfying the equation.

The physical laws governing a phenomenon are often written in the form of differential equations. For example, the formula force = mass times acceleration ($F = ma$) can be written by the equation

$$F\left(x, \frac{dx}{dt,}\ t\right) = m\,\frac{d^2 x}{dt^2}$$

where force is a function of the position of particle x, the velocity of motion $\frac{dx}{dt}$ (a first derivative), and of time t, and where acceleration = $\frac{d^2 x}{dt^2}$ (a second derivative) and m = mass of the particle.

Integration is the reverse of differentiation just as extracting the square root is the reverse of squaring; that is, if one starts with a suitable function and finds its integral and differentiates the integral, the result is the original function. Only those expressions which have been found previously by differentiation can be integrated.

An integral is best thought of as a sum of products of the dependent variable y times small increments in the independent variable dx. One may imagine a curve of y plotted against x. The total area under that curve can be obtained by first dividing the area into a series of parallel, thin strips of width dx so narrow that y is practically constant in the interval dx, then multiplying the length of each strip y by its width dx, and finally summing over all the strips. The area under the curve between any two points on the x axis, x_1 and x_2, is said to be equal to the integral of y with respect to x, from x_1 to x_2 written $\int_{x_1}^{x_2} y\,dx$. In words, the meaning of the symbol $\int$ is: divide the area x into many small bits each of area delta-x, multiply each bit by the corresponding value of y and add these products.

The tall S-shaped part of the symbol is called an *integral sign* and is nothing but an elongated S which stands for summation. The symbol dx is part of the symbol of integration. Between these two parts is written the function to be integrated; it is called the *integrand*. The integrand may be written to show the variable of integration explicitly, as in $y(x)$, which explains that y is a function of x. This symbol should agree with that following the d. The values x_1 and x_2 which appear at the bottom and top of the integral sign $\int$ are called the *lower and upper limits of integration*. These symbols or numbers (x_1 and x_2) designate between which values of the variable of integration one is to total the parts of the integrand. (See Fig. 1.15.)

To designate the integral of a function f of x, mathematics uses the following symbols for two slightly different meanings of integral:

$$\int_a^b f\,dx$$ (read "integral from a to b of f with respect to x.")

$$\int f\,dx$$ (read "integral $f\,dx$" or integral of f with respect to x.")

The first symbol is called the *definite integral* and refers to the area under the curve f, from $x = a$ to $x = b$. (See Fig. 1.15.) The second symbol is called the *indefinite integral* and stands for any function whose derivative with respect to x is f. All such functions are very much alike because they only differ from each other by a constant. For example, the indefinite integrals of the function $\frac{1}{3}x$ include $\frac{1}{6}x^2 + 13$ or $\frac{1}{6}x^2 - 2.36$ or $\frac{1}{6}x^2 + 365.28$, etc., because for every one of these functions, the derivative is exactly the same, namely $\frac{1}{3}x$.

Whereas the object of differential calculus was to find the relation between dy and dx when the relationship between y and x was known, the object of integral calculus is to find the relation between y and x when the relationship between dy and dx is known. Integration makes it possible then to determine the general relation between two variables from a

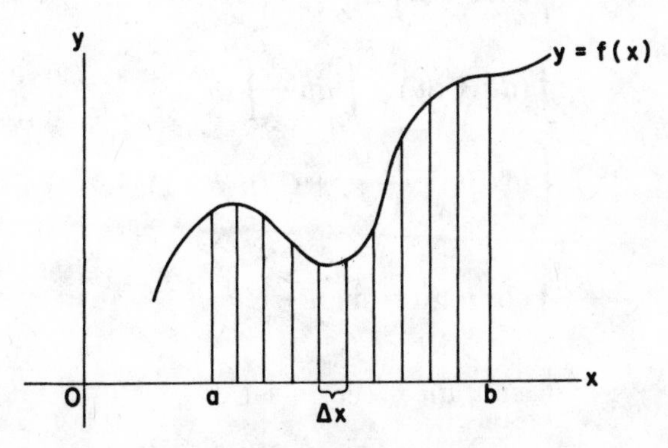

Fig. 1.15. A definite integral. The definite integral from a to b of $f(x)$ with respect to x is written:

$$\int_a^b f(x)\,dx$$

and is read as the integral of $f(x)$ with respect to x between the limits a and b. It is the area between the curve $y = f(x)$, the x-axis, and the lines $x = a$ and $x = b$.

knowledge of the way in which they are varying over very small ranges. For example, if the velocity at two different times is known, it is possible to find the general expression which gives the distance traveled by a falling body as a function of time. (See example number three after derivatives.)

Example: To solve the differential equation

$$\frac{dy}{dx} = 3x^2$$

change to differentials

$$dy = \frac{dy}{dx} \times dx$$

$$dy = 3x^2 \times dx$$

It is known from differentiation that

$$d\,(x^3) = 3x^2\,dx$$

$$dy = 3x^2\,dx = d\,(x^3) = d\,(x^3 + c)$$

Then $y = x^3 + c$

Some rules for integrals where C is a constant and u and v are functions of x:

$$\int du = u + C$$

$$\int a\,du = a \int du$$

$$\int (du + dv) = \int du + \int dv$$

$$\int u^n\,du = \frac{u^{n+1}}{n+1} + C \;(n \neq -1)$$

$$\int \cos u\,du = \sin u + C$$

$$\int \sin u\,du = -\cos u + C$$

Handbooks of mathematical tables give formulas for the evaluation of integrals of various functions. It is not necessary for one to know how all the entries in such tables have been arrived at, just as at a more elementary level one may use tables of logarithms without knowing how the values in these tables were calculated.

Example: Solve the following definite integral:

$$\int_1^2 y\,dx = \left.\frac{x^2}{2}\right|_{x=2} - \left.\frac{x^2}{2}\right|_{x=1}$$

The symbol $\left.\frac{x^2}{2}\right|_{x=2}$ means substitute 2 for x in $\frac{x^2}{2}$.

$$= \frac{2^2}{2} - \frac{1^2}{2}$$

$$= \frac{4}{2} - \frac{1}{2} = \frac{3}{2}$$

PART 2

Mathematical Applications

2.1 VARIABLES AND CONSTANTS

A *variable* is any quantity which has different values under different conditions; it is any quantity which varies. In an investigation, effort should be made to hold all quantities constant except two. The independent variable is given known values and the resulting values of the other variable, the dependent variable, are observed and recorded. From these observations the relationship between the two variables can be computed unless the results have been distorted by a third variable which was not controlled.

For example, if one were plotting a graph showing the settling time of particles of sediment in water as a function of the grain size of the sediment, the independent variable would be the grain size of the sediment and the values of the dependent variable, the settling time, would be observed and recorded. If a third variable such as the density of the liquid medium (e.g., water versus oil) or the height of the water column had not been controlled (i.e., kept the same) the relationship between the two variables would have been distorted.

An *absolute* or *universal constant* is a quantity whose magnitude cannot be changed under any circumstances (e.g., π, Planck's constant, gas constant). A *parameter* is a quantity whose magnitude can change but is not allowed to change during a certain discussion or sequence of mathematical operations, i.e., in the particular problem under consideration. This term is sometimes called an *arbitrary constant.* A parameter actually may be thought of as either a variable that is to be given a fixed value throughout a problem or as a constant which may be replaced by another constant in another problem.

$$V = \pi r^2 h$$

π = a constant equal to 3.1416
r^2 = a variable
h = a variable

$$V = l \times w \times h$$

l, w, and h are all variables.

2.2 FORMULAS

A *formula* is a brief and concise way of stating a rule. A rule of procedure is abbreviated by using single letters to represent words of the rule. In most cases this would be the initial letter of each word, except, of course, in cases where the same initial letter occurs more than once. A formula expresses relationships among variables conveniently so that new facts may be discovered. Its usefulness lies in the fact that formulas express statements that can be applied to many different cases.

Students should think of the formula as a concise language and as a shorthand rule for computation. It is, in fact, a general solution since it is an expression for the dependence of one variable upon another.

The ability to use a formula includes the following elements:

1. The ability to recognize a formula, i.e., to interpret the symbolism. The algebraic symbolism must be retranslated into a verbal statement. This implies among other things familiarity with conventional symbols, such as the use of A for area, V for volume, h for altitude, t for time, and so on. Students should be aware, however, that the letter r may stand for a rate in miles per hour in one formula and for length of the radius of a circle in another.

2. The ability to substitute in a formula. This ability to substitute correctly is largely a matter of practice. However, it should be remembered that intelligent comprehension and interpretation are far more important than manipulation.

3. Units of dimension. The use of units in solving formulas is very important. For additional precaution, many students and scientists follow each factor in a formula by its corresponding unit(s). The following rules should be followed in the use of dimensional units in solving problems:

   a. Quantities expressed in similar dimensions may be added or subtracted.
   b. In equality expressions, both sides of the equation must be equal dimensionally as well as numerically.
   c. Unlike dimensions cannot be canceled or otherwise simplified.

The mathematical translation from one set of units in which a property was measured into a different set of units is known as *conversion*. Conversion of units is most conveniently performed by means of conversion factors. A *conversion factor* is a numerical ratio of units which is equal to 1. Whenever a problem involving conversion of units is given, the first step is always to define the fundamental relation(s) which equals each unit concerned.

Examples and Problems

How many centimeters are there in 1 yard?

Fundamental relations are 1 yard = 3 feet
1 foot = 12 inches
1 inch = 2.54 centimenters

$$1 \text{ yd} = \frac{1 \text{ yd}}{1} \times \frac{3 \text{ ft}}{1 \text{ yd}} \times \frac{12 \text{ in.}}{1 \text{ ft}} \times \frac{2.54 \text{ cm}}{1 \text{ in.}} = 91.44 \text{ cm}$$

Note that each conversion factor is equal to unity and thus the value of the original quantity is not altered.

One ounce equals how many grams?

$$1 \text{ oz} = \frac{1 \text{ oz}}{1} \times \frac{1 \text{ lb}}{16 \text{ oz}} \times \frac{454 \text{ g}}{1 \text{ lb}} = 28.375 \text{ g}$$

How many grams are in 0.75 ounce?

$$0.75 \text{ oz} = \frac{0.75 \text{ oz}}{1} \times \frac{1 \text{ lb}}{16 \text{ oz}} \times \frac{454 \text{ g}}{1 \text{ lb}} = 21.28125 \text{ g} = 21.3 \text{ g}$$

or

$$0.75 \text{ oz} = \frac{0.75 \text{ oz}}{1} \times \frac{28.375 \text{ g}}{1 \text{ oz}} = 21.28125 \text{ g or } 21.3 \text{ g}$$

Some examples of formulas:

$V = \pi r^2 h$ (volume of a cylinder)

$E = IR$ (electricity; volts = amperes times ohms)

$F = \frac{9}{5} C + 32$ (Fahrenheit temperature from Centigrade)

$C = \frac{5}{9} (F - 32)$ (Centigrade temperature from Fahrenheit)

$V = lwh$ (volume of a rectangular solid)

$A = \pi r^2$ (area of a circle)

$P = 4a$ (perimeter of a square)

Use of an example:

What is the volume in milliliters of a column whose height is 90 centimeters and its diameter 2.5 centimeters?

The formula for a cylinder is $V = \pi r^2 h$.

$V = 3.1416\ (2.5\ \text{cm})^2 \times 90\ \text{cm} = 1{,}767.15\ \text{cm}^3 = 1{,}767\ \text{ml}$ (Note that one ml is equal to one cubic cm.)

2.3 CHEMICAL FORMULAS AND EQUATIONS

A symbol represents one atom of an element. A formula represents one molecule of a substance (either an element, oxygen, O_2, or a compound water, H_2O). A number in front of a formula tells how many separate molecules there are of that substance. Thus $7NO_2$ means seven separate molecules of the gas nitrogen, each consisting of one atom of nitrogen joined to two atoms of oxygen by chemical forces. Similarly, $8Fe_2O_3$ means eight molecules of iron oxide (which are found in iron ore), each consisting of five atoms joined together—two of iron, three of oxygen. Sometimes formulas are made just a little more complicated than the examples quoted so far. The formula for alcohol (the active ingredient of intoxicating liquors) is usually written as C_2H_5OH and not C_2H_6O. The reason for this apparent complication is that it gives more information about how the atoms are arranged in the molecule. It tells, for example, that one of the six hydrogen atoms in each molecule of alcohol is different from the rest by being joined to an oxygen atom and not to a carbon atom.

There are differences between the symbols used in chemical equations and those used in mathematics. The position of the formulas in chemical equations is most important. In particular, the reactants in chemical equations are always shown on the left-hand side while the reaction products are entered on the right. Furthermore, the addition sign (+) does not necessarily denote addition in the mathematical sense. On the left-hand side at least it means that the substances react together to form the products shown to the right of the equal sign (=). To overcome any possible confusion, the equal sign is sometimes replaced by an arrow ($\longrightarrow$) to show the direction of the reaction. The symbol for a reversible reaction is ($\rightleftharpoons$).

The very name equation implies that there must be the same number of atoms of each element on the two sides; the equation must balance. This is in accordance with a basic law of chemistry which states that in a chemical reaction matter can neither be created or destroyed. (In nuclear reactions it is possible to convert mass to energy.)

All substances which take part in chemical reactions exist as molecules and molecules of both elements and compounds contain a definite number of atoms. In compounds there is a fixed ratio between the number of atoms in each element present. A different compound is obtained if the ratio is altered.

Thus the only way to make equations balance is to insert whole number multipliers in front of some or all of the formulas. In building up the equation, the best procedure is first to write down all the formulas of all the reactants and all the products. Then, by trial and error, the correct multipliers are found. (See Table 2.1 as an aid.)

Table 2.1. Table of Common Valences

+1	+2	+3	+4	-1	-2	-3
Na	Ca	Al	Sn(ic)	Cl	O	PO_4
K	Mg	B		Br	S	BO_3
Li	Sr	Sc		NO_3	Cr_2O_7	
H	Ba	Bi		OH	SO_4	
NH_4	Zn	Sb		CN	SO_3	
Cu(ous)	Cu(ic)	Cr		ClO_3	CrO_4	
Ag	Fe(ous)	Fe(ic)		ClO_2	CO_3	
Rb	Cd			ClO		
Cs	Pb			NO_2		
	Sn(ous)			I		
	Be			HCO_3		
	Hg(ic)			$C_2H_3O_2$(acetate)		
	Ni			F		
	Co					
	Hg_2(ous)					

Note: Writing molecular formulas when the valences of the constituents are known is really very simple; just remember that the total of positive charges must equal the total of the negative charges.

For example, Calcium Chloride

Ca has a valence of +2 (from table above)
Cl has a valence of −1

Therefore, two chlorides are needed for one calcium.
Thus $CaCl_2$.

Examples and Problems

In building up the equation for the combustion of magnesium ribbon in oxygen to magnesium oxide, the following steps are followed:

$Mg + O_2 \longrightarrow MgO$ — Write the formulas for the reactants and products.

$Mg + O_2 \longrightarrow 2MgO$ — Extra oxygen atom on left side, so double right side.

$2Mg + O_2 \longrightarrow 2MgO$ — Now deficiency of Mg on left side, so double it. Now equation balances.

Given the weight of a compound, determine the weight of a constituent:

A certain solution contains 600 mg of sodium fluoride, NaF. How many milligrams of fluorine are represented in the solution? (Note: First look for the atomic weights of the elements in question in the Appendix of this book or in any other chemistry tables, for example, in the *Handbook of Chemistry and Physics*, The Chemical Rubber Company.)

Na F
23 + 19 = 42

$$\frac{42}{19} = \frac{600 \text{ mg}}{x \text{ mg}}$$

$x = 271.43$ mg

It was found experimentally that a certain quantity of the compound zinc sulfate contained 0.4 moles* of zinc (Zn), 0.4 moles of sulfur (S), and 1.6 moles of oxygen (O). What is the chemical formula?

The formula is $ZnSO_4$ since for every one mole of zinc and sulfur there are four moles of oxygen.

Determine the percentage composition of a substance:

Procedure: Find the molecular weight of the substance.
Divide the atomic weight of each element (or its multiple) by the molecular weight.
Multiply by 100 to give the percentage.

Find the percentage composition of Na_2HPO_4 (anhydrous sodium phosphate).

$$\begin{array}{cccc} Na_2 & H & P & O_4 \\ (2 \times 22.99) + & 1.008 + & 30.97 + & (4 \times 16.00) = \\ 45.98 + & 1.008 + & 30.97 + & 64.00 = 141.958 \end{array}$$

$$\text{percent Na} = \frac{45.98}{141.958} \times 100 = 30.98\%$$

$$\text{percent H} = \frac{1}{141.958} \times 100 = 0.70\%$$

$$\text{percent P} = \frac{30.97}{141.958} \times 100 = 21.82\%$$

$$\text{percent O} = \frac{64}{141.958} \times 100 = 45.08\%$$

Calculate the percent composition of $KClO_3$.

First calculate the molecular weight of each element in the compound.

$$\begin{aligned} 1\ K &= 39.100 \\ 1\ Cl &= 35.457 \\ 3\ O &= 48.000 \\ \hline KClO_3 &= 122.557 \end{aligned}$$

*See Section 2.8 for the definition of a mole.

$$\text{Percent K} = \frac{39.100}{122.557} \times 100 = 31.903\%$$

$$\text{Percent Cl} = \frac{35.457}{122.557} \times 100 = 28.931\%$$

$$\text{Percent O} = \frac{48.000}{122.557} \times 100 = 39.165\%$$

Percent Composition	
	K 31.903%
	Cl 28.931%
	O 39.165%
Check Total	99.999%

Calculate the percent S in SO_2 and SO_3.

Molecular weight $SO_2 = 32.066 + 32.000 = 64.066$
Molecular weight $SO_3 = 32.066 + 48.000 = 80.066$

$$\text{Percent S in } SO_2 = \frac{32.066}{64.066} = 50.051\%$$

$$\text{Percent S in } SO_3 = \frac{32.066}{80.066} = 40.049\%$$

Calculate the percent composition of an alloy containing 13.25 g Bi, 8.00 g Pb, and 3.75 g Sn.

Total weight of alloy = 13.25 g + 8.00 g + 3.75 g = 25.00 g

$$\text{Percent Bi} = \frac{13.25\text{ g}}{25.00\text{ g}} \times 100 = 53.0\%$$

$$\text{Percent Pb} = \frac{8.00\text{ g}}{25.00\text{ g}} \times 100 = 32.0\%$$

$$\text{Percent Sn} = \frac{3.75\text{ g}}{25.00\text{ g}} \times 100 = \frac{15.0\%}{100.0\%}$$

Determine the percentage of water in sodium phosphate, $Na_2HPO_4 \cdot 5H_2O$.

$$\begin{array}{ccccccccc} Na_2 & & H & & P & & O_4 & & 5H_2O \\ (2 \times 22.99) & + & 1.008 & + & 30.97 & + & (4 \times 16.00) & + & (6 \times 18.02) = \\ 45.98 & + & 1.008 & + & 30.97 & + & 64.00 & + & 90.10 = 232.06 \end{array}$$

232.06 = molecular weight of the hydrated sodium phosphate

$$\frac{232.06}{90.1} = \frac{100\%}{x\%}$$

$x = 38.8\%$

The formula of a substance can be determined from its percentage composition by the following method:

Divide the percentage composition values by the atomic weights of the elements involved.

Determine the ratio between these quotients to obtain the relative number of atoms of each kind in the compound.

For example, What is the formula of a compound that contains 32.38% sodium, 22.57% sulfur, and 45.05% oxygen?

$$\text{sodium} = \frac{32.38}{22.99} = 1.408$$

$$\text{sulfur} = \frac{22.57}{32.06} = 0.704$$

$$\text{oxygen} = \frac{40.05}{16.0} = 2.816$$

The ratio is 0.704:1.408:2.816 as 1:2:4.
Therefore the formula is Na_2SO_4.

Calculate the empirical formula of a compound containing 11.19% hydrogen and 88.81% oxygen.

(Note: The empirical formula is the simplest formula of the compound. It represents the actual number of atoms of each type present in the compound.)

In 100 g of the compound there are 11.19 g hydrogen and 88.81 g oxygen.

(Divide each weight by respective gram atomic weight to obtain the number of gram atoms.)

$$\frac{11.19\text{ g}}{1.008\text{ g/gram atom}} = 11.10 \text{ gram atoms for hydrogen}$$

$$\frac{88.81\text{ g}}{16.00\text{ g/gram atom}} = 5.550 \text{ gram atoms for oxygen}$$

The ratio equals 2 gram atoms hydrogen to 1 gram atom oxygen.
Thus the formula is H_2O.

Calculate the empirical formula of a compound containing 43.67% phosphorus and 56.33% oxygen.

Atomic weight phosphorus = 31 rounded off
Atomic weight oxygen = 16

$$\frac{43.67\text{ g}}{31\text{ g/gram atom}} = 1.4 \text{ gram atoms for phosphorus}$$

$$\frac{56.33\text{ g}}{16\text{ g/gram atom}} = 3.5 \text{ gram atoms for oxygen}$$

The ratio equals 1:3.5. This must be doubled to obtain whole numbers. Thus the formula is P_2O_7.

What is the molecular formula of a compound which is 85.63% carbon and 14.37% hydrogen whose molecular weight is 56.108?

(Note: The molecular formula represents the actual number of atoms of each type present in the compound.)

$$\frac{85.63\text{ g}}{12.0\text{ g/gram atom}} = 7.14 \text{ gram atoms for carbon}$$

$$\frac{14.37\text{ g}}{1.0\text{ g/gram atom}} = 14.4 \text{ gram atoms for hydrogen}$$

The ratio is 1:2. Thus the empirical formula is CH_2.

The weight of this formula is $1 \times 12.0 + 2 \times 1.008 = 14.027$.

The given molecular weight (GMW) was 56.108.

$$\frac{56.108}{14.027} = 4.$$ Thus the molecular formula is 4 (CH_2) or C_4H_8.

What is the formula of the hydrate of 5.00 g of the salt magnesium sulfate ($MgSO_4$) that lost 2.56 g of water of crystallization upon heating?

Compound contains 2.44 g of $MgSO_4$ and 2.56 g of H_2O

Molecular weight $MgSO_4$ = 120 g

Molecular weight H_2O = 18 g

$$\frac{2.44\text{ g}}{120\text{ g/GMW}} = 0.0203 \text{ GMW for } MgSO_4$$

$$\frac{2.56\text{ g}}{18.0\text{ g/GMW}} = 0.142 \text{ GMW for } H_2O$$

The ratio is 1:7.

Thus the formula for the hydrate is $MgSO_4 \cdot 7H_2O$.

2.4 GRAPHS

A *graph* is a diagram showing the relationship of two or more variable quantities. A graph represents material in a form that is more readily grasped than is a table of statistical data. Graphs should be introduced whenever a vivid presentation of number relations is desired. Graphs are used extensively for reference, for illustrative purposes, and for analysis. They are used to present facts or tabulated data in a more usable form.

A *mathematical graph* is usually in the form of a line or curve drawn in a frame of reference formed by the two axes, the horizontal x-axis and the vertical y-axis. When the relation of two dependent quantities is to be plotted, one axis is used as a scale to represent the changes in value of one of the given quantities; the other axis measures the relative change of the other. The construction of graphs should enable the student to organize and interpret in terms of trends the data collected.

Picture Graphs

A picture graph involves the use of pictures to represent numerical data. (See Fig. 2.1.) In constructing a picture graph an appropriate picture is used to represent the subject of the graph. The exact number

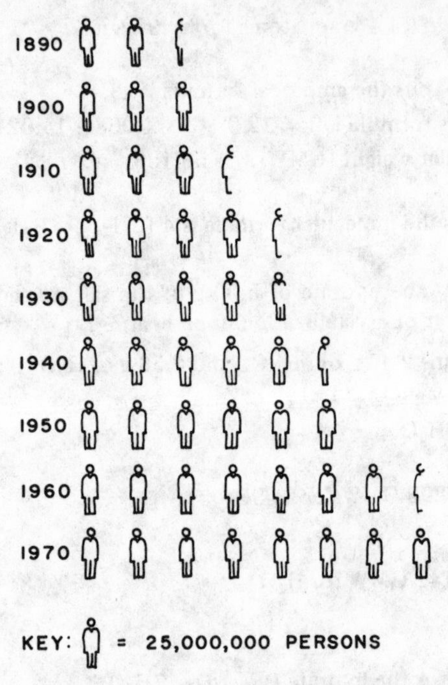

Fig. 2.1. United States population 1890–1970.

which each picture represents should be stated in the graph. The data must be rounded off to the nearest of the representative units and the corresponding number of units drawn. If odd figures need to be represented, only a portion of the picture should be shown. One should realize that picture graphs are less accurate than the data on which they are based.

Bar Graphs

Bar graphs resemble picture graphs in many respects but the use of bars permits a more accurate representation of data. (See Fig. 2.2.) In making bar graphs one should use scaled paper whenever possible. The bars should

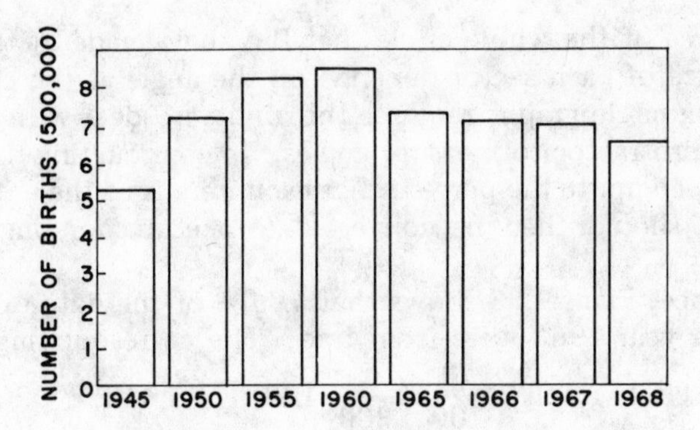

Fig. 2.2. Births in the United States.

be uniform in width. The scale should always be stated and must start at zero so that the relative lengths will be correct. A histogram is a bar graph of a frequency function in which the range of the variable x is divided into intervals and the frequency of each interval is indicated by the areas of vertical bars. If the intervals are equal, the heights serve as an exact measure of relative frequency.

Circle Graphs

Circle graphs are used to show the parts into which a unit is divided. The making of circle graphs depends on recognition of the fact that the area of a section of a circle bounded by two radii and an arc is the same

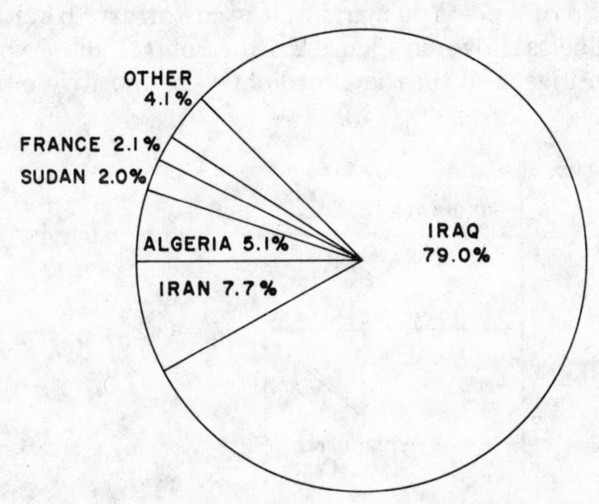

Fig. 2.3. Date exports. World total 347,000 metric tons. Average 1961–1963.

Iraq = 79.0% = 284°	Sudan = 2.0% = 7.20°
Iran = 7.7% = 27.72°	France = 2.1% = 7.56°
Algeria = 5.1% = 18.36°	Other = 4.1% = 14.76°

part of the area of the whole circle that the angle made by the radii is of 360°. The size of each sector depends on the angle at the center of the circle. Making circle graphs requires the ability to deal with proportions and to use a protractor to measure angles. In constructing the graph, the first step is to compute the percent that each item is of the entire amount. Each percent should then be converted to a corresponding number of degrees.

For instance, Figure 2.3 shows that 79.0% of the date exports in the world for the year 1963 were from Iraq. The corresponding number of degrees in the circle graph is $\frac{79}{100} = \frac{x}{360}$; $100x = 28{,}440$; $x = 284.4°$.

Line Graphs

Line graphs are helpful in enabling one to interpolate between the values on which the graph is based to estimate possible future values. In plotting a line graph one needs two scales—a horizontal and a vertical scale. Each scale should be labeled clearly to show what quantity is measured by it. Points should be located before the graph can be drawn. A point is located by means of a pair of numbers; one of these two numbers (x) tells how far to count on the horizontal scale and the other number (y) tells how far to count on the vertical scale. The numbers x and y are known as coordinates of the point; if it is desirable to distinguish between them, x is called the abscissa of the point and y the ordinate. When specifying the position of a point on a graph, it is customary to state the x value before the y value as shown in Figure 2.4 for *Point A* and *Point B*. Not all numbers are positive, and on the coordinate axes positive numbers are to

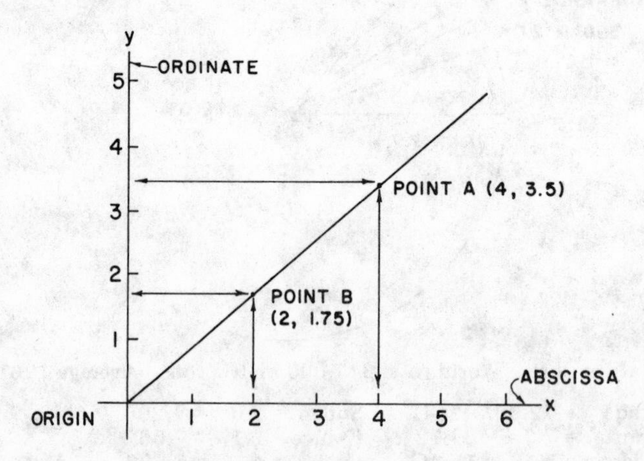

Fig. 2.4. The coordinates of a point.

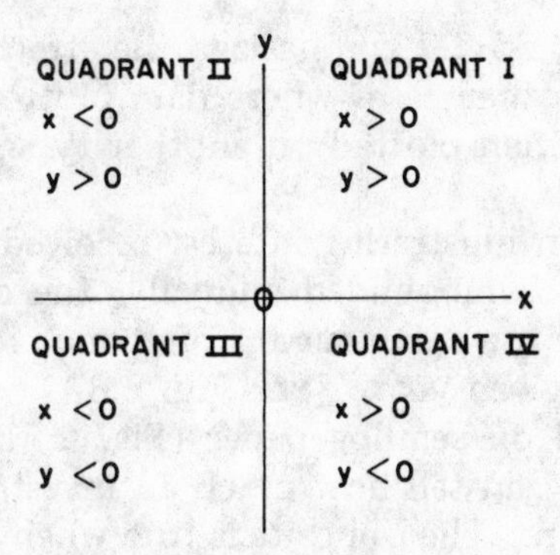

Fig. 2.5. Signs of the coordinates in the four quadrants of the xy plane.

the right or up, and negative numbers are to the left or down. (See Fig. 2.5.) It is through the exaggeration of scale and the failure to indicate scale that graphs can be used to distort the true picture of a situation.

The data as plotted in Figure 2.6 appear to report relatively stable temperatures for the week. On the other hand, the same data plotted on an expanded scale in Figure 2.7 appear to imply a great variation in temperature throughout the week, thus overemphasizing the effect of a 10-degree change in temperature.

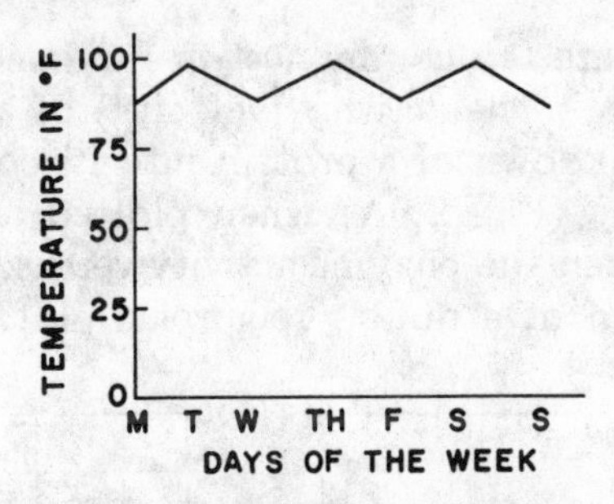

Fig. 2.6. Broken-line graph.

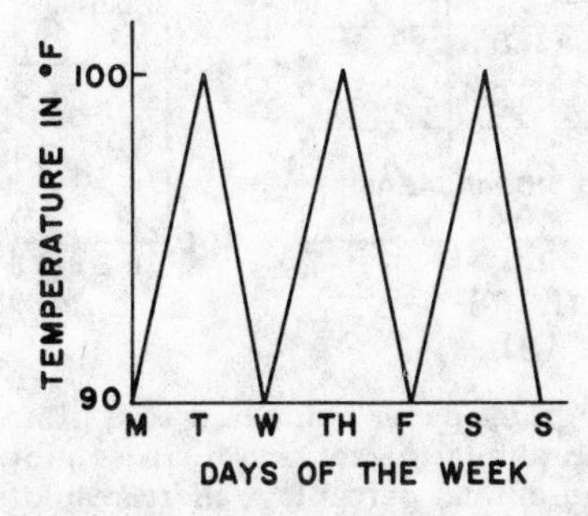

Fig. 2.7. Exaggerated broken-line graph.

While Figures 2.6 and 2.7 emphasized the effect of scale changes on a line graph, there are other cases where data plotted on one type of line graph changes form when plotted on another type, as in the case of the log graph.

For example, when illustrating interest received on $100 in a savings account yielding 10% compounded annually, the data when plotted on regular graph paper show a rising curve from which it is difficult to extract actual amounts for a given year. (See Fig. 2.8A). However, when these same data are plotted on semilog paper (Fig. 2.8B), the curved line becomes linear and one can tell how much an investment would be worth at any time in the future. The opposite is true when an increase by a fixed amount (e.g., $10.00) is plotted. (See dotted lines in Figs. 2.8A and 2.8B).

Thus regular graph paper best shows the amount of change; semilog paper best shows the rate or percentage of change. The fact that a regular percentage increase gives a straight line on semilog paper enables one to use this paper for calculating growth rates or compound interest.

It is of interest to note that determination of the rate of some chemical reactions necessitates logarithmic plots. One example is the semilogarithmic plot of first order type reactions which as in the above example of Figures 2.9A and 2.9B is curved on regular graph paper and a straight line on semilog paper. The slope of this straight line is used in calculating the rate of the reaction.

Since line graphs are the commonest in fundamental science, another example will be given. The activity (velocity) of an enzyme catalyzing, for example, the breakdown of a protein into its component amino acids may represent a curve (Fig. 2.9A) when plotted on regular graph paper and a straight line when the enzyme activity versus protein concentration is plotted in the form of a double reciprocal plot. In this case, the ad-

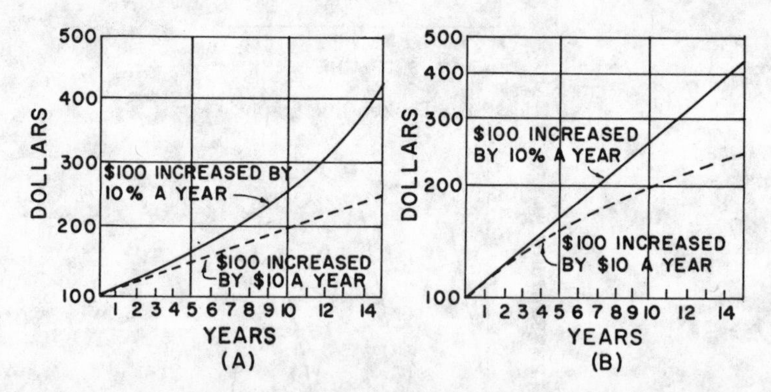

Fig. 2.8. Plot on regular graph paper compared to a plot on semilog paper. (A) On regular paper an annual percentage increase plots as a curved line. (B) On semilog paper an annual percentage increase plots as a straight line. Dotted lines represent increase by a fixed amount (see text).

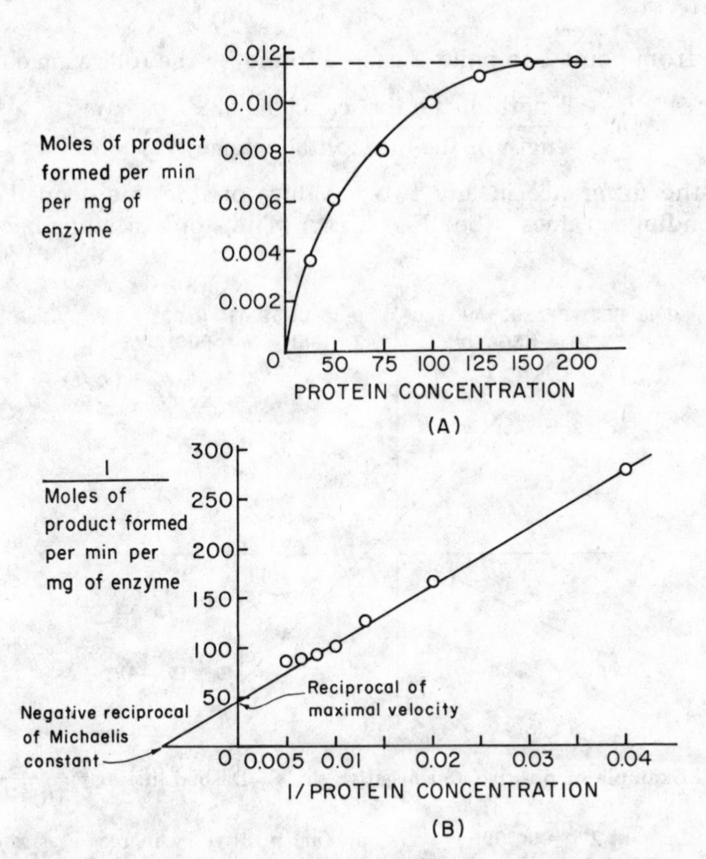

Fig. 2.9. Plots of enzyme velocity dependence on substrate concentration. In Figure 2.9a the substrate is protein which is being hydrolyzed (broken down) by a proteolytic enzyme. The data in Figure 2.9a were replotted in the form of a double reciprocal plot to obtain Figure 2.9b. The kinetic constants can best be obtained from the linear plot of Figure 2.9b.

vantage of the double reciprocal plot is in evaluating kinetic constants of the enzyme reaction. For instance, the enzyme maximal activity, when using the curve, can be determined only as an estimate by extrapolation as shown by the dotted line in Figure 2.9A; however, when the graph is in the form of the double reciprocal plot, the exact maximal activity is determined by reading off the number on the y-axis where the straight line and the y-axis intersect (Fig. 2.9B). This number is then converted to an actual value by finding its reciprocal.

2.5 SLOPE

The ratio of the vertical increase to the corresponding horizontal increase of a line is defined as the slope of a line. The slope (m) of a line segment from point 1 to point 2 can be found by the following equation:

$$m = \frac{\text{length of the vertical change}}{\text{length of the horizontal change}} = \frac{y_2 - y_1}{x_2 - x_1}$$

that is, the difference in any two y values on the line divided by their corresponding x values. (See Fig. 2.10.) The slope of a line becomes in-

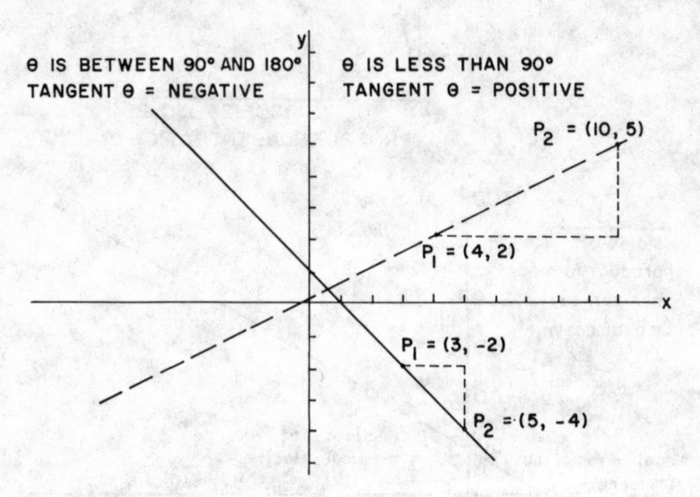

Fig. 2.10. Example of positive and negative slope. Dashed line $m = \frac{5-2}{10-4} = \frac{3}{6} = \frac{1}{2}$; tangent $27° = 0.5095 = \frac{1}{2}$; slope ($m$) positive; y increases as x increases. Solid line $m = \frac{-4-(-2)}{5-3} = \frac{-4+2}{2} = \frac{-2}{+2} = -1$; tangent $135° = -1$; slope (m) negative; y decreases as x increases.

finite as its inclination angle approaches zero. A straight line in a plane is represented by an equation of the form $y = mx + b$. This is the slope-intercept equation where m = the slope of the line and b = the y-intercept.

The ratio of the change in y over the change in x also gives the tangent $\left(\text{tangent} = \frac{\text{ordinate}}{\text{abscissa}}\right)$ of the angle θ made by the graph line with the x axis; $m = \tan\theta$. Thus the slope of a line may also be determined by measuring the angle between the graph line and the x axis with a protractor and looking up the tangent of the line in the tables. An acute angle (less than 90°) has a positive tangent, and an obtuse angle (90° to 180°) has a negative tangent; thus the sign of the slope agrees with the sign of the tangent.

The distance (s) between two points on a line is given by the equation

$$s = \sqrt{(x_2 - x_1)^2 + (y_2 - y_1)^2}$$

This relation follows the rule of geometry* that the square of the hypotenuse of a right-angled triangle is equal to the sum of squares of the other two sides. Figure 2.11 makes the reason clear.

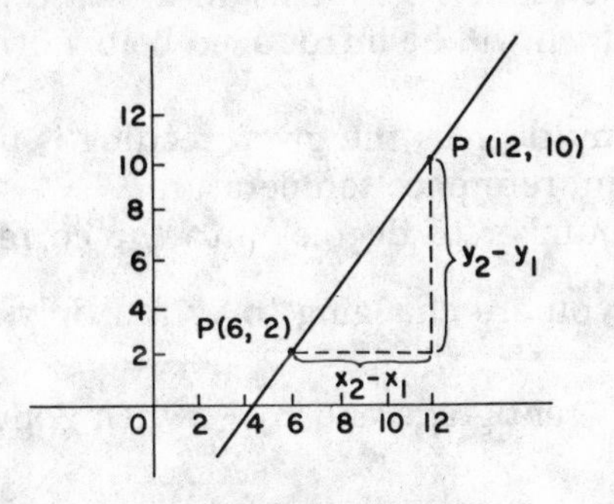

Fig. 2.11. The distance s between two points.

$$s = \sqrt{(12 - 6)^2 + (10 - 2)^2} = \sqrt{6^2 + 8^2} = \sqrt{36 + 64} = \sqrt{100} = 10$$

It is immaterial which point is considered x_1, y_1 or x_2, y_2 for the answer is the same if the points are interchanged in the formula. Also note that the square of a negative or positive number is always positive.

2.6 TEMPERATURE SCALES

Temperature determines the direction of heat flow between an object and its surroundings: heat flows from matter at a higher temperature to matter of a lower temperature. In measuring temperature by thermometers, the primary references are the boiling and freezing points of pure water. The interval between these two points is divided into a convenient number of units (degrees) in terms of which temperatures are measured.

The two temperature scales most frequently used are *Fahrenheit* and *centigrade* (now officially named *Celsius*). Another scale much used in scientific work is the *Kelvin* (or absolute) scale.

In the Fahrenheit scale the temperature interval between the freezing and boiling points of water is divided into 180 equal divisions. The zero of the scale is placed 32 of these divisions below the freezing point of water. The centigrade scale makes use of the same freezing point and boiling point of water for its reference temperatures, but it has 100 divisions in this range (thus the name centi-grade) and starts its numbering from zero at the freezing point. Since 100 divisions on the centigrade scale occupy the same space as 180 divisions between freezing and boiling on the Fahrenheit scale, the relative size of the divisions on the two scales

*Pythagorean theorem.

is $\frac{100}{180}$, or when simplified, $\frac{5}{9}$. It should be noted that the centigrade divisions are the larger of the two, and thus in a given temperature interval there will be fewer centigrade divisions included.

To translate a Fahrenheit temperature into a centigrade temperature there is a formula which will be introduced below or alternatively a three-step process:

1. Find how many degrees the given reading is upward or downward from the freezing point reference temperature.

2. Convert this number of degrees into the corresponding number of degrees in the scale you are changing to by multiplying it by $\frac{5}{9}$ when converting Fahrenheit to centigrade and by $\frac{9}{5}$ when converting centigrade to Fahrenheit.

3. Count this new number of degrees upward or downward from the freezing point on the scale you are changing to and thus determine the final reading on the new scale.

The following problem is an illustration of the use of this method of thinking:

What temperature on the centigrade scale corresponds to 98.6°F (body heat)?

First: Sketch the two temperature scales as shown in Figure 2.12 and mark the given temperature at its appropriate location.

Second: The given temperature 98.6°F can be seen to be 66.6° above the freezing point.

Third: Since the centigrade degrees are larger in size there will be only $\frac{5}{9}$ as many of them in the interval represented by 66.6° Fahrenheit. Multiplying 66.6 by $\frac{5}{9}$ one gets 37° centigrade.

Fourth: Counting 37° centigrade upward from the freezing point on the other scale (which is zero) one gets the final centigrade reading of 37° centigrade.

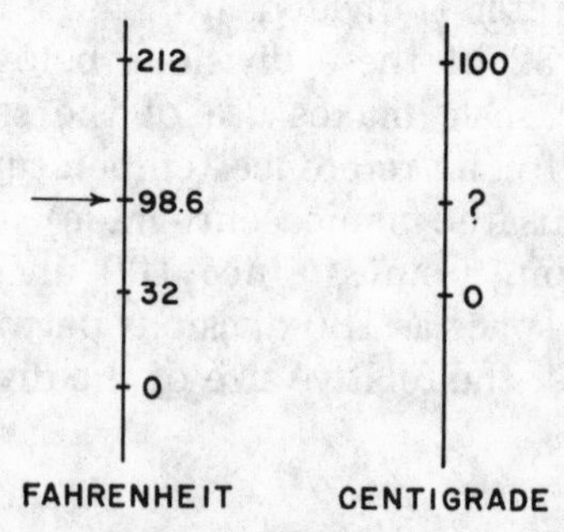

Fig. 2.12. Sketch of temperature scales illustrating problems.

Thus, if a temperature on the centigrade scale is to be converted to the Fahrenheit scale, the equivalent number of divisions on the Fahrenheit scale above the lower fixed point is calculated first by multiplying the centigrade reading by $\frac{9}{5}$. But since the lower fixed point has a value of 32° on the Fahrenheit scale, 32 must be added to the result of the first calculation.

Conversely, if a reading in Fahrenheit degrees is to be converted to the centigrade scale, 32 must first be subtracted from the original reading. By this means the number of divisions by which the reading differs from the level of the lower fixed point is found, and it is this latter figure which is multiplied by $\frac{5}{9}$. The result of this calculation gives the temperature on the centigrade scale.

Therefore, to convert from one system to the other, use the following formulas:

$$F^\circ = \frac{9}{5}\, C^\circ + 32$$

$$C^\circ = \frac{5}{9}\,(F^\circ - 32)$$

Examples: Convert 50°C to Fahrenheit.

$$\left(\frac{9}{5} \times 50^\circ\right) + 32^\circ =$$

$$\left(\frac{450}{5}\right) + 32^\circ = 90 + 32 = 122^\circ F$$

Convert −12°C to Fahrenheit.

$$\left(\frac{9}{5} \times -12^\circ\right) + 32 =$$

$$\left(\frac{-108}{5}\right) + 32^\circ = -21.6 + 32 = 10.4^\circ F$$

Convert 182°F to centigrade.

$$\frac{5}{9}\,(182^\circ - 32^\circ) =$$

$$\frac{5}{9}\,(150^\circ) = \frac{750}{9} = 83.3^\circ C$$

Convert −80°F to centigrade.

$$\frac{5}{9}\,(-80^\circ - 32^\circ) = \frac{5}{9}\,(-112^\circ) = \frac{-560}{9} = -62.2^\circ C$$

To create confidence that this system is simple, an approximation can be made that $\frac{5}{9} \cong \frac{5}{10} = \frac{1}{2}$ and $\frac{9}{5} \cong \frac{10}{5} = 2$. This approximation can be illustrated as follows:

20°C is equal to ? It is equal to double 20 plus 32 ≅ 72° F.

20° F is equal to ? It is equal to 20 minus 32 keeping the sign (here it is minus 12) divided by 2: - 12/2 ≅ - 6° C.

The Kelvin scale is similar to the centigrade scale in that the size of divisions is the same in both scales. The Kelvin scale has its zero point at -273°C. To consider the reason for this choice of a zero point one must consider the molecular motion concept of temperature. Every substance contains a certain amount of heat—even ice. This simply means that the molecules of any substance (the smallest possible parts of that substance) are constantly in motion. When a substance is heated, its molecules are given the energy to move faster. The hotter the substance the faster its molecules move, and the cooler the substance the slower its molecules move. If you continue this cooling idea, there must come a point where there is no heat and where all molecule movement absolutely ceases. This point is known as *absolute zero*, for it is obviously the lowest temperature that can be attained since temperature is a measurement of the degree of heat (i.e., the motions of the molecules or atoms that compose it).

Since the speed of motion of the molecules determines the temperature of an object, there is no logical upper limit to the temperature conceivable. In the opposite direction, however, the molecules of an object

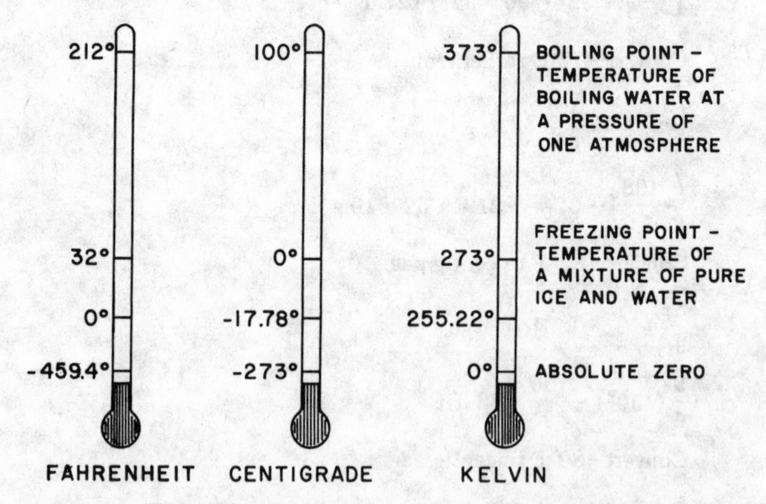

Fig. 2.13. Comparison of temperature scales.

may be slowed down only to the point where motion ceases. When using the centigrade scale, it has been shown that this point lies at about $-273°C$ ($-273.16°$). If one chooses this for the beginning of the Kelvin temperature scale, one never has to deal with negative temperatures. One may convert centigrade readings into Kelvin by adding the number 273. (See Fig. 2.13.)

Examples and Problems

20°C = ?° Kelvin

$20° + 273° = 293°$ Kelvin

- - - -

50°F = ?° Kelvin

This is actually two problems because first one must convert from F to C and then from C to Kelvin.

To find C

$$\frac{5}{9}(50° - 32) = \frac{5}{9}(18) = \frac{90}{9} = 10°C$$

and then $10° + 273° = 283°$ Kelvin
so that 50°F = 283° Kelvin

If F = 59°, C = ?

$C = 5/9\ (F - 32)$
$C = 5/9\ (59 - 32)$
$C = 5/9\ (27)$
$C = 135/9 = 15°C$

If C = 80°, F = ?

$$F = \frac{9C}{5} + 32$$

$$F = \frac{9(80)}{5} + 32$$

$$F = \frac{720}{5} + 32 = 144 + 32 = 176°F$$

If F = 50°, C = ?

$C = 5/9\ (F - 32)$
$C = 5/9\ (50 - 32)$
$C = 5/9\ (18) = 90/9 = 10°C$

Which is warmer, 85°F or 30°C?

85°F = ?°C 30°C = ?°F

$$C = \frac{5}{9}(85 - 32)$$

$$= \frac{5}{9}(53)$$

$$= \frac{265}{9} = 29.44°C$$

$$F = \frac{9(30)}{5} + 32$$

$$= \frac{270}{5} + 32$$

$$= 54 + 32 = 86°F$$

85°F = 29.44°C
86°F = 30°C
30°C is warmer than 85°F

During an experiment in a laboratory a temperature of 1000°C was reached. What would the reading on a Fahrenheit scale be?

$$F = \frac{9(1000)}{5} + 32$$

$$= \frac{9000}{5} + 32$$

$$= 1800 + 32 = 1832°F$$

If a European friend explained that an oven should be set at 235°C, what Fahrenheit oven reading should be used?

$$F = \frac{9(235)}{5} + 32 = \frac{2115}{5} + 32 = 423 + 32 = 455°F$$

If C = −40°, F = ?°

$$F = \frac{9(-40)}{5} + 32$$

$$= \frac{-360}{5} + 32 = -72 + 32 = -40°F$$

Therefore, −40°C = 40°F and this is the only place where the temperature readings are identical.

2.7 pH

Aqueous solutions contain positively charged hydrogen ions (or protons, H^+) and negatively charged hydroxyl ions (OH^-). In water these are entirely derived from the ionization (breaking up) of water molecules to H^+ and OH^-. In pure water at 25°C this ionization process results in a hydrogen-ion concentration of 1×10^{-7} moles per liter (or 6.023×10^{16} H^+ per liter). The pH scale is a mathematical means of designating the actual concentration of H^+ (and thus OH^-) ions in any aqueous solution;

it is a measure of the hydrogen-ion concentration. Solutions of pH 7.0 are neutral, those above pH 7.0 are basic, and those below pH 7.0 are acidic. The lower the pH the more acidic is the solution. The scale on pH meters is from zero (0) to 14. (See Table 2.2.)

Table 2.2. The pH Scale

[H+] (M)		*pH*		*[OH⁻] (M)*
1.0	strongly acidic	0		10^{-14}
0.1		1		10^{-13}
0.01		2		10^{-12}
0.001		3	weakly acidic	10^{-11}
0.0001		4		10^{-10}
0.00001		5		10^{-9}
10^{-6}		6		10^{-8}
10^{-7}		7	neutral	10^{-7}
10^{-8}	weakly basic	8		10^{-6}
10^{-9}		9		10^{-5}
10^{-10}		10		0.0001
10^{-11}		11		0.001
10^{-12}		12	strongly basic	0.01
10^{-13}		13		0.1
10^{-14}		14		1.0

Note: As the pH of a solution decreases, its [H+] increases, i.e., its acidity increases, and vice versa. A tenfold change in [H+] is represented by a pH difference of one unit.

When the hydrogen-ion concentration of solutions is expressed quantitatively, it varies from a value of nearly 1 for a normal solution of a strong acid to about 1×10^{-14} for a normal solution of a strong alkali. Consequently, there is a variation of about 10,000,000,000,000 in the numerical values within these two limits. The use of ordinary notation for handling a number of such magnitude in computations that involve hydrogen-ion concentration is impractical.

To simplify the statement of hydrogen-ion concentration, it is convenient to use logarithmic notation, with the mantissa usually rounded off to one or two places of decimals. (See Table 2.3.) It has become customary, therefore, to speak of the hydrogen-ion concentration of a given solution in terms of its pH value which is defined as the logarithm of the reciprocal of the hydrogen-ion value. This relationship between a pH value and the hydrogen-ion concentration (in g ion/liter) can be explained in various ways.

1. As stated before, the pH of a solution equals the logarithm to the base 10 of the reciprocal of its hydrogen-ion value.

Table 2.3. pH of Some Fluids

Fluid	*pH*
Seawater	5.5
Blood plasma	7.4
Intracellular fluids	
Muscle	6.1
Liver	6.9
Gastric juice	1.2–3.0
Saliva	6.4–6.9
Cow's milk	6.6
Urine	5–8
Tomato juice	4.3
Grapefruit juice	3.2
Coke	2.8
Lemon juice	2.3

$$pH = \log \frac{1}{[H^+]}$$

2. And since the logarithm of a reciprocal equals the negative logarithm of a number, this equation may also be stated: the pH of a solution is the negative of the logarithm to the base 10 of its hydrogen-ion value.

$$pH = -\log [H^+]$$

3. If the $[H^+]$ is written as a power of 10, the corresponding pH value is the index of this exponential term without its negative sign, e.g., a $[H^+]$ of $10^{-3.72}$ g ion/liter is equivalent to a pH value of 3.72.

It is especially important to note that the pH* scale is logarithmic, not arithmetic. To say that two solutions differ in pH by one pH unit means that one solution has ten times the hydrogen ion concentration of the other.

Buffer

A buffer is a solution of a weak acid and its salt capable of resisting to a considerable extent changes in pH upon addition of an acid or a base. (See Appendix A.1 for the definition of pK.) The pH meters are standardized with buffers of a given pH. The primary use of buffers, however, is to prevent changes in pH in a reaction releasing H^+ or OH^- ions and thus making the pH a constant value. The buffering capacity of a solution is defined as the mole equivalents of $[H^+]$ or $[OH^-]$ that are required to change one liter of 1 M buffer by 1.0 pH unit. (Molarity M is defined in the next section.)

*In general, when a quantity is preceded by the letter p, the combination means -1 times the logarithm of the quantity following the letter p → as pK_{ion} means $-\log K_{ion}$ and pH means $-\log [H^+]$.

The pH of 20 ml of 2 M buffer, pH 7.8, was decreased to pH 5.8 when 4.2 ml of 2.6 N HCl were added. What is the capacity of the buffer at pH 7.8?

4.2 ml of 2.6N HCl contain $\frac{4.2}{1000} \times 2.6$ equivalents of H^+ = 10.92 milliequivalents of H^+.

1000 ml of buffer would require $\frac{1000}{20} \times 10.92 \times 10^{-3} = 0.546$ which is the buffer capacity.

Note: The buffer capacity at constant pH is directly proportional to the buffer concentration. In the above problem, the same buffer, pH 7.8, at 1 M concentration (instead of 2 M) would have half the buffer capacity, i.e., 0.273 equivalents of H^+ per liter.

Examples and Problems

If the hydrogen-ion concentration is 0.000001 moles per liter (i.e., 1×10^{-6}) what is the pH?

$$pH = \log \frac{1}{[H^+]} = \log \frac{1}{10^{-6}} = \log \frac{10^6}{1} = 6.00 = pH$$

If the hydrogen-ion concentration of a solution is 6×10^{-5}, what is the pH?

$$pH = \log \frac{1}{6 \times 10^{-5}} = \log \frac{10^5}{6} = \log 10^5 - \log 6 = 5 - 0.77 = 4.23 = pH$$

Note that the hydrogen-ion concentration in this problem is 60 times that of the above (i.e., 60 times more acidic), yet when looking at the pH values of 4.23 *versus* 6.00 the difference seems small.

What is the hydrogen-ion concentration in moles/liter of a solution if the pH is 5.4?

$pH = -\log [H^+] = 5.4$ or $\log [H^+] = -5.4$

$-5.4 = (-6 + 0.6)$

$[H^+] = \text{antilog} (+0.6 - 6) = \text{antilog} (0.6) \times \text{antilog} (-6) = 4.0 \times 10^{-6}$ moles/liter

What is the pH of a 0.25 M solution of HCl?

HCl is completely ionized so the hydrogen-ion concentration is 0.25M.

$pH = -\log [H^+] = -\log 0.25 = -\log 2.5 \times 10^{-1}$

$\log 2.5 = 0.40$

$pH = -(\log 2.5 \times 10^{-1}) = -(\log 2.5 + \log 10^{-1})$

$= -(0.4 - 1) = 0.6 = pH$

What is the pH of a solution whose $[H^+]$ equals 6.5×10^{-10}?

$$pH = \log \frac{1}{6.5 \times 10^{-10}} = \log \frac{10^{10}}{6.5}$$

$$= \log 10^{10} - \log 6.5 = 10 - 0.81 = 9.19 = pH$$

Calculate the pH of a 0.20 M solution of acetic acid which is 2.6 percent ionized.

$$[H^+] = 2.6\% \text{ of } 0.20 \text{ moles/liter}$$
$$= 0.026 \times 0.20 \text{ moles/liter} = 5.2 \times 10^{-3} \text{ moles/liter}$$

$$pH = \log \frac{1}{5.2 \times 10^{-3}} = \log \frac{10^3}{5.2} = \log 10^3 - \log 5.2$$

$$= \log 10^3 - \log 5.2 = 3.00 - 0.72 = 2.28 = pH$$

A solution has a pH of 6.35. What is the hydrogen-ion concentration expressed in moles per liter?

$$pH = -\log [H^+] = 6.35$$
$$\log [H^+] = -6.35$$
$$= -7.00 + 0.65$$
$$[H^+] = \text{antilog } (-7.00 + 0.65)$$
$$= \text{antilog } -7.00 \times \text{antilog } 0.65$$
$$= 10^{-7} \times 4.5$$
$$= 4.5 \times 10^{-7} \text{ moles/liter}$$

2.8 SOLUTIONS

The term *solution* may be applied to any homogeneous mixture of two or more substances. A solution, even though it is homogeneous, is not a compound, but is a mixture because its composition is variable. For instance, salt dissolved in water makes a salt solution. However, one can add more salt and get a stronger salt solution or add more water and get a weaker salt solution, but both are still salt solutions.

Ordinarily one thinks of solutions as being liquid, but there are also other types. Air is a solution and is an example of gases dissolved in gases. One can also have solids dissolved in liquids, etc. Gaseous solutions are not too important because each gas behaves as if it alone were present and solid solutions are too complicated for this text, thus this section will be devoted to the discussion of liquid solutions.

For some solutions it is convenient to designate one material as the solvent and the other material or materials as solutes. The solvent may be defined as the substance whose physical state (e.g., solid, liquid, or gas) is preserved when the solution is formed or the substance present in the larger amount. It is the substance that dissolves another substance. The solute is that part of a solution which is considered to be dissolved in the other, the solvent. In the salt solution mentioned above, salt is the solute and water is the solvent. An aqueous solution is one in which water is the solvent.

The concentration of a solution may be expressed in various ways:

Molarity

Molarity is the number of moles [i.e., the molecular weight or formula weight in grams called gram molecular weight* (GMW)] of solute in one

*Value of GMW of a compound = sum of (atomic weight of every element × number of atoms). See Appendix A.31 for atomic weights.

liter of solution. The usual symbol for molarity is M. All solutions of the same molarity contain in the same volume of solution the same number of solute molecules. [1 M sucrose solution = one mole solid solute (342 grams) plus enough water to make the total volume of the solution 1000 ml or one liter.]

$$M = \frac{\text{grams of solute}}{\text{liter of solution} \times \text{GMW of solute}}$$

or

$$M = \frac{\text{grams of solute} \times 1000}{\text{milliliters of solution} \times \text{GMW of solute}}$$

Examples and Problems

To make 250 ml of 0.05 M magnesium chloride ($MgCl_2$), first look in the atomic weight table (see Appendix A.31 and find Mg = 24.31 and Cl = 35.45; thus 24.31 + 2(35.45) = 95.21, the gram molecular weight of $MgCl_2$.

Substituting in the above formula,

$$0.05\ M = \frac{\text{grams of } MgCl_2 \times 1000 \text{ ml}}{250 \text{ ml} \times 95.21 \text{ GMW}}$$

$$1000 \times \text{grams of } MgCl_2 = 0.05 \times 250 \times 95.21 = 1190.125$$

$$\text{grams of } MgCl_2 = \frac{1190.125}{1000} = 1.19025$$

So one would weigh on the analytical balance 1.190 g of $MgCl_2$, place it in a 250-ml volumetric flask, add distilled water to dissolve this solute, and make up the volume to the 250 ml mark on the flask.

How many grams of KCl are required to make 1.00 liter of a 4.00 M solution?

The GMW of KCl is 39.1 g + 35.5 g = 74.6 g.

Thus 74.6 g KCl per liter = a 1.00 M solution.

Thus 4 × 74.6 or 298.4 g KCl per liter of solution is equivalent to a 4.00 M solution.

How many grams of $MgCl_2$ are required to make 10.00 liters of a 1.5 M solution?

95.21 g $MgCl_2$ per 1 liter = a 1.00 M solution

1.5 × 95.21 g or 142.82 g $MgCl_2$ per 1 liter = a 1.5 M solution.

Thus 10 × 142.82 or 1428.2 g $MgCl_2$ per 10 liters = a 1.5 M solution.

How many grams of $BaCl_2$ are required to make 6 liters of a 1.2 M solution?

GMW $BaCl_2$ = 137 g + 2 × 35.5 g = 208 g

208 g $BaCl_2$ per 1 liter = a 1 M solution

6 × 208 g or 1248 g $BaCl_2$ per 6 liters = a 1 M solution
Thus 1.2 × 1248 g or 1497.6 g $BaCl_2$ per 6 liters = a 1.2 M solution.

How many grams $FeSO_4$ are needed to make 500 ml of a 0.5 M solution?

GMW $FeSO_4$ = 55.8 g + 32.0 g + 4 × 16.0 g = 151.8 g
151.8 g $FeSO_4$ per 1 liter = a 1 M solution
0.5 × 151.8 or 75.9 g $FeSO_4$ per 1 liter = a 0.5 M solution

Thus $\frac{500}{1000}$ × 75.9 g or 37.95 g $FeSO_4$ per 500 ml = a 0.5 M solution.

What would be the molarity if 3.2 g NaOH were dissolved to make 250 ml of solution?

3.2 g NaOH per 250 ml of solution is equivalent to $\frac{3.2}{250}$ g NaOH per 1 ml of solution.

Thus $\frac{3.2}{250}$ × 1000 or 12.8 g NaOH per 1 liter of solution.

GMW NaOH = 40.

$\frac{12.8 \text{ g NaOH}}{40 \text{ g/mole}}$ = 0.32 moles NaOH per liter of solution makes it a 0.32 molar solution since molarity is defined as moles per liter of solution.

What would be the molarity if 10.0 g Na_2CO_3 (GMW 106 g) were dissolved to make 400 ml of solution?

10 g Na_2CO_3 per 400 ml of solution is equivalent to $\frac{10.0}{400.0}$ g Na_2CO_3 per 1 ml of solution.

Thus $\frac{10.0}{400.0}$ g × 1000 or 25 g Na_2CO_3 per 1 liter of solution.

Since GMW = 106, $\frac{25 \text{ g}}{106 \text{ g/mole}}$ = 0.24 moles or 0.24 molar.

How many grams of NaCl are there in 2 liters of a 3.0 M solution?

Molarity × liters × weight = grams

$1 = \frac{g}{M \times w}$? grams = 2 liters × $\frac{3 \text{ M NaCl}}{1 \text{ liter}} \times \frac{58.5 \text{ g NaCl}}{1 \text{ mole NaCl}}$

= 351 g NaCl

How would you make 500 ml of a 2 M solution of NaCl?

Since 500 ml is one-half of 1 liter, it would take one-half of the GMW of NaCl to make a 1 M solution. To make a 2 M solution, it would take then the full GMW which is 58.5 g of NaCl and water added until the total volume of the solution was 500 ml. The resulting solution is 2 M in NaCl.

A bottle of commercial sulfuric acid (H_2SO_4) is labeled 90% sulfuric acid: density 1.8 g/ml. What is the molarity of this solution? Note: Formula weight H_2SO_4 = 98 g.

One must find the number of moles in 1 liter of this solution.

$$1 \text{ liter} \times \frac{1000 \text{ ml}}{1} \times \frac{1.8 \text{ g}}{1 \text{ ml}} \times \frac{90 \text{ g } H_2SO_4}{100 \text{ g soln.}} \times \frac{1 \text{ mole}}{98 \text{ g}}$$

$$= \frac{1 \times 1000 \times 1.8 \times 90}{100 \times 98} = 16.43 \text{ moles.}$$

The concentration is thus 16.43 M.

A liter of solution of NaCl was labeled 1 M. A. Is this solution isotonic, hypotonic, or hypertonic? B. Make a half liter of isotonic solution from a stock of this 1 M solution.

(Note: An *isotonic solution* is one in which the salt concentration is equal to 0.85% and if the salt used is NaCl then it is also called physiological saline. *Hypotonic* and *hypertonic* refer to salt concentrations less than 0.85% and greater than 0.85%, respectively.)

A. 1 M is made by dissolving 58.5 g of NaCl in 1 liter (58.5 is the formula weight of NaCl).

58.5 g per 1000 ml is the same as 5.85 g per 100 ml, which is 5.85%.

Therefore, the answer to the first part of the question is that a concentration of 5.85% salt represents a hypertonic solution, i.e., a solution whose concentration is greater than 0.85%.

B. To make one-half liter of isotonic solution from 1 M, one should pour 100 ml of the 1 M salt solution into a graduated cylinder and then make up the volume to the 688 ml mark using distilled deionized water. After mixing, a volume of 500 ml, i.e., one-half liter, is taken and labeled isotonic or physiological saline solution.

$$x \text{ ml} = \frac{100 \text{ ml} \times 5.85\%}{0.85\%} = \frac{585}{0.85\%} = 688.23 \text{ ml}$$

Gram Equivalent Weight

An equivalent weight of a substance participating in a neutralization reaction is that weight equivalent in reacting power to an atom of hydrogen or mathematically the mass of a substance numerically equal to the ratio of its molecular weight divided by its valence. A milliequivalent weight is one-thousandth of the equivalent weight. A gram equivalent weight is the equivalent weight expressed in grams and is therefore that weight equivalent in reacting power to a gram atom (1.008 g) of hydrogen.

Monobasic acids, monoacidic bases, and salts with monovalent ions have equivalent weights and molecular weights of the same value. Acids

Table 2.4. Equivalent Weight of a Few Common Elements

Element	*Equivalent Weight (grams)*
Hydrogen	1.008
Oxygen	8.000
Sulfur	16.000
Carbon	3.000
Bromine	79.920
Aluminum	9.000
Calcium	20.040
Zinc	32.700
Sodium	23.000
Potassium	39.100
Lead	103.610
Chlorine	35.460

Equivalent Weights of Some Radicals*

Radical	*Formula*	*Equivalent Weight (grams)*
Sulfate	SO_4	48.0
Nitrate	NO_3	62.0
Ammonium	NH_4	18.0
Carbonate	CO_3	30.0
Chromate	CrO_4	58.0
Chlorate	ClO_3	83.5
Hydroxide	OH	17.0

Note: Equivalent weight = $\frac{\text{molecular weight}}{\text{valence}}$

*A radical is found in an inorganic compound containing more than two elements whose chemical behavior is such that the compound may be considered as being composed of only two groups, one of which may be an element and the other a complex group containing more than one element. For example, $CaSO_4$ behaves chemically as though it were made of two groups, calcium (Ca) and sulfate (SO_4). For compounds containing such radicals, one equivalent of the compound will contain one equivalent of the element and one equivalent of the radical. For example, the equivalent weight of sodium sulfate is 71.0 g.

like sulfuric acid $H_2^+ (SO_4)^{--}$ (valence = 2), bases like barium hydroxide $Ba^{++} (OH)_2^-$ (valence = 2), and salts like calcium chloride $Ca^{++}Cl_2^-$ (valence = 2) have equivalent weights that are one-half the value of their molecular weights. The equivalent weight of the salt ferric nitrate $Fe^{+++} (NO_3)_3^-$ (valence = 3) is one-third the molecular weight. (Note: Valence (or oxidation state) is the number of equivalents of an element or compound present in one mole of the compound.) (See Table 2.4.)

Examples and Problems

What are the equivalent weights of barium and oxygen in barium oxide, BaO, and in barium peroxide, BaO_2?

$$\text{BaO, equivalent weight of Ba} = \frac{137.36}{2} = 68.68$$

$$\text{equivalent weight of O} = \frac{16.0}{2} = 8.00$$

$$\text{BaO}_2\text{, equivalent weight of Ba} = \frac{137.36}{2} = 68.68$$

$$\text{equivalent weight of O} = \frac{16.0}{1} = 16.00$$ (because oxygen has an oxidation number of −1 in peroxides).

Ten grams of a metal react with an acid to yield 0.080 g hydrogen. What is the equivalent weight of the metal?

(Note: To find the equivalent weight of the metal, one must find what weight of it will yield one equivalent weight of hydrogen, i.e., 1.008 g).

$$\frac{10.0 \text{ g}}{0.08 \text{ g}} \times 1.008 \text{ g} = 126 \text{ g} = \text{equivalent weight of the metal}$$

In oxidation-reduction reactions or complex-formation titration, additional calculations are necessary, therefore, chemists prefer the use of the molarity (described above) over normality when expressing concentration of salts.

Normality

The normality of a solution is the number of gram equivalents of solute in 1 liter of solution, or gram milliequivalents in a ml of solution. Normality is commonly used in expressing concentration of acids and bases while molarity is used for expressing the concentration of salts. The normality of a solution is its relation to the normal solution and the symbol N is used as the abbreviation for the word *normal.*

$$N = \frac{\text{grams of solute}}{\text{liter of solution} \times \text{equivalent weight}}$$

or

$$N = \frac{\text{grams of solute} \times 1000}{\text{milliliter of solution} \times \text{equivalent weight}}$$

Examples and Problems

To make 500 ml of 0.002 N barium hydroxide $Ba(OH)_2$, first find the equivalent weight of this base which is its gram molecular weight divided by two (171.38/2 = 85.69).

Substituting in the above formula,

$$0.002\,N = \frac{\text{grams of barium hydroxide} \times 1000}{500 \text{ ml} \times 85.69 \text{ eq wt}}$$

grams of barium hydroxide × 1000 = 0.002 × 500 × 85.69

$$\text{grams of barium hydroxide} = \frac{85.69}{1000} = 0.08569 \text{ g}$$

which one must weigh and dissolve in a volume of 500 ml of water (equivalent weight/1000 ml)

How many grams of NaOH are required to make 7.0 liters of a 4.0 N solution?

Note: The equivalent weight of a base is the gram molecular weight divided by the number of hydroxyl groups (OH^-) present.

The equivalent weight of NaOH is 40 g

$$4.0\,N = \frac{\text{grams of solute}}{7 \times 40}$$

grams of solute = 4.0 × 7 × 40 = 1120 g

What would be the normality of a solution containing 4 g NaOH per 200 ml?

4 g NaOH per 200 ml is equivalent to $\frac{4}{200}$ g or 0.02 g per 1 ml of solution.

Therefore, per liter of solution there are 0.02 × 1000 ml, or 20 g NaOH/liter.

Since the equivalent weight is 40 g, then 20 g NaOH is 0.5 equivalent weights (20 g/40 g), which is 0.5 N.

What is the normality of concentrated H_2SO_4, specific gravity 1.86, 99% by weight H_2SO_4?

Assume there is 1 liter of solution, the density of which is 1.86 g/ml.

Therefore, since mass = density × volume
= 1.86 g/ml × 1000 ml or 1860 g.

The weight of the solution is 1860 g.

99%, or 1841.4 g, of this weight is pure H_2SO_4.

The equivalent weight of H_2SO_4 is 98/2 or 49 g.

Thus the normality is $\frac{1841.4\text{g}}{49 \text{ g/eq wt}} = 37.58$.

There are 37.58 equivalents per liter of solution, which is 37.58 N.

It is not necessary to calculate both the molarity and the normality of a given solution since they both depend on the same formula weight. The normality is always a whole number multiple of the molarity because one formula weight must contain a whole number of hydrogen (H^+ or OH) equivalents. For example, KOH has a molecular weight and an equivalent weight of 56 so that the molar and normal strengths are always the same. For example, in a 0.5-liter solution containing 140g of KOH the molarity and normality are calculated as follows:

$$\text{Molarity} = \frac{140 \text{ g}}{0.5 \text{ liter} \times 56 \text{ GMW}} = 5 \text{ M}$$

$$\text{Normality} = \frac{140 \text{ g}}{0.5 \text{ liter} \times 56 \text{ eq wt}} = 5\ N$$

$CaCO_3$ has a molecular weight of 100 g and an equivalent weight of 50 g so that there are always twice as many equivalent weights as molecular weights in any given amount. Its normality, therefore, is always twice its molarity. Thus 6 M $CaCO_3$ is 12 N $CaCO_3$.

The concentration of a solution is given as 2.0 g of NaCl per liter of solution. Convert this to units of molarity and normality.

1 mole NaCl = 58.5 g NaCl

$$\text{Molarity} = \frac{2.0 \text{ g NaCl}}{1 \text{ liter} \times 58.5 \text{ g}} = 0.0342 \text{ M}$$

$$\text{Normality} = \frac{2.0 \text{ g NaCl}}{1 \text{ liter} \times 58.5 \text{ g}} = 0.0342\ N$$

because the molecular and equivalent weights of NaCl are the same.

One milliequivalent of any acid will neutralize one milliequivalent of any base, thus the following relation has been established:

$$\text{ml of acid} \times N \text{ of acid} = \text{ml of base} \times N \text{ of base}$$

How many milliliters of 2.0 N hydrochloric acid HCl can be neutralized by 20.0 ml of 1.5 N potassium hydroxide KOH?

Substituting in the above formula,

x ml of HCl × 2.0 N = 20 ml of KOH × 1.5 N

$$x \text{ ml of HCl} = \frac{20 \times 1.5}{2.0} = 15 \text{ ml}$$

If sulfuric acid H_2SO_4 was used in this example in place of HCl, the answer would still be 15 because ml × N of any acid equals ml × N of any base.

A 70-ml sample of NaOH solution reacts completely with 50 ml of 0.1 N H_2SO_4. What is the normality of the base?

$$70 \text{ ml} \times N = 50 \text{ ml of } 0.1\,N$$

$$N = \frac{50 \text{ ml} \times 0.1\,N}{70 \text{ ml}} = 0.07\,N$$

How many ml of 0.2 N KOH are required to titrate 50 ml of 2.5 N HNO_3?

$$\text{ml} \times 0.2 = 50 \text{ ml} \times 2.5\,N$$

$$\text{ml} = \frac{50 \times 2.5}{0.2} = \frac{125}{0.2} = 625 \text{ ml}$$

Percent Composition

There are three common ways of denoting percent composition. The weight percent is the number of grams of solute in 100 g of solution. Thus a 10% solution of sodium chloride is prepared by dissolving 10 g of salt in enough water to make 100 g.

$$\text{weight percent} = \frac{\text{weight of solute}}{\text{weight of solution}} \times 100$$

Example: 60 percent nitric acid HNO_3 means that the reagent contains 60 g of HNO_3 per 100 g of the concentrated solution.

What is the percentage of sulfur in a mixture containing 100 pounds of sulfur and 50 pounds of magnesium?

$$\text{Weight \% sulfur} = \frac{100 \text{ lb sulfur} \times 100}{150 \text{ lb mixture}} = 66.6\%$$

Weight percent has the great advantage of being independent of temperature.

$$\text{volume percent} = \frac{\text{volume of solute}}{\text{volume of solution}} \times 100$$

Example: 30% ethanol refers to the presence of 30 ml of ethanol in every 100 ml of liquid

$$\text{weight-volume percent} = \frac{\text{weight of solute in grams}}{\text{volume of solution in ml}} \times 100$$

Example: 15% sodium acetate solution refers to a solution which was prepared by dissolving 15 g of sodium acetate in water and then diluting to 100 ml. The expression of percent in this manner is most frequently used in chemistry laboratories.

The denominator in all the above cases refers to the solution and not to the solvent alone.

Mole Fraction and Mole Percent

A mole of a substance is its molecular weight expressed in grams. This is the most useful unit since one mole of every compound contains 6.023×10^{23} molecules (the Avogadro number). A millimole and a micromole represent one-thousandth and one-millionth of a mole, respectively.

The equation 1 mole = 6.023×10^{23} molecules gives one a means of translating from moles to number of molecules, and since moles are related to units of grams, all three sets of units can be interconverted.

Examples and Problems

How many molecules are there in 200 g of water (H_2O)?

$$1 \text{ mole } H_2O = 18 \text{ g}$$
$$1 \text{ mole } H_2O = 6.023 \times 10^{23} \text{ molecules}$$
$$200 \text{ g } H_2O = 200 \text{ g } H_2O \times \frac{1 \text{ mole}}{18 \text{ g}} \times \frac{6.023 \text{ molecules}}{1 \text{ mole}}$$
$$= 6.69 \times 10^{24} \text{ molecules } H_2O$$

What is the mass of 1 atom of copper (Cu)?

$$1 \text{ atom Cu} = \frac{1 \text{ atom Cu}}{1} \times \frac{1 \text{ mole Cu}}{6.023 \times 10^{23} \text{ atoms Cu}} \times \frac{64 \text{ g Cu}}{1 \text{ mole Cu}}$$
$$= 10.63 \times 10^{-23} \text{ g Cu}$$

How many moles of H_2SO_4 are there in 196 g?

$$1 \text{ mole } H_2SO_4 = 98 \text{ g}$$
$$\text{Thus } 196 \text{ g } H_2SO_4 = 196 \text{ g} \times \frac{1 \text{ mole}}{98 \text{ g}} = 2 \text{ moles of } H_2SO_4$$

How many grams of $CaCl_2$ are there in 50 moles?

$$1 \text{ mole } CaCl_2 = 111 \text{ g } CaCl_2$$
$$\text{Thus } 50 \text{ moles } CaCl_2 = 50 \text{ moles } CaCl_2 \times \frac{111 \text{ g}}{1 \text{ mole}} = 5{,}550 \text{ g } CaCl_2$$

How many moles are there in 120 g of NaCl?

$$120 \text{ g NaCl} = 120 \text{ g NaCl} \times \frac{1 \text{ mole}}{58.5 \text{ g}} = 2.05 \text{ moles NaCl}$$

Concentrations of solutions are sometimes expressed in terms of the mole fraction of the solute or solvent present. The *mole fraction* of the solute is $\frac{n}{n+N}$ where n is the number of moles of solute present and N is the number of moles of solvent. Mole fractions of solute and solvent should add up to one. *Mole percent* equals mole fraction times one hundred.

A solution contains 4 moles of salt in 50 moles of water. What is the mole fraction of the salt? of the water?

$$\text{Mole fraction of salt} = \frac{4}{4+50} = \frac{4}{54} = \frac{2}{27}$$

$$\text{Mole fraction of water} = \frac{50}{4+50} = \frac{50}{54} = \frac{25}{27}$$

Molality

The molality of a solution is the number of moles of solute in 1000 g of solvent. The usual symbol of molality is m. All solutions of the same molality contain in the same mass of solvent the same number of solute molecules. Thus 1 m sucrose solution = one mole solid solute (342 g) added to 1000 g of water.

$$m = \frac{\text{grams of solute} \times 1000}{\text{grams of solvent} \times \text{molecular weight of solute}}$$

This expression of the concentration of a solution is not very common.

Ionic Strength

Ionic strength, usually represented by the symbol u, is half the sum of the molar concentrations of the ions multiplied by the valence squared. It is a measure of the intensity of the electrical field due to ions in a solution. Thus, if a solution contains a number of different ionic species (1, 2, and 3) and their respective molarities are M_1, M_2, and M_3, in gram ions per 100 g of solvent, and Z_1, Z_2, and Z_3 are the corresponding valences (i.e., the charges carried by the ions), then the ionic strength is given by

$$u = \frac{1}{2}(M_1 Z_1^2 + M_2 Z_2^2 + M_3 Z_3^2)$$

the sum being taken for all the ions present.

Examples and Problems

Calculate the ionic strength of 0.5 M NaCl and 0.5 M $MgCl_2$ solutions.

0.5 M NaCl $\quad u = \frac{1}{2}(0.5 \times 1^2 + 0.5 \times 1^2) = 0.5$ M

Note: Molarity equals ionic strength in salts which have monovalent charge.

$$0.5\ M\ MgCl_2 \quad u = \frac{1}{2}(0.5 \times 2^2 + 2\,(0.5 \times 1^2)) = 1.5\ M$$

Calculate the ionic strength of 0.1 M Na_2HPO_4 assuming complete ionization of the salt to give

$$Na_2HPO_4 \longrightarrow 2Na^+ + HPO_4^{--}$$

$$u = \frac{2^*\,[Na^+]\,(ZNa^+)^2 + [HPO_4^{--}]\,(ZHPO_4^{--})^2}{2}$$

$$= \frac{2\,[0.1]\,(1)^2 + [0.1]\,(-2)^2}{2}$$

$$= \frac{0.2 + 0.4}{2}$$

= 0.3 M (note that the units of ionic strength are those of concentration).

The reporting of concentrations per unit weight of solvent or solution (molality, weight percent, mole fraction) has the advantage that these values will be independent of temperature. The density of a liquid changes with temperature; therefore, any concentration scale based on volume will vary slightly with the variation of temperature, and a solution which is 1.00 molar at 5°C will be only about 0.96 molar at 95°C. For dilute aqueous solutions, the molarity and the molality are approximately the same, since a liter of dilute solution does contain about 1000 g of water (density = 1.0 at 20°C). For concentrated solutions, as well as solvents other than water (which have densities different than 1.0), this is not likely to be true. An extreme case is that of solutions in mercury, for which a 0.1 molal solution (moles in 1000 g) is approximately 1.3 molar (moles per 1 liter).

Saturated Solutions

A saturated solution at some given temperature is one in which the molecules of the solute in solution are in equilibrium with undissolved molecules. A supersaturated solution is one in which more solute is in solution than is present in a saturated solution of the same substances at the same temperature and pressure. This is accomplished by heating the solution, adding more solute, and then carefully cooling. The solute that went into solution at the higher temperature does not appreciably precipitate out of solution upon cooling. The cool solution is then considered a saturated solution.

---

*Please note that brackets [ ] are used to denote concentrations.

The method of percent saturation has the advantage that weighing is not needed. For example, if one dissolves a solute of ammonium sulfate $(NH_4)_2SO_4$ in a certain quantity of water until no more dissolves, the solution at that temperature is considered 100% saturated in ammonium sulfate. This is not the same as 100 g per 100 ml. In fact, in the case of ammonium sulfate, a salt which has a great solubility in water, a solution of it is 100% saturated when it contains 76.14 g in 100 ml at 23°C. If a liter solution is desired whose percent saturation in ammonium sulfate is 50%, one simply takes 500 ml of the saturated solution and mixes it with 500 ml of water. If, on the other hand, a 25% saturation is desired, the following formula must be followed:

$$\% \text{ saturation in decimal notation} = \frac{x}{V + x}$$

where V = the volume in ml of solution to be treated and
x = ml of the saturated solution to be added

For example, if one has 1000 ml of water and wants to make its percent saturation in sucrose 25%, apply the formula:

25% = 0.25 in decimal notation

$$0.25 = \frac{x}{1000 + x}$$

$$x = 250 + 0.25x$$

$$0.75x = 250$$

$$x = \frac{250}{0.75} = 333.3$$ is the number of milliliters of saturated solution of sucrose one must add to the 1000 ml of water to obtain 25% saturation

To increase the percent saturation in the 1333.3 ml of solution to 40%

$$0.40 = \frac{x}{1000 + x}$$

$$x = 400 + 0.4x$$

$$0.60x = 400$$

$$x = \frac{400}{0.60} = 666.6 \text{ ml}$$

Since 333.3 ml were added already to make the 25% saturated solution, 666.6 - 333.3 = 333.3 ml must be added to the 1333.3 ml to increase the concentration of sucrose in this solution from 25% to 40% saturation.

Aliquot

An aliquot is defined as any part that is contained a whole number of times in a quantity. It is simply a portion of a whole.

Dilution and Dilution Factor

A dilute solution is one which contains a small proportion of solute. When a chemist makes a 1 liter solution of, for example, 1 M KCl, he weighs 56.1 g, the GMW of this salt, dissolves it in a liter volumetric flask, and adds water to the liter mark. This stock solution can then be used for making various dilutions.

$$\text{final molarity desired} = \frac{\text{aliquot of stock soln.} \times \text{molarity of stock soln.}}{\text{volume in milliliters of solution desired}}$$

Case 1: If a liter of 0.001 M solution of KCl is desired, one can take 1 ml aliquot of the 1 M stock and add it to 999 ml of water. Case 2: If only 10 ml of 0.001 M KCl is desired, one takes 0.01 ml and adds to it 9.99 ml of water, and so on. Case 3: If a 100 ml solution of 0.000001 M, i.e., 1×10^{-6} M, is desired from the above 1 M stock, one does not have to measure 0.0001 ml and make it up to 100 ml, but instead makes first a 0.001 M solution as in Cases 1 and 2 and then takes 0.1 ml of that and makes it up to 100 because $\frac{0.1 \times 0.001}{100} = 0.000001$ M, the desired concentration.

The concentrations of weak solutions are very frequently expressed in terms of dilution factor (i.e., one part per so many parts).

Examples and Problems

Determination of dilution factor.

What is the dilution factor needed to make a final concentration of 0.0001 M KCl from 0.05 M KCl?

Using the relation $\text{dilution factor} = \frac{\text{initial concentration}}{\text{final concentration}}$

$$\frac{0.0500}{0.0001} = \text{dilution factor} = 500 \text{ times}$$

Thus one should take 1 part of the 0.05 M KCl and make it up to 500 by adding 499 parts of water, or whatever solvent it is to be diluted in.

If one has a 1 M solution and then takes 1 ml of it and dilutes it with water to 8 ml and then takes from the resulting solution 0.2 ml and dilutes it again with water to 5 ml, what is the final dilution?

$$1 \times \frac{1}{8} \times \frac{0.2}{5} = \frac{0.2}{40} = \frac{2}{400} = \frac{1}{200}$$

The answer is 1 to 200 or 1:200.

A laboratory bottle is labeled 5.0 M KCl. How would one make from this 100 ml of a 2.0 M KCl solution?

$$0.2 \text{ M KCl} = \frac{x \text{ ml} \times 5.0 \text{ M}}{1000 \text{ ml}}$$

$$200 \text{ M ml} = x \text{ ml} \times 5.0 \text{ M}$$

$$x \text{ ml} = \frac{200 \not{M} \text{ ml}}{5.0 \not{M}} = 40 \text{ ml}$$

So take 40 ml of the 5.0 M KCl stock solution and add 60 ml of H_2O to make 100 ml of 2.0 M KCl.

A blood sample was diluted 1 to 15 with physiological saline solution for preparation of a platelets count. The blood platelets were counted using a 0.4 cu mm chamber. If the number of platelet cells found was 8000, what is the platelet count per 1 cu mm of the patient's blood?

By the use of proportion, one finds that if there are
8000 cells in 0.4 cu mm, there will be
8000 × 2.5 = 20,000 cells per 1 cu mm
Since the blood was diluted 1:15,
20,000 × 15 = 300,000 platelets per 1.0 cu mm of the patient's blood.

Since all percentages are a ratio of parts per hundred, dilution factor is merely another way of expressing the concentration of solutions in terms of percent (and less frequently of the percent of mixtures of solids). For example, 5% means 5 parts per 100 or 1:20 which is read 1 to 20. The dilution factor and percent concentration are proportional and can easily be interconverted by the use of proportion and cross multiplication as illustrated in the following examples:

Given the concentration in percent, determine the dilution factor.

Example: Change 0.08% to a dilution factor.
Since dilution factor is defined as one part per x parts,

$$\frac{0.08\%}{100\%} = \frac{1 \text{ part}}{x \text{ parts}}$$

$x = 1250$, i.e., the dilution factor = 1:1250.

Given the dilution factor, determine the concentration in terms of percent.

Example: Change 1:8000 to percent concentration.

$$\frac{8000 \text{ parts}}{1 \text{ part}} = \frac{100\%}{x\%}$$

$x = 0.0125\%$

Calculation of the dilution factor or percent concentration of a solution of a known concentration and volume.

Examples: What is the dilution factor of a solution that contains 8 mg of a substance per ml of solution?

8 mg = 0.008 g

$$\frac{0.008 \text{ g}}{1 \text{ g}} = \frac{1 \text{ ml}}{x \text{ ml}}$$

Since x = 125 ml, the dilution factor = 1:125.

What is the final concentration in terms of percent of 800 ml of a 10% solution which was diluted to 2500 ml?

$$\frac{2500 \text{ ml}}{800 \text{ ml}} = \frac{10\%}{x}$$

$x = 3.2\%$

What is the final dilution factor of a homogeneous protein solution if 10 ml of a 1:50 dilution of the solution is diluted to 250 ml?

1:50 = 2.0%

$$\frac{250 \text{ ml}}{10 \text{ ml}} = \frac{2.0\%}{x}$$

Since $x = 0.08\% = 0.08$ per 100, i.e., $\frac{0.08}{100} = 1250$,

the dilution factor is 1:1250.

What is the dilution factor of a solution of a compound if 100 ml contain 25 g?

100 ml of water weigh 100 g

$$\frac{100 \text{ g}}{25 \text{ g}} = \frac{100\%}{x}$$

$x = 25\%$

Preparation of a solution of a desired concentration (expressed in terms of dilution factor) from another solution of known concentration.

Examples: To make 10 liters of 1:1000 solution of a certain compound, how many milliters of a 1:20 stock solution are used?

10 liters = 10,000 ml

1:20 = 5.0% 1:1000 = 0.1%

$$\frac{5.0\%}{0.1\%} = \frac{10{,}000 \text{ ml}}{x}$$

x = 200 ml are needed

So take 200 ml of the stock solution and add 9800 ml solvent to make the volume 10 liters.

How much water should be added to 250 ml of a 1:1000 solution of a substance to make a final concentration of 1:2000?

1:1000 = 0.1% 1:2000 = 0.05%

$$\frac{0.05\%}{0.1\%} = \frac{250 \text{ ml}}{x \text{ ml}}$$

x = 500 ml of 0.05% solution

500 ml − 250 ml = 250 ml

Determine the relative amounts of solutions of different concentrations that should be used to make a mixture of a desired concentration.

Examples: What will be the resulting protein concentration if one mixes 0.5 ml of a 2.5 mg/ml solution with 0.5 ml of a 5.0 mg/ml solution?

$$\begin{array}{rcrcr} 2.5 & \times & 0.5 & = & 1.25 \\ 5.0 & \times & \underline{0.5} & = & \underline{2.5} \\ & & 1.0 & & 3.75 \end{array}$$

$$\frac{3.75}{1} = 3.75 \text{ mg/ml}$$

What is the resulting concentration of a mixture of 100 ml of 0.1 M compound and 50 ml of 0.4 M of the same compound?

$$\begin{array}{rcrcr} 0.1 & \times & 100 & = & 10 \\ 0.4 & \times & \underline{50} & = & \underline{20} \\ & & 150 & & 30 \end{array}$$

$$\frac{30}{150} = 0.2 \text{ M}$$

What is the resulting concentration of a mixture of 200 ml of a 14% sugar solution and 75 ml of a 5% sugar solution?

$$\begin{array}{rcrcr} 14 & \times & 200 & = & 2800 \\ 5 & \times & \underline{75} & = & \underline{375} \\ & & 275 & & 3175 \end{array}$$

$$\frac{3175}{275} = 11.55\%$$

In what proportion should alcohols of 70% and 20% strengths be mixed to make 50% alcohol?

Let x = proportion of 70% alcohol

Let y = proportion of 20% alcohol

$0.70x + 0.20y = 0.50(x + y)$ or

$70x + 20y = 50x + 50y$

$70x - 50x = 50y - 20y$

$x(70 - 50) = y(50 - 20)$

$$\frac{x}{y} = \frac{50 - 20}{70 - 50} = \frac{30 \text{ parts of } 70\%}{20 \text{ parts of } 20\%}$$

Given a liter of 95% alcohol, how much water should be mixed to this liter to make 70% alcohol?

$$\frac{70\%}{95\%} = \frac{1000 \text{ ml}}{x}$$

$x = 1357$ Therefore use 1000 ml of 95% alcohol and enough water to make 1357 ml.

What is the percentage of alcohol in a mixture of 1500 ml of 20% alcohol, 500 ml of 70% alcohol, and 1000 ml of 95% alcohol?

$$\begin{aligned} 20 \times 1500 &= 30{,}000 \\ 70 \times 500 &= 35{,}000 \\ 95 \times \underline{1000} &= \underline{95{,}000} \\ 3000 &\quad 160{,}000 \end{aligned}$$

$$\frac{160{,}000}{3{,}000} = 53.3\%$$

Determine the volume of a concentrated solution required to prepare a desired quantity of a diluted solution.

Examples: A chemical supplier distributes enzyme concentrates. One sample was labeled enzyme concentration = 4 mg/ml; specific activity = 2000 units per mg. What dilution factor must one make to obtain a sample containing 50 units in 0.2 ml of solvent?

Dilution factor is 1 part to x parts. Therefore by proportion

$$\frac{50}{2000} = \frac{1}{x} \quad x = \frac{2000 \times 1}{50} = 40.$$ So the dilution factor is 1:40.

By proportion again, each 40 ml of solution contains 1 ml of enzyme. In 0.2 ml of solution, how much enzyme should there be?

$$\frac{40}{1} = \frac{0.2}{x} \quad x = \frac{1 \times 0.2}{40} = 0.005.$$ Therefore one would take 0.005 ml of enzyme and add 0.195 ml of solvent.

How many milliliters of 20% brine solution having a specific gravity of 1.1478 are required to make 500 ml of diluted brine at 6%?

$500 \times 0.06 = 30$ g of salt in 500 ml of 6% brine

$$\frac{20\%}{100\%} = \frac{30\text{ g}}{x}$$

$x = 150$ g of 20% brine

150 g of water measure 150 ml

$$\frac{150\text{ ml}}{1.1478} = 130.68\text{ ml}$$

Determine the dilution factor needed, knowing that a specific concentration of a substance is required to give a proper reading in a specific instrument (e.g., spectrophotometers, radioactive counters, polarimeters, etc.).

Examples: A heme protein under certain experimental conditions exhibited a maximum light absorption at 500 nm. When a solution of this protein was read in the spectrophotometer, it gave an absorbancy of 0.75. What dilution factor should one make to get an absorbancy of 0.05?

$$\frac{0.05}{0.75} = \frac{1}{x} \qquad x = \frac{0.75 \times 1}{0.05} = 15.$$ So the dilution factor is 1:15.

Given a similar heme protein with a protein concentration of 10 mg/ml, and an absorbancy at 500 nm of 0.9, to what concentration must this protein be diluted to give an *OD* of 0.15? (*OD* is defined in the next section.)

$$\frac{10}{0.9} = \frac{x}{0.15} \qquad x = \frac{10 \times 0.15}{0.9} = 1.667\text{ mg/ml}$$

In determining the catalytic efficiency of some crystalline enzyme preparation, it was found that it must be diluted to a concentration of 0.10 mg/ml to get a reading in the range of 200–400 on some instrument. What dilution must be made to get this value if this enzyme preparation was concentrated to 50 mg/ml? In this case the range 200–400 can be assumed to equal a unit of 1. Therefore by the use of proportion,

$$\frac{0.1}{1} = \frac{50}{x} \qquad x = \frac{1 \times 50}{0.1} = 500.$$

Hence the dilution is 1 to 500, i.e., 1:500 or 499 ml of solvent must be added to 1 volume of the enzyme. In practice one takes 0.01 ml of the enzyme and adds 4.99 ml of solution.

Conversion of percent wet weight to percent dry weight.

Example: If some moist compound contains 8% solute and 40% water, what will be the percentage of solute after the compound is dried?

100 g of moist compound would contain 40 g of water and would therefore weigh 60 g after drying.

$$\frac{60\%}{100\%} = \frac{8.0\%}{x}$$

$$x = 13.33\%$$

Making stock solutions of various substances in terms of percent concentration.

Examples: How many g of glucose are required to prepare 5000 ml of a 5% solution?

volume in ml × percent expressed as a decimal = grams of solute

5% = 0.05

5000 × 0.05 = 250 g of solute

How many milliliters of a 6% stock solution can be made from 60 g of a compound?

$$\frac{6\%}{100\%} = \frac{60\text{g}}{x\text{ g}}$$

$x = 1000$ g, weight of the solution if it were water
volume in ml = 1000 ml

Conversion of percent concentration to grams per milliter to molarity.

Example: Given a 1% solution of a protein whose molecular weight is 180,000 daltons, i.e., grams/mole, and a 1% solution of sodium chloride whose molecular weight is 58.5, what is the concentration in g/ml and the molarity?

For the protein, the 1% is 1 g per 100 ml of water = 0.01 g/ml = 10 mg/ml

$$\text{Molarity (M)} = \frac{\text{grams of solute} \times 1000}{\text{ml of solution} \times \text{GMW of solute}}$$

$$M = \frac{0.01 \times 1000}{1 \times 180{,}000} = 0.000055555 = 5.56 \times 10^{-5}\ M$$

For the salt, the 1% is also 1 g per 100 ml of water or 0.01 g/ml

$$M = \frac{0.01 \times 1000}{1 \times 58.5} = 0.171 = 1.71 \times 10^{-1}\ M$$

2.9 DETERMINATIONS OF SOLUTE CONCENTRATION BY OPTICAL DENSITY MEASUREMENTS

Beer's Law

The fraction of the incident light absorbed by a solution at a given wavelength is related to the thickness of the absorbing layer and to the concentration of the solution. This is the essence of the Beer-Lambert law, best known as Beer's law.

For a solution having a characteristic absorption coefficient equal to k, a constant, this law can be expressed by the following relation:

$$\frac{I}{I_0} = e^{-kCl}$$

Converting to logarithms to the base 10, inverting I/I_0 to remove the minus sign and replacing k by ϵ (the molar extinction coefficient), the following is obtained:

$$2.303 \log \frac{I_0}{I} = \epsilon Cl = OD$$

where $\frac{I_0}{I}$ = the ratio of the incident light to the emergent light

OD = the fraction of light absorbed and stands for optical density of the solution (also called absorbancy in American literature and extinction E, not extinction coefficient ϵ, in European literature)

ϵ = the extinction coefficient specific for the absorbing material and the wavelength of the radiation. The molar absorption coefficient must be known if one is to determine the concentration of a substance in solution. This coefficient is sometimes referred to as the absorbancy index, not absorbancy, in many of the European journals.

C = the concentration of the substance in solution

l = the length of the light path through a spectrophotometer cell or cuvette. Most cuvettes are made with one centimeter light path. In these cases, the formula simplifies to $OD = \epsilon C$

Before any study of concentration can be made, one must know the extinction coefficient (ϵ) of the substance to be studied. To find the extinction coefficient (ϵ) of a substance by use of the equation $\epsilon = OD/C$, one needs to know two things:

1. The concentration of the substance. This can be done by weighing a certain amount of the substance in question and dissolving it in a certain volume of solvent which does not absorb light at the same wavelength as the solute (i.e., the substance being studied).

2. The optical density (OD) of the solution of the substance measured in a one cm path length cell in a spectrophotometer against the reference solvent.

Because $OD = 2.303 \log \frac{I_0}{I}$ is a dimensionless number, when the light path length l is in centimeters and C is in terms of weight per unit volume (e.g., mole per liter), the ϵ which represents the area per unit of absorbing substance has the dimensions of $cm^2 \times mole^{-1}$ or $cm^2/mole$. Therefore, the practical aspect of Beer's Law is that once one has determined the ϵ for a given substance in a particular solvent, any concentration of that substance can be calculated from optical density measurements.

Example: What is the concentration of a protein solution whose OD reading in a 1 centimeter cuvette at 280 nanometers is 0.50 if its $\epsilon_{1\%}$ (i.e., the extinction coefficient for 1% of this protein solution) is 13.5.

First note that 1% is 1 g per 100 ml or 1000 mg per 100 ml, which is 10 mg/ml.

Thus for 1 mg/ml the value of ϵ is $\frac{13.5}{10} = 1.35$.

Now applying the formula $OD = \epsilon Cl$

$0.5 = 1.35 \times C \times 1$

$C = \frac{0.50}{1.35} = 0.37$ mg/ml is the concentration of that protein.

Most solutions obey Beer's Law if dilute,* but deviations are observed in concentrated solutions (e.g., at $OD > 1.0$); however, the more general law of Lambert (which states that equal fractions of the incident light are absorbed by successive layers of equal thickness of the light-absorbing substance) is always obeyed. For highly concentrated solutions with very high extinction coefficients, *transmittance* rather than OD is measured in the spectrophotometer. Nearly all spectrophotometers have an OD and a transmittance (T) scale represented by $\%T$. (See Fig. 2.14.)

The relation between OD and transmittance is as follows:

$$OD = \log \frac{1}{T} = -\log T$$

Conversion from $\%T$ to OD is done by simply converting the value $\%T$ to decimal notation and then substituting the value in the formula.

Example: $12.7\% = \frac{12.7}{100} = 0.127.$

$$OD = -\log 0.127 = -(1.1038) = 0.8962$$

*An OD of 0.4 to 0.5 gives a minimum of stray light in the spectrophotometer.

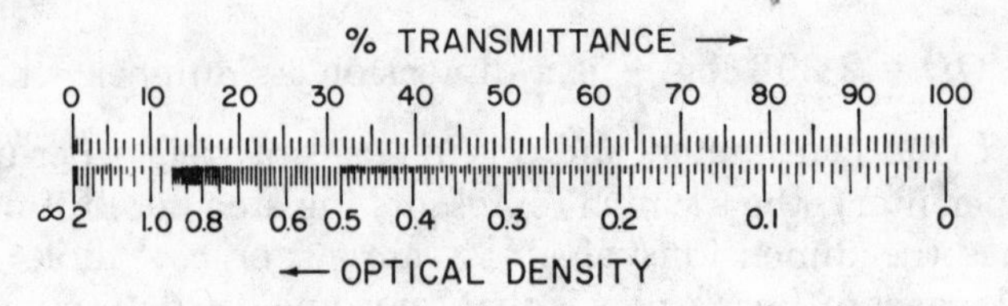

Fig. 2.14. Spectrophotometer scale.

If one has two solutions of the same substance but of different concentrations and measured their optical density at a certain wavelength in a cuvette of one centimeter path length, then by the use of proportions the relation between their optical densities and concentrations (within a range where Beer's Law is obeyed) can be determined as follows:

$$\frac{\text{concentration of standard}}{OD \text{ of standard}} = \frac{\text{concentration of unknown}}{OD \text{ of unknown}}$$

Thus,

$$\text{concentration of unknown} = \frac{OD \text{ of the solution of unknown concentration}}{OD \text{ of the same solution of known concentration (i.e., the standard)}} \times \text{concentration of the standard}$$

Example: A given bovine serum albumin solution used as a standard had a concentration of 0.7 mg/ml and an *OD* at 280 nm of 0.96 in a buffer. An unknown portion of this bovine serum albumin solution was diluted with the same buffer. What is the concentration of the resulting solution if its *OD* at 280 nm is 0.32?

$$\text{concentration of unknown} = \frac{0.32}{0.96} \times 0.8 = 0.27 \text{ mg/ml}$$

Determination of Concentration Using Empirical Formulas

Chemists working with substances of unknown extinction coefficients or of unknown purity have reworked to the use of approximate methods for determining concentrations. For example, the concentration of many protein solutions is determined by optical density measurement at one (ex. at 280 nm) or two wavelengths (ex. at 280 and 260 nm or 215 and 225 nm). The concentration of a protein in a heterogeneous sample may also be obtained by determining the amount of a specific protein constituent, as for example N_2 or Fe, whose percent in the protein is known.

In accurate spectrophotometric measurements of the concentration of solutions of pure substances, the *OD* value is divided by a predetermined extinction coefficient ϵ of the substance in question. However, when ϵ is not known, the biophysical chemist employs a formula which encompasses a certain factor. This factor is a function of the optical density characteristics of the solute in relation to that at the specific wavelength used in the measurement.

Examples of some factors used in the determination of protein concentration.

Protein concentration in μg/ml = *OD* at 215 nm − *OD* at 225 nm × the factor 144 = μg protein/ml

Protein concentration mg/ml = Factor calculated for the *OD* reading of $\frac{280 \text{ nm}}{260 \text{ nm}} \times OD$ 280 nm

Protein concentration mg/ml = Nitrogen value × the factor 6.25

Example: Determine the molar concentration and the gram percent of a plasma protein solution diluted 1 to 10,000 if it gave an *OD* of 0.8 and 0.3 at 215 nm, and 225 nm, respectively. This protein is known to have a molecular weight of 36,000 g/mole.

The concentration using the above formula in μg/ml = (0.8 − 0.3) × 144 = 0.5 × 144 = 72 × 10,000 = 720,000 μg/ml

720,000 μg/ml = 720 mg/ml
= 0.72 g/ml
= 0.0072 g/100 ml
= 0.0072 gram percent

1 M of this solution is GMW per liter, i.e., 36,000 g/1000 ml

and we have 0.0072 g/100 ml which is 0.00072 g per 1000 ml.

$$\text{Therefore the molarity} = \frac{0.00072}{36,000} = \frac{2}{1000,000,000} = 2 \times 10^{-8} \text{M} = 200\ \mu\text{M}$$

This concentration of this substance seems very small compared to expressing it in μg/ml (720,000 μg/ml versus 200 μM) because of the high molecular weight of that substance.

2.10 AVOGADRO NUMBER

The methods by which the volume of a given weight of gas—or the weight of a given volume—are calculated depend on *Avogadro's Law* which states that under the same conditions of temperature and pressure, equal volumes of all gases contain the same number of molecules.

When the molecular weight of a compound is taken to indicate a number of grams, the expression is called the *mole* or *gram molecular weight** of that compound. In all gases one mole has a common volume, 22.4 liters, and contains 6.023×10^{23} tiny particles called molecules under standard conditions of temperature and pressure (STP)—that is, 0°C and a barometric pressure of 760 mm of mercury. (See Fig. 2.15.)

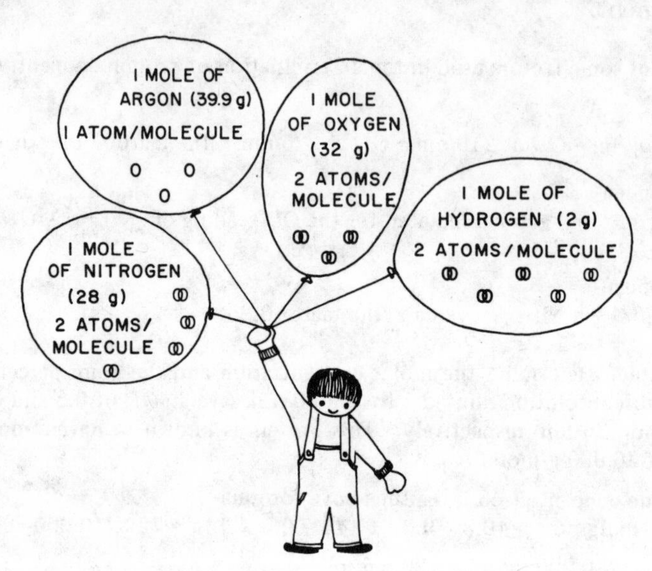

Fig. 2.15. Each balloon contains 1 mole of a gas with a volume of 22.4 liters and 6.025×10^{23} molecules.

Examples and Problems

Given the weight of a gas under standard conditions of temperature and pressure, determine its volume.

Example: What is the volume (STP) of 9.5 g of hydrogen (H_2)?

Molecular weight of hydrogen = $2 \times 1 = 2$
2 g of hydrogen measure 22.4 liters

$$\frac{2\ \text{g}}{9.5\ \text{g}} = \frac{22.4\ \text{liter}}{x\ \text{liter}} \qquad x = 106.4\ \text{liters}$$

Determine the weight of a compound required to make a specified volume of a gas.

Example: How many grams of zinc are required to make 150 liters (STP) of hydrogen?

*Thus the unit of molecular weight (MW) is g/mole.

$Zn + H_2SO_4 = ZnSO_4 + H_2$

65 2

65 g of zinc will produce 2 g of hydrogen which measure 22.4 liters

$$\frac{22.4 \text{ liters}}{150 \text{ liters}} = \frac{65 \text{ g}}{x \text{ g}} \qquad x = 435.3 \text{ g}$$

Determine the volume of a gas when correction for temperature and pressure must be made.

Example: The volume of a gas measured at 800 mm is 400 ml. What is its volume at 760 mm, if the temperature remains constant?

$$\frac{760 \text{ mm}}{800 \text{ mm}} = \frac{400 \text{ ml}}{x \text{ ml}}$$

$$x = \frac{400 \times 800}{760} = 421.05 \text{ ml}$$

2.11 GAS LAWS

The gas laws express the effects of temperature and pressure on the volume of a given mass of gas. The volume of a gas is sensitive to changes in temperature and in external pressure. An ideal or perfect gas is one for which the volume varies in direct proportion to the absolute temperature under constant pressure (*Charles's law*)—a given mass of gas will increase in volume by $\frac{1}{273}$ of its volume at 0°C for every 1°C rise in temperature. The volume of an ideal gas varies in inverse proportion to the applied pressure at constant temperature (*Boyle's law*)—e.g., by doubling the pressure on 2 liters of gas the volume will decrease until 1 liter remains. Thus, the product of P and V of a given mass of gas is constant at a fixed temperature. As the pressure goes up, the volume of a gas goes down, and as the pressure goes down, the volume goes up.

$$PV = \text{constant}$$

Therefore,

$$\frac{V_1}{V_2} = \frac{P_2}{P_1} \quad \text{or} \quad P_1 V_1 = P_2 V_2 \qquad \text{(See Fig. 2.16.)}$$

Additionally, at constant pressure,

$$\frac{V_1}{V_2} = \frac{T_1}{T_2} \quad \text{or} \quad V_1 T_2 = V_2 T_1 \qquad \text{(See Fig. 2.16.)}$$

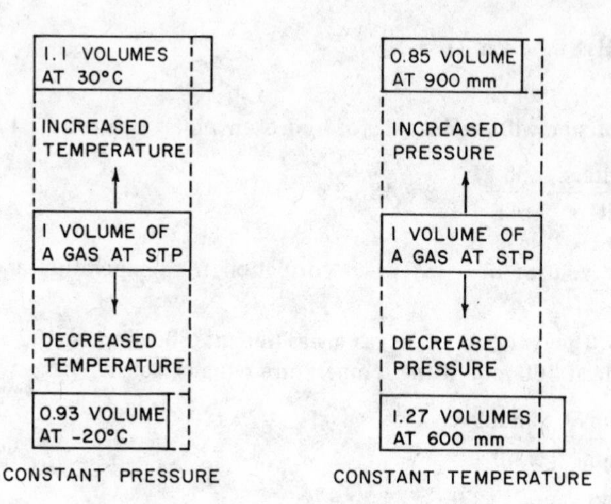

Fig. 2.16. Effect of change in temperature and pressure on a gas.

When Boyle's and Charles's laws are combined, it is possible to solve problems where both temperature and pressure change at the same time. The combined laws may be stated as follows:

$$\frac{P_1 V_1}{T_1} = \frac{P_2 V_2}{T_2}$$

The value of these laws is their usefulness for calculating, for example, the volume of a gas of any required pressure, if the volume at another pressure is known.

Example: At 273°K and 600 mm pressure, a given mass of gas occupies 500 ml. If the pressure is increased to 800 mm, what must the temperature be to give a volume of 600 ml?

Substitute the data in the combined formula and solve for T_2.

$$\frac{600 \text{ mm} \times 500 \text{ ml}}{273^\circ\text{K}} = \frac{800 \text{ mm} \times 600 \text{ ml}}{T_2}$$

$$T_2 = \frac{800 \text{ mm} \times 600 \text{ ml} \times 273^\circ\text{K}}{600 \text{ mm} \times 500 \text{ ml}} = \frac{218{,}400}{500} = 436.8^\circ\text{K}$$

The weight of 22.4 liters of a gas at STP is called gram molecular weight of the gas (GMW) and this volume is called the gram molecular volume (GMV). This number was selected because it is the volume occupied by 32 g of O_2 at STP and because O_2 is the standard for comparison in the calculation of molecular weights. The number of molecules actually present in this GMV is 6.023×10^{23}, which is called Avogadro's number or N (see previous section).

Examples and Problems

5.00 g of a gas occupy 500 ml at STP. What is the GMW of the gas?

500 ml = 5.00 g at STP

Therefore 1 ml weighs $\frac{5.0 \text{ g}}{500 \text{ ml}}$ and

22.4 liters or 22,400 ml weigh 22,400 ml $\times \frac{5.0 \text{ g}}{500 \text{ ml}} = 224$ g

224 g is the GMW of the gas since it is the weight which occupies 22.4 liters at STP.

How much will 600 ml of O_2 weigh at STP?

22,400 ml of O_2 weigh 32.0 g

1 ml weighs $\frac{32.0 \text{ g}}{22{,}400 \text{ ml}}$

600 ml weigh $\frac{32}{22{,}400} \times 600 = 0.86$ g

4.8 g of a gas occupy 250 ml at 25°C and 720 mm pressure. What is the GMW of the gas?

Volume at STP = 250 ml $\times \frac{273}{298} \times \frac{720}{760} = 216.972$

4.8 g occupies 216.972 ml at STP

Thus 1 ml weighs $\frac{4.8 \text{ g}}{216.972 \text{ ml}}$ and

then $22{,}400 \times \frac{4.8 \text{ g}}{216.972 \text{ ml}} = 495.547$ g which is the GMW of the gas.

Proportional variation implies that one quantity is equal to a constant numerical multiplier times the second quantity. If this constant is set equal to nR (where n is the number of moles of gas and R is called the gas constant), *the ideal gas law or equation becomes* $PV = nRT$. Therefore, R, the molar gas constant (which has the same value for all gases), has the dimensions of pressure times volume divided by number of moles times temperature.

$$\text{pressure} = \text{force per unit of area} = \text{force} \times \text{length}^{-2}$$

$$\text{volume} = \text{length}^3$$

The product of force and length $\left(\frac{\text{force}}{\text{length}^2} \times \frac{\text{length}^3}{1} = \text{force} \times \text{length}\right)$ is energy, and energy may be stated in various ways; the three most common are (1) liter-atmosphere, (2) ergs, and (3) calories.

One mole ($n = 1$) of an ideal gas occupies 22.414 liters at standard temperature and pressure [STP = 1 atmosphere pressure (760 mm of mercury) and 273.15° Kelvin]. Therefore,

$$R = \frac{1 \text{ atm} \times 22.414}{1 \text{ mole} \times 273.15°\text{K}} = 0.082054 \frac{\text{liter-atm}}{°\text{K mole}}$$

An erg is the work done by a force of 1 dyne operating over a distance of 1 cm, and a calorie is 4.184×10^7 ergs. Using the respective conversion factors

$$R = \frac{8.314 \times 10^7 \text{ ergs}}{°\text{K mole}} \qquad R = \frac{1.987 \text{ calories*}}{°\text{K mole}}$$

It is not always possible to collect a gas in the pure state. That is, it may be mixed with other gases. *Dalton's law* states that in a mixture of gases, each gas exerts a pressure independent of others and proportional to the relative amount by volume of that gas in the mixture. Thus a mixture of gases which is 80% by volume N_2 and 20% by volume O_2 at the total atmospheric pressure of 760 mm Hg will have 80% of 760 mm or 608 mm pressure due to N_2 and 20% of 760 mm or 152 mm pressure due to the O_2. Each gas exerts its own pressure as if it alone were present, since in the gaseous state the molecules are so far apart that there is plenty of room for other types of atoms to exist without interaction between them. Thus, if we collect a gas over water, there will be a mixture of two gases—the one collected plus some water vapor. If the concentration or the amount by volume of the water vapor is known, the pressure which is called the partial pressure can be calculated. It would be difficult to measure the amount of water vapor present in a sample collected over water. Fortunately, there is another means of obtaining the vapor pressure of the water. Since the pressure due to the water vapor is dependent only upon the temperature and not upon the size of the container, if the temperature is known, the partial pressure due to the water vapor can be found by referring to a table.

Example: The total pressure of gases (O_2, CO_2, N_2, and H_2O) in the air cells of the lungs (alveoli) equals 760 mm Hg; thus, if any three of the partial pressures of these four gases are known, the fourth one can be obtained by subtraction.

Suppose one must determine the PO_2 (partial pressure of oxygen) in a patient

*It is apparent that in general R has the units of energy degree^{-1} mole^{-1} and the unit that changes is energy.

whose arterial PCO_2 (partial pressure of carbon dioxide) is found to be 40 mm Hg. Assume the percent of nitrogen in the inspired air to be a constant at 80% and the PH_2O (partial pressure of water vapor) equal to 47 mm Hg at 37°C.

Total pressure = 760 mm Hg at STP = $PCO_2 + PO_2 + PN_2 + PH_2O$

Thus PO_2 = 760 mm Hg − PH_2O − PN_2 − PCO_2

The partial pressure due to water vapor must first be subtracted from the total pressure of 760 mm Hg.

$$760 - 47 = 713 \text{ mm Hg} = PO_2 + PN_2 + PCO_2$$

Since nitrogen is 80% of the total gases present (O_2, CO_2, N_2)

$$\therefore PN_2 = \frac{80}{100}(PO_2 + PCO_2 + PN_2)$$

$$\frac{80}{100} \times 713 = 570.4 \text{ mm Hg is } PN_2$$

$$\text{Thus } PO_2 = 760 - 47_{H_2O} - 570.4_{N_2} - 40_{CO_2}$$

$$= 760 - 657.4$$

$$PO_2 = 102.6 \text{ mm Hg}$$

The pressure of carbon dioxide (P_{CO_2}) in arterial blood also may be determined by an indirect calculation of the pressure of CO_2 from a nomogram (Fig. 2.17) based on the Henderson-Hasselbalch equation. (See Appendix A.1.) This equation is based on three unknowns: the blood pH (the normal pH values for blood are 7.35 to 7.45), the plasma bicarbonate concentration, and the partial pressure of the CO_2 in the blood. If any two are known, the third may be calculated. A straight line drawn through any two known points on any two scales crosses the other two scales at points indicating simultaneously occurring values.

Examples and Problems

Using the ideal gas law, $PV = nRT$, the volume of a mole of gas at a certain temperature and pressure can be calculated. What is the volume of 1 mole of gas at 0°C (273.16°K) and 1 atmosphere (atm) pressure?

$$V = \frac{nRT}{P} = \frac{(1 \text{ mole})(0.082054/\text{liter atm})(273.16°\text{K})}{(1 \text{ atm})(°\text{K mole})}$$

$$= 22.414 \text{ liters}$$

Note: The conditions of 0°C and 1 atmosphere are known as standard condition. The volume of a mole of gas (22.414 liters) under these conditions is known as the gram molecular volume.

A sample of gas occupying 225 cm^3 at STP was found to weigh 0.925 g. What is its molecular weight?

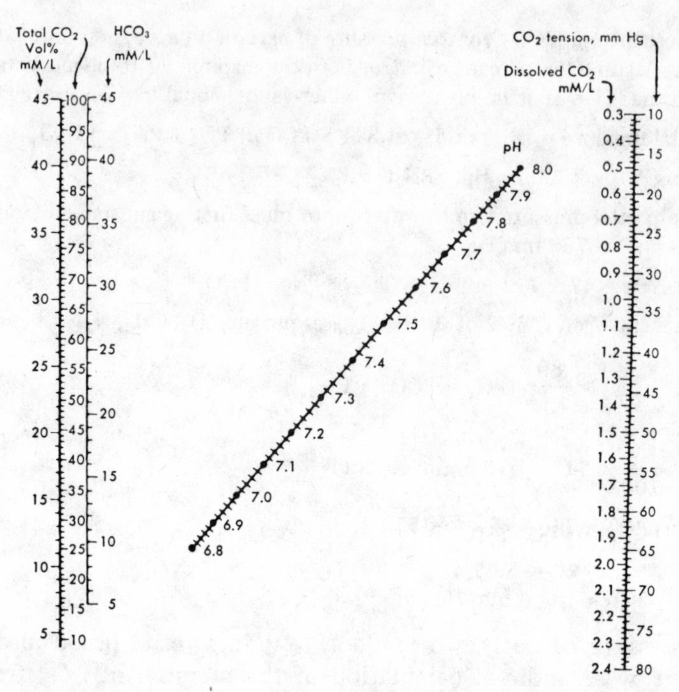

Fig. 2.17. Nomogram for calculating P_{CO_2}. What is the P_{CO_2} value for a patient with an arterial pH of 7.30 and a total CO_2 of 30 mM/liter? Draw a line from the total CO_2 column reading 30 through the pH column intersecting the 7.30 point. The line intersecting the mm Hg column gives a value of 60. Thus the patient's P_{CO_2} is 60 mm Hg. (From F. C. McLean, *Physiol. Rev.* 18:495.)

To find the molecular weight, one must find the weight in grams of 1 mole of the gas.

$$1 \text{ mole gas} = 1 \text{ mole gas} \times \frac{22.414 \text{ STP}}{1 \text{ mole gas}} \times \frac{1000 \text{ cm}^3}{1 \text{ liter}} \times \frac{0.925 \text{ g}}{225 \text{ cm}^3 \text{ STP}}$$

$$= 92.146 \text{ g} = \text{molecular weight}$$

How many grams of ammonia are there in 500 cm^3 STP of NH_3 gas?

$$500 \text{ cm}^3 \text{ STP NH}_3 = 500 \text{ cm}^3 \times \frac{1 \text{ liter}}{1000 \text{ cm}^3} \times \frac{1 \text{ mole NH}_3}{22.414 \text{ STP}} \times \frac{17.032 \text{ g NH}_3}{1 \text{ mole}}$$

$$= \frac{500 \times 17.032}{1000 \times 22.414} = 0.38 \text{ g NH}_3$$

A liter of gas at 10°C is heated at constant pressure to 25°C. What volume will it occupy at 25°C?

$$V_2 = \frac{V_1 T_2}{T_1}$$

$$V_2 = \frac{1 \text{ liter} \times 298^\circ\text{K}}{283^\circ\text{K}} = 1.053 \text{ liters}$$

A flask at constant pressure holds 505 cm^3 of air at 17°C. What volume will the air have at 27°C?

$$V_2 = \frac{505 \text{ cm}^3 \times 300^\circ\text{K}}{280^\circ\text{K}} = \frac{151{,}500 \text{ cm}^3}{280} = 541.07 \text{ cm}^3$$

A football is inflated to a pressure of 20 lb/in.^{2} when the temperature is 25°C. What will the pressure be at 5°C?

$$P_2 = \frac{20 \text{ lb/in.}^2 \times 278^\circ\text{K}}{298^\circ\text{K}} = \frac{5560}{298} = 18.66 \text{ lb/in.}^2$$

How many moles of a gas occupy 1.75 liters at 730 mm pressure and at 25°C?

$$\text{Volume at STP} = 1.75 \times \frac{730}{760} \times \frac{273}{298} = \frac{348{,}757.5}{226{,}480.0} = 1.54 \text{ liters}$$

$$\text{moles} = \frac{1.54 \text{ liters}}{22.414 \text{ liters mole}^{-1}} = 0.0687 \text{ moles}$$

If the vapor pressure of H_2O at 25°C is 23.8, what is the number of moles in 17.3 liters of O_2 collected over H_2O at 720 mm of pressure at 25°C?

$$\text{Total pressure} = P_{O_2} + P_{H_2O}$$

$$720 \text{ mm} = P_{O_2} + 23.8$$

$$P_{O_2} = 720 \text{ mm} - 23.8 = 696.2 \text{ mm}$$

$$\text{Volume at STP} = 17.3 \times \frac{696.2}{760} \times \frac{273}{298} = \frac{3{,}288{,}082.98}{226{,}480.00} = 14.518 \text{ liters}$$

$$\frac{14.518 \text{ liters}}{22.414 \text{ liters mole}^{-1}} = 0.6477 \text{ moles}$$

2.12 OHM'S LAW

The total energy of an electric current as it passes through a solution is distributed among three factors: (1) current strength (I), which is measured in amperes; (2) its potential (E—electromotive force or EMF), which is measured in volts; and (3) the resistance (R) of the conducting medium, which is measured in ohms. The relationship of these three factors is expressed by the Ohm's law, $I = \frac{E}{R}$, and therefore, $E = IR$ and $R = \frac{E}{I}$. *Ohm's law* states that the current strength (I) is directly proportional to the applied electromotive force (E) and inversely proportional to the resistance (R). In electrical measurements, often two of the three variables (current, voltage, or resistance) are known. The third unknown variable can be found through a ratio equation.

Examples and Problems

A 4-ohm resistance is connected to a cell whose voltage is 12 volts. What current will result in the circuit?

$$I = \frac{E}{R} = \frac{12}{4} = 3 \text{ amperes}$$

An electric coffee pot has a resistance of 48 ohms. When it is used on a 128-volt line, how much current does it take?

$$I = \frac{E}{R} = \frac{120}{48} = 2.5 \text{ amperes}$$

How much current will be produced when the resistance is 20 ohms and the voltage is 220?

$$I = \frac{E}{R} = \frac{220}{20} = 11 \text{ amperes}$$

What voltage is necessary to produce a current of 8 amperes when the resistance is 64 ohms?

$$E = IR = 8(64) = 512 \text{ volts}$$

What is the resistance of a 10-ampere lamp on a 220-volt circuit?

$$R = \frac{E}{I} = \frac{220}{10} = 22 \text{ ohms}$$

What is the resistance of a 220-volt circuit that uses a current of 0.5 ampere?

$$R = \frac{E}{I} = \frac{220}{0.5} = 440 \text{ ohms}$$

2.13 DENSITY, MASS, AND VOLUME

Density is the ratio of the mass of a substance to its volume, i.e., it is the mass of a substance divided by its volume. This is simply a mathematical definition of density.

It should be pointed out that there is a definite distinction between the mass and the weight of a substance. The *mass* of a body may be described as that property which remains constant when various forces are applied to it. Mass, like weight, is measured in pounds or kilograms. Mass is the amount of matter in the object and is essentially constant, regardless of its location or state of motion relative to an observer (unless it is moving with a speed close to the speed of light). But the weight of an object varies. The *weight* of a body is the gravitational force with which the body is attracted toward the earth. The weight of an object is the force of gravitational attraction acting on its mass. Weight varies as the gravitational force varies. At the center of the earth where gravity is zero the weight would be zero but the mass would be the same as it was at the earth's surface. On the planet Jupiter, which exerts a gravitational force much greater than the earth's, the weight of the object would be correspondingly greater, but again the mass would be the same.

There are many possible units for density, such as lb/ft^3, g/cc, kg/liter and g/ml.

Examples and Problems

The density of a certain object is 4.00 g/ml and it weighs 1.00 kg. What is its volume?

$D = M/V$ and therefore $V = M/D$

However, the density is in g/ml and the mass in kg, so change the kg to g by multiplying by 1000 to get 1000 g.

$$V = \frac{M}{D} = \frac{1000 \text{ g}}{4.00 \text{ g/ml}} = 250 \text{ ml}$$

By carrying the units all the way through the problem and by canceling in the proper places, the unit ml is arrived at which is the unit of volume desired. If volume was desired and the answer came out in g or ml/g or some other units, one would immediately know something was wrong because the answer must come out in the units of the thing one is looking for.

A cubical box is 50 cm on a side and weighs 5000 g. What is its density?

Volume = $l \times w \times h$ = 50 cm × 50 cm × 50 cm = 125,000 cc

$$D = M/V = \frac{5000 \text{ g}}{125,000 \text{ cc}} = 0.04 \text{ g/cc}$$

A container full of mercury weighs 10,000 g. The empty container weighs 6000 g. The density of mercury is 13.6 g/cc. What is the volume of the box?

Mass of mercury plus container	10,000 g
Mass of container	6,000 g
Mass of mercury	4,000 g

$D = M/V$

$$13.6 \text{ g/cc} = \frac{4000 \text{ g}}{V}$$

$$V = \frac{4000 \text{ g}}{13.6 \text{ g/cc}} = 294.12 \text{ cc}$$

Volume of mercury = 294.12 cc
Inner volume of box = 294.12 cc

An object has a mass of 1000 g and a volume of 500 cc. What is its density?

$D = M/V = 1000 \text{ g}/500 \text{ cc} = 2.0 \text{ g/cc}$

A steel ball has a density of 9 g/ml and a volume of 15 liters. What does it weigh?

15 liters = 15,000 ml

$$D = \frac{M}{V} \qquad 9 \text{ g/ml} = \frac{M}{15{,}000 \text{ ml}}$$

$M = 15{,}000 \text{ ml} \times 9 \text{ g/ml} = 135{,}000 \text{ g}$

The density of water is 62.4 lb/ft³ at 20°C. What is the volume in liters occupied by 1600 g of water?

$$V = \frac{M}{D} = \frac{1600 \text{ g}}{62.4 \text{ lb/ft}^3} = \frac{1600 \text{ g} - \text{ft}^3}{62.4 \text{ lb}}$$

$$V = \frac{1600 \text{ g ft}^3}{62.4 \text{ lb}} \times \frac{1 \text{ lb}}{454 \text{ g}} \times \left(\frac{12 \text{ in.}}{1 \text{ ft}}\right)^3 \times \left(\frac{2.54 \text{ cm}}{1 \text{ in.}}\right)^3 \times \frac{1 \text{ liter}}{1000 \text{ cm}^3}$$

$$= \frac{1600 \times 1728 \times 16.4 \text{ liters}}{62.4 \times 454 \times 1000} = \frac{45{,}342{,}720 \text{ liters}}{28{,}329{,}600} = 1.6 \text{ liters}$$

The density of a gram of alcohol at 20°C is 0.79 g/cm³. What is the mass of 125 cm³ of alcohol?

$M = DV = 0.79 \text{ g/cm}^3 \times 125 \text{ cm}^3 = 98.75 \text{ g.}$

2.14 VISCOSITY

The *viscosity* of a fluid is a measure of the resistance of the fluid to flow. In the process of flow, the molecules comprising the fluid move past one another and viscosity arises from frictional effects of relative motion. (See Fig. 2.18.) Both liquids and gases are viscous, but the molecules of a gas are so nearly independent of one another that the viscosity of a gas is of a much smaller order of magnitude than the viscosity of a liquid.

The unit of viscosity is the *poise* which requires a force of 1 dyne* to cause two parallel planes of 1 square centimeter area 1 centimeter apart in the liquid to slide past one another at a relative velocity of 1 centimeter

*One dyne is a force which, acting on a mass of 1 gram for 1 second, increases its velocity by 1 centimeter per second.

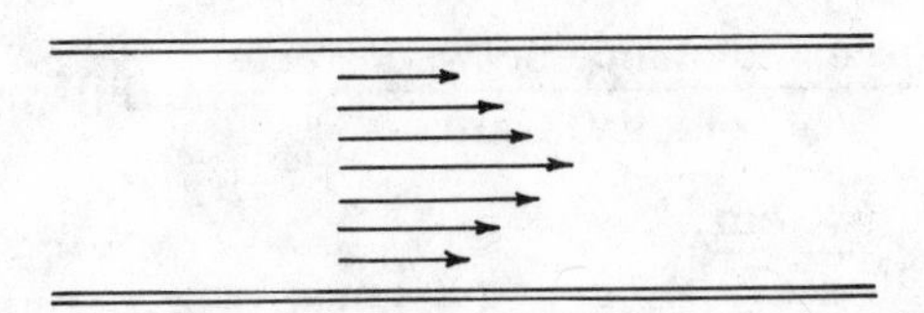

Fig. 2.18. Fluid friction (viscosity). For relatively slow motion, most liquids flow as if they were divided into sheets flowing one over another with velocities proportional to the distance (d) away from that part of a liquid in contact with a surface where friction is greatest and laminar flow is slowest.

per second. Most liquids have viscosities which are only a small fraction of a poise, so that *viscosity coefficients* are frequently given as multiples of a *centipoise*, one one-hundredth of a poise, or of a *millipoise*, one one-thousandth of a poise. The viscosity of water at 20° C is approximately 1 centipoise or 10 millipoises.

The Greek letter η (eta) represents the coefficient of viscosity or simply viscosity. It is small for liquids which flow readily like water and large for liquids like glycerin and molasses. (See Table 2.5.) The coefficient of viscosity is markedly dependent on temperature, increasing for gases and decreasing for liquids as the temperature is increased. The viscosity of liquids decreases about 2% for each degree rise in temperature (C°). Viscosity is also increased or decreased by the addition of solutes. For example, the viscosity of water is increased by the addition of a quantity of sugar.

The coefficient of viscosity is defined as $\eta = \dfrac{fd}{uA}$ in which

f = force (dynes) required to maintain a constant difference between the velocities of two parallel layers moving in the same direction
d = distance (cm) between two layers
u = difference in velocity (cm/sec) of the two parallel layers
A = area (cm^2) of surface contact of the two layers

Table 2.5. Viscosities of Liquids

Substance	*20°*	*30°*	*40°*	*50°*	*60°*
Water	1.009	0.800	0.654	0.549	0.470
Carbon tetrachloride	0.969	0.843	0.739	0.651	0.585
Methyl alcohol	0.593	0.515	0.449	0.395	0.349
Ethyl alcohol	1.200	1.003	0.831	0.701	0.591
Benzene	0.647	0.561	0.492	0.436	0.389
Mercury	1.550	1.50	1.45	1.41	1.37

Note: Coefficient of viscosity in centipoises at various temperatures.

Thus viscosity $= \dfrac{\text{force} \times \text{distance between faces}}{\text{velocity} \times \text{area}}$ and its units are

$$\eta = \frac{\text{dynes} \times \text{cm}}{\text{cm/sec} \times \text{cm}^2} = \frac{\dfrac{\text{g} \times \text{cm}}{\text{sec}^2} \times \text{cm}}{\text{cm/sec} \times \text{cm}^2} = \frac{\text{g} \times \text{cm}}{\text{sec}^2} \times \frac{\text{cm}}{1} \times \frac{1}{\text{cm}^2} \times \frac{\text{sec}}{\text{cm}}$$

$$= \frac{\text{g}}{\text{sec} \times \text{cm}} = \text{poise}$$

The viscosity of a liquid can be most conveniently measured by observing the flow of liquid through a capillary. If equal volumes of two liquids flow through the same capillary under the same driving pressure, the following is the ratio of viscosity coefficients: $\dfrac{\eta_1}{\eta_2} = \dfrac{d_1 t_1}{d_2 t_2}$ in which η_1 and η_2 are the coefficients of viscosity of the respective liquids, d_1 and d_2 are the relative densities, and t_1 and t_2 are the relative times of flow. Thus, if the viscosity of a reference liquid is known (usually water), the relative viscosity of another liquid can be determined from density and time flow data.

Example: In a fermentation process, a standard stock of fermentation concentrate whose viscosity was 1.07 centipoises was accidentally diluted with water. This solution had a relative density of 1.28 and a flow time of 0.54 minutes through a specified distance on a calibrated viscosometer. What is the viscosity of the resulting dilute solution if its density was found to be 1.16 and its flow time is 0.48 minutes?

Assume η_1 is the standard solution and η_2 is unknown.

The formula for viscosity coefficient is $\dfrac{\eta_1}{\eta_2} = \dfrac{d_1 t_1}{d_2 t_2}$.

Therefore $\dfrac{1.07}{x} = \dfrac{1.28 \times 0.54}{1.16 \times 0.48}$

$$\frac{1.07}{x} = \frac{0.6912}{0.5568}$$

$$x = \frac{1.07 \times 0.5568}{0.6912} = \frac{0.5958}{0.6912} = 0.862 \text{ centipoises}$$

2.15 SPECIFIC GRAVITY

Specific gravity is the ratio of the density of a substance to the density of another substance taken as the standard—usually air or water. The formula may be written

$$\frac{\text{specific gravity}}{\text{(at a specified temperature)}} = \frac{\text{mass of a known volume}}{\text{mass of an equal volume}}$$

For solids and liquids the standard is usually distilled water at 4°C (39°F) and for gases it is usually air at 20°C (68°F) and at the standard atmospheric pressure of 760 mm of mercury.

The numerical value of specific gravity is the same in all systems of measurement since the value is a ratio between like quantities and thus has no units. Density, however, has units. In the metric system they are g/ml and in the English system they are lb/cu ft. The specific gravity of water is always 1, but the density of water is 1 g/milliliter or 62½ lb/cu ft.

Examples: Find the specific gravity of concentrated sulfuric acid at 20°C given

1 ml of H_2SO_4 = 1.84 g
1 ml of H_2O = 1 g

Therefore specific gravity $= \frac{1.84}{1.00} = 1.84$

The specific gravity of a certain solution is 1.97. What is the weight of 250 ml?

1.97 × 250 = 492.5 g

A 10% solution of table salt has a specific gravity of 1.07. If 500 ml of this solution were evaporated to dryness, what weight of salt would remain?

1.07 × 500 = 535.0 (total weight of solution)
535.0 × 0.10 = 53.5 g (salt remaining)

The specific gravity of a solid heavier than and insoluble in water may be calculated by simply dividing the weight of the solid in air by the weight of water it displaces when immersed in it. The weight of water displaced (apparent loss of weight in water) is equal to the weight of an equal volume of water.*

Example: A piece of glass weighs 52 g in air and 32 g when immersed in water. What is its specific gravity?

52 g − 32 g = 20 g of displaced water
(weight of an equal volume of water)

Specific gravity of glass $= \frac{52}{20} = 2.6$

To calculate the weight of a given volume or the volume of a given weight of a liquid, its specific gravity must be known. Because of the

**Archimedes' principle:* a solid body immersed in liquid is buoyed up by a force equal to the weight of liquid it displaces.

simple relationship between the units in the metric system, such problems are easily solved when only metric quantities are involved but are most complex when units of the English system are used.

Examples and Problems

Given the volume and specific gravity of a liquid, calculate the weight of the liquid.
Weight of liquid = weight of equal volume of water × specific gravity of liquid

What is the weight of 5200 ml of alcohol having a specific gravity of 0.820?
5200 ml of water weighs 5200 g
5200 g × 0.820 = 4264 g

Given the weight and specific gravity of a liquid, calculate the volume of the liquid.

$$\text{Volume of liquid} = \frac{\text{volume of equal weight of water}}{\text{specific gravity of liquid}}$$

What is the volume in pints of 25 lb of glycerin having a specific gravity of 1.25?

1 oz = 28.35 g
1 lb = 453.6 g = 454 rounded off
25 lb = 454 g × 25 = 11,350 g
11,350 g of water measure 11,350 ml

$$\frac{11{,}350 \text{ ml}}{1.25} = 9080 \text{ ml} \qquad \text{Again, } 28.35 \text{ g} = 1 \text{ oz}$$

$$\text{Therefore } \frac{9080 \text{ g}}{28.35 \text{ g/oz}} = 320.28 \text{ oz}$$

$$\frac{320.28 \text{ oz}}{16 \text{ oz/pint}} = 20.0175 \text{ pints}$$

What is the molarity of a bottle of glacial acetic acid reagent given its specific gravity = 1.05?

Molecular weight of acetic acid = 60.05.
Recalling that molarity is the GMW per 1000 ml,

$$\text{Molarity of the acid} = \frac{\text{Specific gravity} \times 1000}{60.05} = \frac{1.05 \times 1000}{60.05} = 17.5 \text{ M}$$

2.16 RELATIVE HUMIDITY

Atmospheric air is a mixture of gases, consisting of about 80% nitrogen, 18% oxygen, and small amounts of carbon dioxide, water vapor, and other gases. The mass of water vapor per unit volume is called the *absolute humidity*. The total pressure exerted by the atmosphere is the sum of the pressures exerted by its component gases. These pressures are called partial pressures of the components. It is found that the partial pressure of each of the component gases of a gas mixture is very nearly the same as would be the actual pressure of that component alone if it occupied the same volume as does the mixture, a fact known as *Dalton's law*. That is, each of the gases of a gas mixture behaves independently of the others. The partial pressure of water vapor in the atmosphere is ordinarily a few millimeters of mercury.

If the concentration of water vapor, or the absolute humidity, is such that the partial pressure equals the vapor pressure, the vapor is said to be saturated. If the partial pressure is less than the vapor pressure, the vapor is unsaturated. The ratio of the partial pressure to the vapor pressure at the same temperature is called the *relative humidity*, and is usually expressed as a percentage.

$$\text{Relative humidity (\%)} = 100 \times \frac{\text{partial pressure of water vapor}}{\text{vapor pressure at same temperature}}$$

The relative humidity is 100% if the vapor is saturated and zero if no water vapor at all is present.

Example: The partial pressure of water vapor in the atmosphere is 8 mm and the temperature is 20°C. Find the relative humidity.

From Table 2.6, the vapor pressure at 20°C is 17.5 mm.

$$\text{Therefore relative humidity} = \frac{8}{17.5} \times 100 = 46\%$$

Table 2.6. Vapor Pressure of Water

t(°C)	*Vapor Pressure*	t(°F)
0	4.58	32
5	6.51	41
10	8.94	50
15	12.67	59
20	17.5	68
40	55.1	104
60	149.0	140
80	355.0	176
100	760.0	212
120	1490.0	248
140	2710.0	284
160	4630.0	320
180	7510.0	356
200	11650.0	392
220	17390.0	428

Since the water vapor in the atmosphere is saturated when its partial pressure equals the vapor pressure at the air temperature, saturation can be brought about either by increasing the water vapor content or by lowering the temperature. For example, let the partial pressure of water vapor be 8 mm when the air temperature is 20°C as in the preceding example. Saturation or 100% relative humidity could be attained either by introducing enough water vapor (keeping the temperature constant) to increase the partial pressure to 17.5 mm or by lowering the temperature to 9.1°C at which, by interpolation from Table 2.6, the vapor pressure is 8 mm.

If the temperature were to be lowered below 9.1°C, the vapor pressure would be less than 8 mm. The partial pressure would then be higher than the vapor pressure and enough vapor would condense to reduce the partial pressure to the vapor pressure at the lower temperature. It is this process which brings about the formation of clouds, fog, and rain. The phenomenon is also of frequent occurrence at night when the earth's surface becomes cooled by radiation. The condensed moisture is called dew. If the vapor pressure is so low that the temperature must fall below 0°C before saturation exists, the vapor condenses into ice crystals in the form of frost.

2.17 REFRACTIVE INDEX

A wave is said to be refracted when its velocity is changed as a result of the wave's passing from one medium to another. The ratio of the velocity of a wave in a first medium to that in a second medium is the refractive index of the second medium with respect to the first, for that particular type of wave. This change in velocity results in a bending of the wave, or a change in direction, if the incident wave strikes the surface of the second medium in any but a perpendicular direction.

If a beam of light travels as shown in Figure 2.19 from air or a vacuum

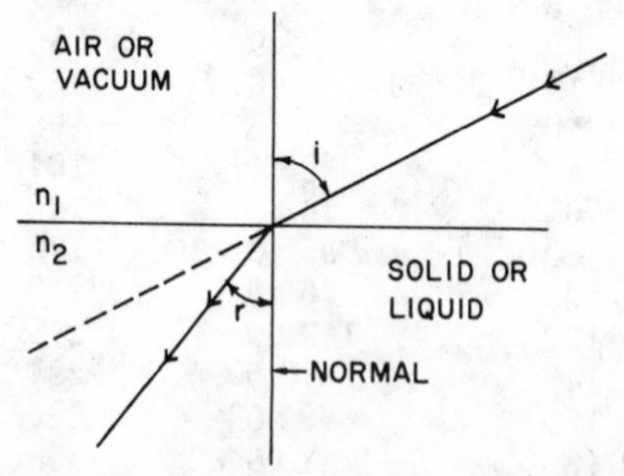

Fig. 2.19. Regular refraction of light. $n_2 > n_1$. (1) When light passes from one medium to a denser medium (optically speaking), the ray is bent toward the normal (as in the figure). (2) When light passes from one medium to a less dense medium, the ray is bent away from the normal.

into a more dense medium, then i is called the *angle of incidence* and r is the *angle of refraction*. The refractive index n of the medium is then defined as the ratio of the sine of the angle of incidence to the sine of the angle of refraction. By measuring the angles of incidence and refraction, the refractive index can be determined, its value depending on the wavelength of the light. A subscript to the letter n indicates the particular type of light used for the measurement of the refractive index (e.g., D-line of sodium or α - , β - , or γ - lines of the Balmer hydrogen spectrum).

Examples and Problems

The velocity of light in space is ≅ 300,000 km/sec. What is the index of refraction of a substance in which the velocity of light is 175 km/sec?

$$\frac{300{,}000 \text{ km/sec}}{175{,}000 \text{ km/sec}} = 1.714$$

A ray of light entering water from air makes an incident angle whose sine is 0.36. What is the sine of the angle of refraction?

$$1.33 \text{ (index of refraction for water)} = \frac{0.36}{x}$$

$$1.33x = 0.36$$

$$x = \frac{0.36}{1.33} = 0.27$$

2.18 MAP SCALES

Scale means the size of the representation of an object with respect to its actual size. A map or diagram must be drawn to scale to be accurate. For example, the scale of a map shows how much of the actual earth's surface is represented by a given measurement on the map. A large-scale map usually covers only a small region and shows most of the details of an area, such as roads or small rivers. On a large-scale map, the measurement scale used represents more accurately the actual earth's surface. A small-scale map leaves out many details and covers a much larger area, such as the world.

Scales can be expressed in three different ways.

Graphic Scale

Scale is shown graphically by means of a straight line on which distances have been marked off. Each mark usually represents a certain number of miles on the earth's surface. Graphic scale, as 0 ——— 10 miles, has the advantage that it remains true even if the map or diagram is photographed larger or smaller.

Words and Figures

A scale is often expressed as so many units on the paper equaling so many units of the actual object. An example is the inch to the mile scale as 1 in. = 8 mi; 1 in. on the map equals 8 mi on the earth's surface.

Numerical Scale or Representative Fraction

The most common method of expressing scale is to write a representative fraction. For example, a scale might be written as 1:62,500 or 1/62,500. This means that one unit of measurement on the map represents 62,500 of the same units on the surface of the earth. The advantage of this method is that the scale is expressed by the fraction regardless of what measurement system is used, the English or the metric. Below is an example of this method.

Examples and Problems

If a globe representing the earth has an equatorial radius of 45 cm, what is its scale compared to the earth?

The equatorial radius of the earth is 6378 km. If 45 cm on the globe equals 6378 km of the earth, then 1 cm on the globe represents 141.73 (or approximately 142 km) on the earth. Expressed in scale form, this is 1 cm/142 km.

What is the ratio if 1 in. on a map stands for 1 mi?

1 mi = 5280 ft

5280 ft × 12 in. = 63,360 in.

$$\text{ratio} = \frac{1}{63{,}360}$$

What is the ratio if $\frac{1}{2}$ in. on a map stands for 1 mi?

63,360 in. = 1 mi

Therefore, there are 125,720 half-in. in 1 mi.

$$\text{ratio} = \frac{1}{125{,}720}$$

A map has a ratio of $\frac{1}{80{,}000}$. What distance on the earth does 1 in. on the map stand for?

80,000 in. or 1.26 mi.

2.19 CIRCLE AND SPHERE

A *circle* is a closed curve lying in a plane and constructed so that all points are equally distant from a fixed point in the plane. The point in the plane from which all points on the circle (Fig. 2.20) are equally distant is called the *center of the circle*. Any of the equal straight-line segments which extend from the center of a circle to the circle itself is called a *radius*. A straight line through the center of a circle, terminated at each end by the circle, is called a *diameter*. A diameter is equal in length to two radii. Any portion of a circle is called an *arc*. The length of a circle (the distance around the space enclosed) is the *circumference* of the circle.

Important formulas relating to circles are:

1. Equation of a circle in the plane—in rectangular Cartesian coordinates $(x - h)^2 + (y - k)^2 = r^2$, where r is the radius of the circle and the center is at the point (h,k). When the center is at the origin, the equation becomes $x^2 + y^2 = r^2$.

2. The circumference of a circle is equal to its diameter multiplied by pi (π) or to twice the radius multiplied by π. $C = \pi d = 2\pi r$.

3. The area of a circle is equal to the square of its diameter divided by four and then multiplied by π or the square of its radius $\left(\frac{d}{2}\right)$ multiplied by π. $A = \pi \frac{d^2}{4} = \pi r^2 = 0.7854\ d^2$.

A *sphere* is a round body all points on the surface of which are equidistant from its center. The terms center, radius, and diameter apply to a sphere as in the case of a circle.

Important formulas relating to spheres are

1. Area of a sphere: $A = 4\pi r^2 = \pi d^2$
2. Volume of a sphere: $V = \frac{4}{3}\pi r^3 = 0.5238\ (d^3)$

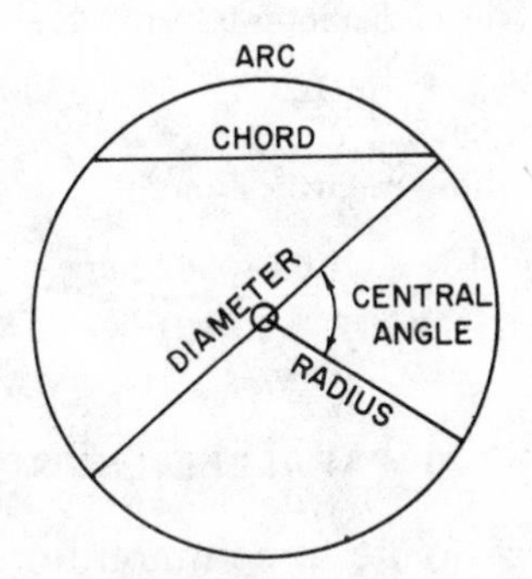

Fig. 2.20. Diagram of a circle.

Examples and Problems

Find the area of a circle whose radius is 7 in.

$$A = \pi r^2 \quad \text{or} \quad 22/7\, r^2$$

$$A = \frac{22}{\not 7} \times \frac{(\not 7)(7)}{1} = 154 \text{ sq in.}$$

Find the circumference of a circle whose diameter is 4 in.

$$C = \pi d = 3.14 \times 4 = 12.56 \text{ in.}$$

Find the circumference of a circle whose radius is 4 in.

$$C = 2\pi r = 2(3.14) \times 4 = 25.12 \text{ in.}$$

Find the area of a circle whose diameter is 4 in.

$$A = \pi \frac{d^2}{4} = \frac{22}{7} \times \frac{4^2}{4} = \frac{22}{7} \times \frac{16}{4} = \frac{352}{28} = 12.57 \text{ sq in.}$$

Find the area of a circle whose radius is 4 in.

$$A = \pi r^2 = \frac{22}{7} \times 4^2 = \frac{22}{7} \times 16 = \frac{352}{7} = 50.28 \text{ sq in.}$$

Find the area of a sphere whose diameter is 4 in.

$$A = \pi d^2 = \frac{22}{7} \times 16 = \frac{352}{7} = 50.28 \text{ sq in.}$$

Find the area of a sphere whose radius is 4 in.

$$A = 4\pi r^2 = 4\left(\frac{22}{7}\right) \times 4^2 = \frac{4 \times 22 \times 16}{7} = \frac{1408}{7} = 201.14 \text{ sq in.}$$

Find the volume of a sphere whose diameter is 4 in.

$$V = 0.5238\,(d^3) = 0.5238 \times 4^3 = 0.5238 \times 64 = 33.5232 \text{ cu in.}$$

Find the volume of a sphere whose radius is 4 in.

$$V = \frac{4}{3}\pi r^3 = \frac{4}{3}\left(\frac{22}{7}\right) \times 4^3 = \frac{4 \times 22 \times 64}{3 \times 7} = \frac{5632}{21} = 268.19 \text{ cu in.}$$

2.20 PARALLEL LINES

Parallel lines have no points in common no matter how far they are extended. They are always equidistant from each other. Two parallel

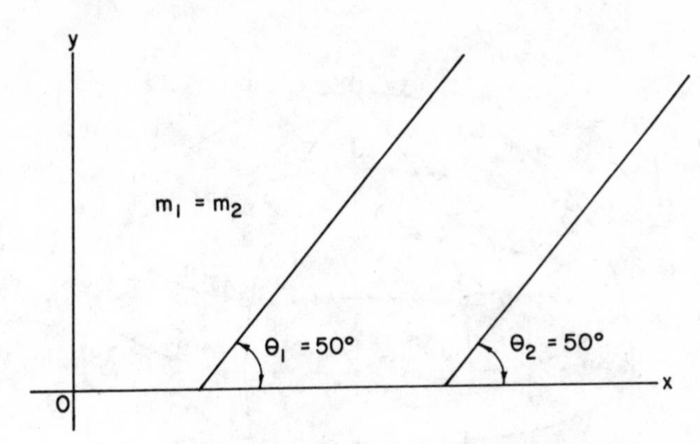

Fig. 2.21. Parallel lines have equal slopes.

lines have equal angles of inclination and thus have equal slopes ($m_1 = m_2$). (See Fig. 2.21.) The concept of parallel lines is needed to understand latitude and earth measurement.

2.21 LATITUDE AND LONGITUDE

It is not possible to go far into this subject until one has some acquaintance with the geometrical idea of parallel lines and some knowledge of the geometry of the circle.

Degrees of arc are angular intervals that geographers and navigators use to figure distances more conveniently on the earth. There are two kinds of degrees used to describe position on the earth's surface, degrees of latitude and degrees of longitude.

Latitude describes the position of a point on the earth's surface in relation to the equator. Imagine a series of lines running around the earth parallel to the equator. These are lines of latitude. (See Fig. 2.22.) Since these small circles are parallel to the great circle of the equator, they are known as parallels of latitude. Every point on the earth's surface can be located on such a line. It shows the position of that spot in relation to the equator. All points on the earth's surface that have the same latitude lie on the same imaginary circle called a parallel of latitude.

Latitude is the measurement of the length of arc between the equator and a given point along a north-south line called a meridian. The latitude of a point is measured *along* the meridian that runs through it. Latitude is measured in degrees and fractions thereof. A point on the equator has a latitude of zero degrees. The North Pole has a latitude of 90° North and the South Pole has a latitude of 90° South.

Latitude is one of the two coordinates that locate any point on the earth. The other coordinate is longitude. Latitude, together with longi-

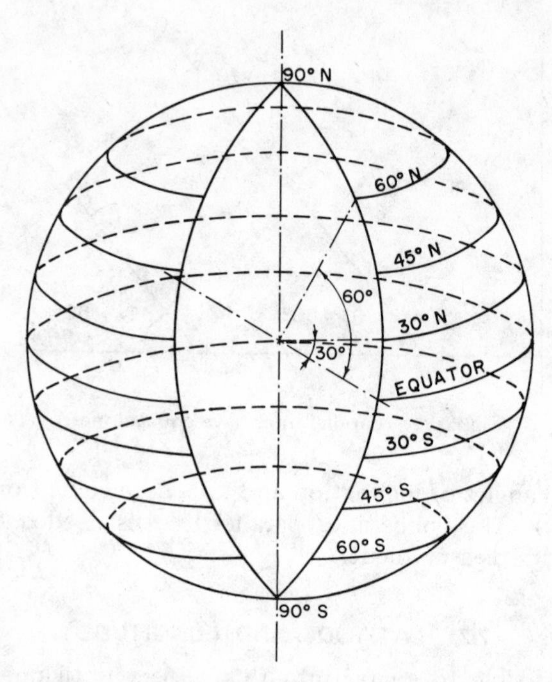

Fig. 2.22. Lines (or parallels) of latitude circle the earth parallel to the equator. They are marked off at one-degree intervals north and south up to 90° (the North and South Poles). Thus the 30° N line of latitude makes an angle of 30° with the equator at the middle of the earth. The distance of a degree of latitude is 69 miles (111 km) except near the poles where, due to a slight flattening of the earth, it is a little more.

tude, forms a convenient grid for determining the exact location of any place on the earth's surface.

Longitude provides a means of determining position directly east or west on the earth's surface with respect to a fixed point of origin. The great circle of 0° longitude is known as the prime meridian and was made to pass arbitrarily through Greenwich, near London, England, and both geographic poles. Longitudes are measured both east and west from the prime meridian through 180° until they meet in a common, *irregular* line over the Pacific Ocean known as the *International Date Line.* (The 180° prime meridian is irregular to avoid splitting countries. This is explained further under the section on Time.)

Like latitude, longitude is measured in degrees and fractions thereof. More technically, longitude is the amount of arc created by drawing a line from the center of the earth to the intersection of the equator and the prime meridian, and another line from the center of the earth to any other point on the equator. (See Fig. 2.23.)

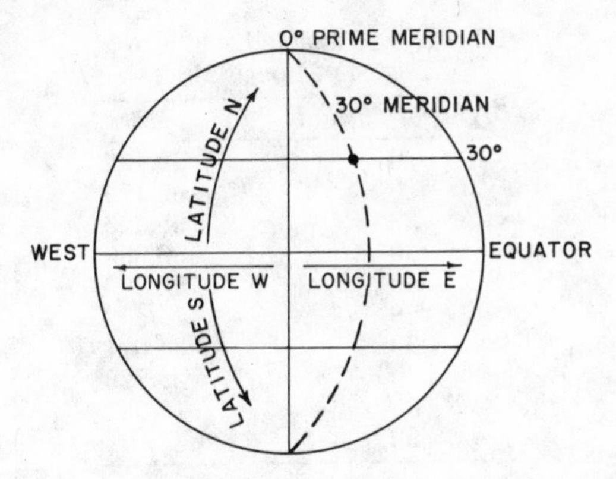

Fig. 2.23. Measurement of latitude and longitude. Point is located at 30° N latitude, 30° E longitude.

The circles made by the meridians around the earth are all equal. The circles made by the parallels get smaller as they approach the poles. At latitude 60°, the circumference of the latitude circle is one-half the circumference of the equatorial circle. The distance between two parallels of latitude 1° apart is about 69 miles. The length of 1° of latitude varies from 68.7 miles near the equator to 69.4 miles near the poles. The variation results from the fact that the earth is not a perfect sphere. The equatorial diameter of the earth is approximately 27 miles greater than its polar diameter. Since all meridians of longitude must pass through the two poles, any group of meridians drawn on a globe of the earth must converge at the two poles. The greatest linear separation of 1° of longitude is at the equator. (See Fig. 2.24.) The length of a degree of longitude for various latitudes is shown in Table 2.7. Thus degrees of latitude are about equal

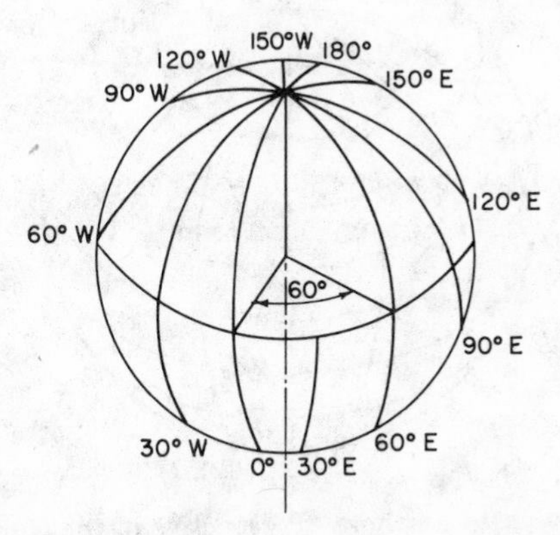

Fig. 2.24. View of earth with cutout section indicating 60° E longitude.

Table 2.7. Variable Lengths of a Degree of Longitude

Degree of Latitude	*Miles*
0°	69.172
10°	68.129
20°	65.026
30°	59.956
40°	53.063
50°	44.552
60°	34.674
70°	23.729
80°	12.051
90°	0

everywhere but the number of miles in a degree of longitude becomes less as one approaches the poles. (See Fig. 2.25.)

An example which illustrates a practical use of this information is that involving the determination of distance between two locations. For instance, what is the distance in kilometers between Kuwait (29° N latitude) and the North Pole? Since the acceptable average length for a degree of latitude is 111 kilometers and Kuwait is 29° from the North Pole (90° − 29° = 61°), then 61° × 111 km/degree or 6771 km is the most direct distance along a meridian of longitude from Kuwait to the North Pole.

Another Example: How far is Jerusalem from the North Pole?

Jerusalem = 31° 46′ North Latitude
90° − 31° 46′ = 58° 14′

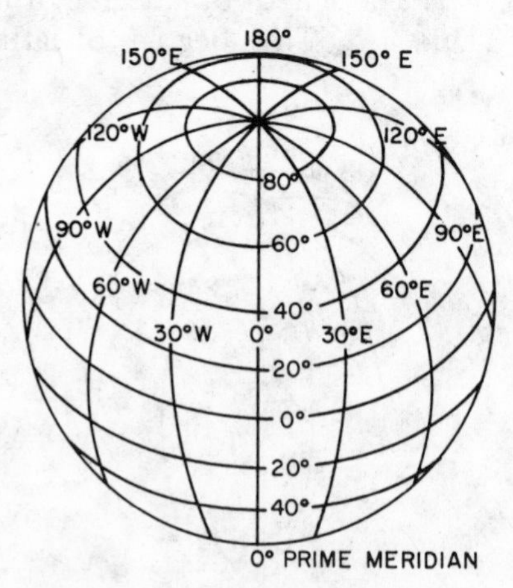

Fig. 2.25. Ordinate system of a sphere. Perspective of the globe with grid formed by parallels of latitude and meridians of longitude.

$$58^\circ \times 111 \text{ km/degree} = 6438 \text{ km}$$

$$\frac{14}{60} \times \frac{111}{1} = \frac{1554}{60} = 25.9 \text{ km}$$

$$6438 + 25.9 \text{ km} = 6463.9 \text{ km from the North Pole}$$

2.22 TIME

Our fundamental time standard is based on the counterclockwise rotation of the earth on its own axis and on its yearly revolution around the sun in the same direction. The basic time unit that is generally used is the *mean solar day* or the average length of time throughout the year between two successive crossings of the meridian by the sun. The unit of time in the metric system is the second which is $\frac{1}{86{,}400}$ of a mean solar day. Because the earth spins on its axis (an imaginary line drawn between the North and South Poles) a different part of its surface is constantly being presented to the sun's rays. It spins counterclockwise, so when the sun appears to be at its highest point in the sky at one place (noon), a place to the east has already passed beneath the overhead sun; it has already experienced noon and the time there is after midday. But a place to the west will not have yet passed beneath the overhead sun and the time there will be before midday. (See Fig. 2.26.)

Local time is based on the apparent overhead passage of the sun with respect to any particular locality. The local time anywhere is noon when the sun is directly above the longitudinal meridian of the observer. Hence, local time is not the same at any two meridians. In making one complete revolution the earth turns through 360° and, as it takes about 24 hours to do so, there is a difference of four minutes between each degree. Thus clocks in one town should show a different time from clocks in another town a few miles to the east or west. When it is 5:00 P.M. local time in

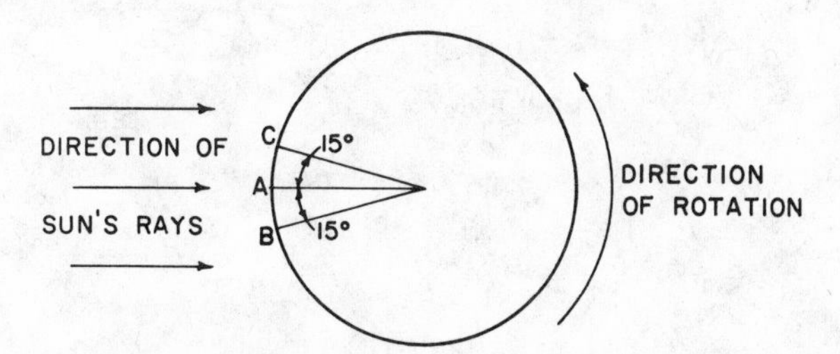

Fig. 2.26. It is noon at point *A*. Point *B*, 15° to the east, has already passed beneath the overhead sun; it is 1:00 P.M. there. Point *C*, 15° to the west, has not yet passed beneath the overhead sun; it is 11:00 A.M. there.

Ames, Iowa, for example, it is only about 4:48 local time in Sioux City, Iowa, about 160 miles to the west. However, this is never practiced and local time is never strictly obeyed.

To avoid the confusion of using local time, standard time is generally used. This is based on the division of the globe into 24 time zones numbered westward from the Greenwich meridian (0° longitude). Each kind of standard time is simply the local civil time of a chosen standard meridian. Each time belt uses the time of the meridian that passes about centrally through it. These meridians are chosen at multiples of 15° from Greenwich so that their times differ exactly by whole hours from that of Greenwich Civil Time. In passing from one time belt to another, one must set his watch ahead if he travels east, or back if he travels west, to agree with the time of that region. (See Fig. 2.27.) The time zones are modified in places to avoid splitting countries, as is the International Date Line.

Figure 2.28 is helpful in understanding the International Date Line, using the concepts of longitude and time zones.

Knowing that the earth rotates 15° an hour with time decreasing from east to west, one should ask himself:

1. In which direction is the unknown time point from the known time point? If the unknown point lies to the east of the known point, the time of the unknown point will be later, and vice versa.

2. How many degrees difference in longitude is there from the known time point to that of the unknown time point?

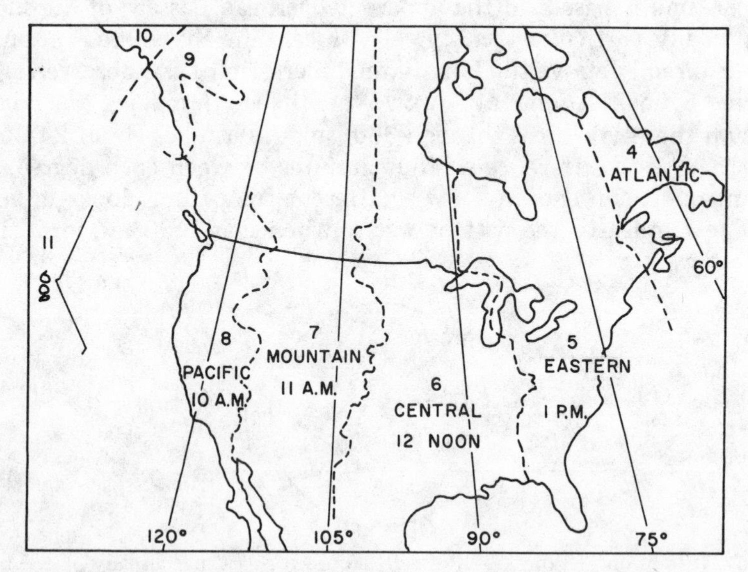

Fig. 2.27. United States standard time belts. Figures indicate the number of hours earlier than Greenwich civil time.

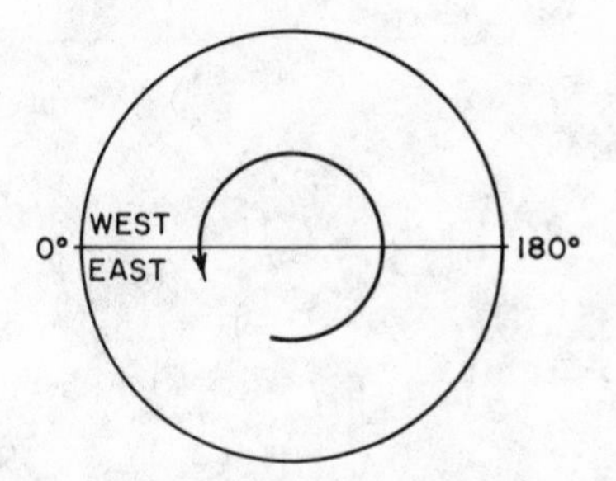

Fig. 2.28. Drawing representing earth seen in polar projection. *Arrow* represents the earth's rotation. *Line* represents the 0–180° meridian.

3. With the above information, one can determine the time between the two points. What is the hour difference in time between the known time point and the unknown time point? To obtain this answer, divide the number of degrees found in step two by 15°.

4. Does the difference in time between the two points involve the hour of midnight?

In Figure 2.29, if one approached the 180° meridian from the east, he would call it noon October 2; if he came from the west, he would call it noon October 1. On crossing the line, each traveler would have to change his date to agree with the time kept in the territory he had entered. The one who travels eastward (i.e., west to east) advances from noon of October 1 to noon of October 2, while the one who travels westward (i.e., east to west) and who has already spent half of October 2, reverts to October 1 and has the afternoon and evening of that day, as well as the morning of October 2, to spend all over again.

The International Date Line does not coincide everywhere with the 180° meridian. (See Fig. 2.30.) This zig-zag line is drawn to avoid differ-

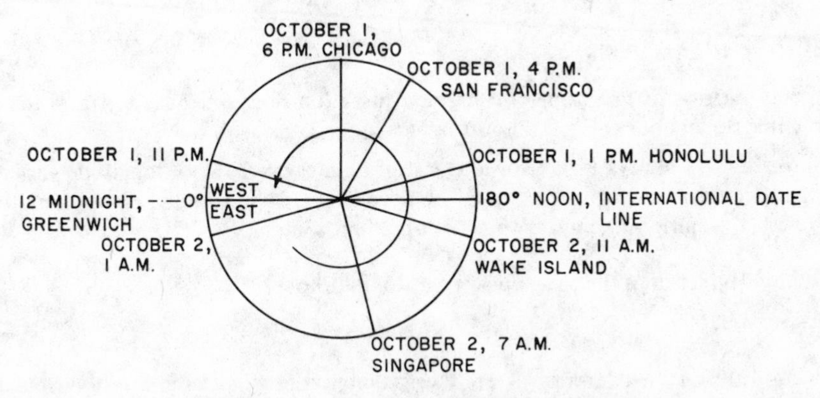

Fig. 2.29. The international date line.

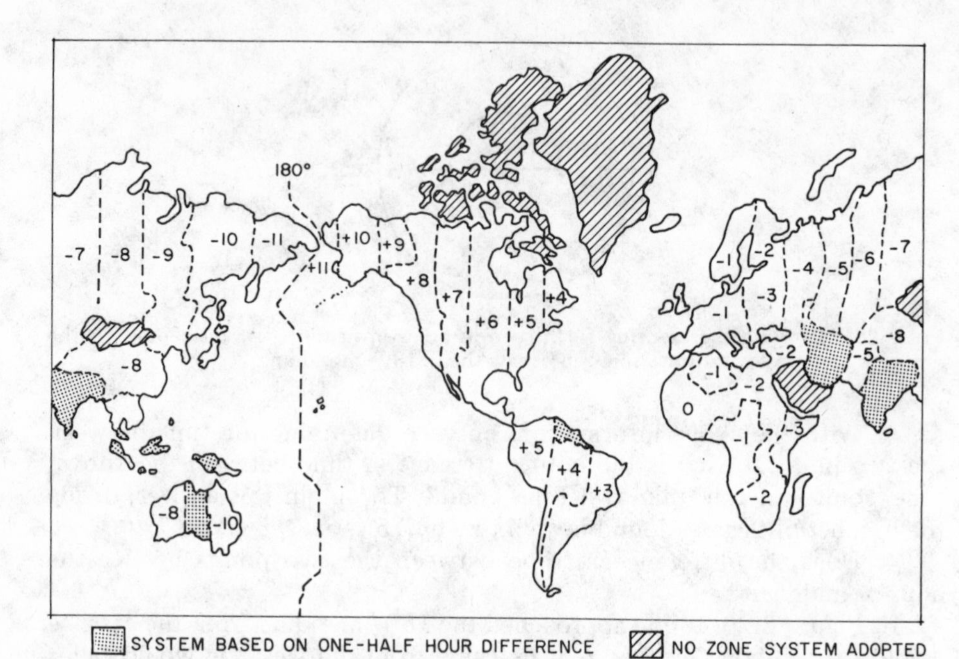

Fig. 2.30. The twenty-four world standard time zones. The middle of the zero time zone passes through Greenwich with its east and west limits 7°30′ on each side. Each 15° zone east and west of the initial zone represents 1 hour of time. The number of hours that must be added to or subtracted from local time to give Greenwich civil time is indicated on the map for each zone.

ences of date at places that are closely related geographically (i.e., to avoid splitting the Soviet Union and the Aleutian Islands and islands in the Pacific). The line detours to the west in the Aleutian Islands and to the east at the Bering Strait and in the South Pacific.

Examples and Problems

How many degrees of longitude and in what direction does one have to travel for the sun to come up an hour earlier? 2 hours later?

Since the sun always rises from the east, if one traveled 15° of longitude east, the sun would appear to come up an hour earlier. If one traveled 30° of longitude west, the sun would appear to come up 2 hours later.

How many degrees does the sun appear to move in 1 hour?

Approximately 15°.

If it were 4:00 P.M. Wednesday at 165° west longitude, what would be the time and day at 165° east longitude?

Thursday, 2:00 P.M.

If a plane flew 500 miles an hour, it would take about 7 hours to go from New York to London. Five time zones are crossed during this trip. If the plane leaves New York at 7:00 A.M., what time is it in London? (12:00 noon) What time would it be in London when the plane landed? (7:00 P.M.) If the plane left London at noon, what time would it be in New York? (7:00 A.M.) What time would it be in New York when the plane landed? (2:00 P.M.)

Using standard time, when it is noon in Miami, Florida, what time is it in Chicago? (11:00 A.M.) Denver? (10:00 A.M.) San Francisco? (9:00 A.M.) New York? (12:00 noon) Lansing, Michigan? (12:00 noon)

If it is 6:00 P.M. in New York, is it 5:00 or 7:00 in Chicago?

(5:00)

If it is 6:00 P.M. in Chicago, is it 5:00 or 7:00 in New York?

(7:00)

2.23 MOTION

Speed is a rate of change of position which may be in varying directions. *Velocity* is the rate of change of position in one particular direction; it is the length traversed per unit time. It is expressed in miles per hour, feet per second, centimeters per second, etc. Velocity has direction as well as quantity. (See Fig. 2.31.)

Acceleration is the rate of change of velocity per unit of time. If the speed of a moving object changes, the object accelerates. If a car at one instant is moving at 35 mph and 1 second later it is moving at 40 mph, its acceleration is 5 mph per sec. If it continues to gain 5 mph every second, it is uniform acceleration. If it slows down, it has a negative acceleration. Acceleration may involve a change in speed or direction, or both.

An object dropped near the surface of the earth (a freely falling object under the influence of earth gravity) falls with a uniform acceleration of about 980.7 cm per sec per sec (or 32 feet/sec^2). This means that each second the object falls the velocity increases 980.7 cm per sec; at the end of every second the object is traveling 980.7 cm per sec faster. (See

Fig. 2.31. Speed versus velocity. Both cars travel from *A* to *B* in 1 hour. They have the same average velocity (30 mph westward), but since the black car travels farther, its speed is greater (40 mph).

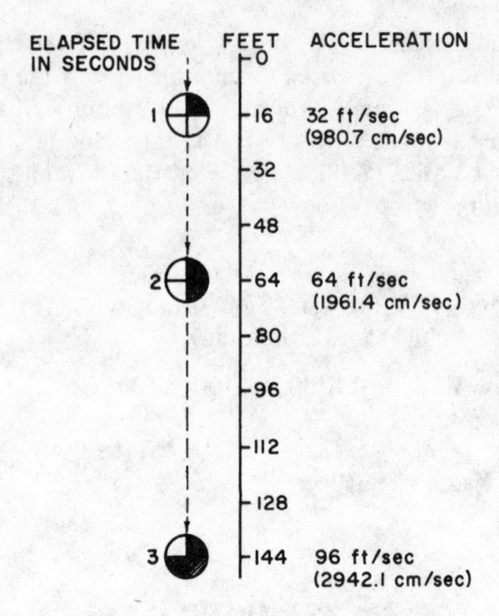

Fig. 2.32. Acceleration of a falling body.

Fig. 2.32.) This law of falling bodies was discovered by Galileo (1564–1642). He established that the distance fallen increases proportionally to the square of the time. This fact is expressed by the formula:

$$s = \frac{gt^2}{2}$$

where s = distance
g = acceleration due to gravity (980.7 cm/sec^2 or 32 ft/sec^2) and
t = time.

Examples and Problems

A car changes its velocity from 20 meters per second to 30 meters per second in a period of 5 seconds. What is its acceleration?

$$a = \frac{\Delta v}{\text{time}} = \frac{30 \text{ m/sec} - 20 \text{ m/sec}}{5 \text{ sec}} = 2 \text{ m/sec}^2$$

Δv = change in velocity

A boy standing on a bridge dropped a stone into the water below. If it took 3 seconds for the stone to reach the water, how high was the bridge?

192 ft

How far does a free falling body travel during the eighth second it falls?

256 ft

How far would a body have to fall to acquire the speed of an automobile that is traveling 60 mph?

$$5280 \text{ ft per mi} \times 60 \text{ mi} = 320{,}800 \text{ ft per hr}$$
$$= 5{,}346.66 \text{ ft per min}$$
$$= 88.99 \text{ ft per sec}$$

Thus a speed of ~89 ft per sec would be reached after $2\frac{3}{4}$ sec of falling.

A falling body has fallen 512 ft from rest. What is its velocity? What length of time has it fallen?

$$512 = \frac{32 \text{ ft/sec}^2 (t^2)}{2}$$

$$1024 = 32 \text{ ft/sec}^2 \, t^2$$

$$t^2 = \frac{1024 \text{ ft}}{32 \text{ ft/sec}^2} = 32 \text{ sec}^2$$

$t = \sqrt{32 \text{ sec}^2} = \sim 5.65$ sec (length of time body has fallen)

During the fifth second, the velocity is 160 ft/sec;

During the sixth second, the velocity is 192 ft/sec;

Thus at 5.65 seconds the velocity is 180.8 ft/sec.

2.24 FORCE

Force is that which is capable of producing an acceleration of a mass, It may be understood as synonymous with the concepts of push and pull. When a rock rests on a table, the force which it exerts on the table is equal to its own weight. Every time a force is exerted, it is balanced by a force exactly equal to it in the opposite direction.

A simple way of showing the working of this law is to fasten the hooks of two spring balances together. (See Fig. 2.33.) When the ends of the balances are pulled apart, *both* of them register the same pull and the springs are pulled in opposite directions. Exactly the same effect occurs if one of the balances is fixed to the wall and the end of the other is pulled. Both will record equal and opposite forces.

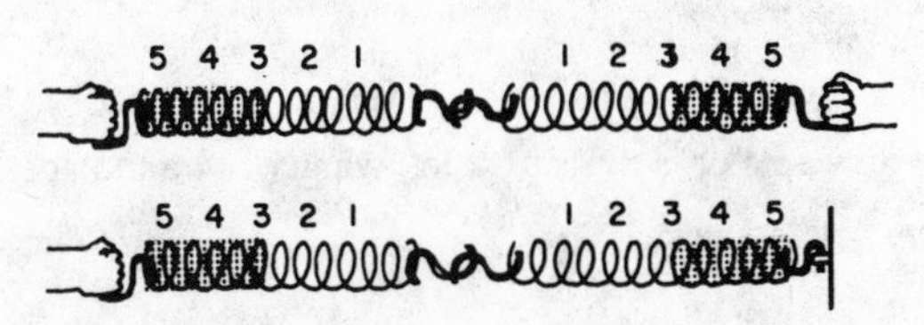

Fig. 2.33. Spring balance showing equal and opposite forces.

A force is required to compress a spring, to lift a weight, or throw a ball. If many forces are acting on an object at the same time, their effect (the resultant force) is the same as a single net force.

In the special but common case in which the net force is zero, the object is said to be in equilibrium. If it is at rest, it will remain at rest, and if it is in motion, it will remain in motion at the same speed and the same direction. The action of an unbalanced net force causes an object to accelerate, i.e., to change its speed or direction, or both. The amount of acceleration is proportional to the size of the force and inversely proportional to the mass of the object:

$$\text{acceleration} = \text{force/mass}$$

The fact that acceleration is inversely proportional to the mass of the object simply means that the same force would accelerate a light object faster than a heavy one. For example, a man can throw a baseball much farther and faster than a cannon ball ten times its weight.

Force is therefore equal to mass multiplied by acceleration

$$\text{force} = \text{mass} \times \text{acceleration}$$

and the following combinations of units may be used:

$$\text{force (in newtons)} = m \text{ (in kilograms)} \times a \text{ (in meters/sec}^2\text{)}$$
$$\text{force (in dynes)} = m \text{ (in grams)} \times a \text{ (in cm/sec}^2\text{)}$$
$$\text{force (in pounds)} = m \text{ (in slugs)} \times a \text{ (in feet/sec}^2\text{)}$$

The basic unit of force is the *newton*, which is the force required to accelerate a mass of one kilogram one meter per second per second. One newton equals 100,000 dynes. The *dyne* is defined as the force when acting upon a mass of one gram for one second will change its velocity by one centimeter per second. Thus dynes equal $\frac{\text{g} \times \text{cm}}{\text{sec}^2}$. In the English-speaking countries, forces are measured in pounds.

Examples and Problems

What force when acting on a mass of 6 kg will produce an acceleration of 5 meters per second?

$$F = ma \qquad F = 6 \text{ kg} \times 5 \text{ m/sec}^2 = 30 \text{ newtons}$$

If a force of 12 newtons acts on a mass of 2 kg, what will be the acceleration?

$$F = ma \qquad a = \frac{F}{m} = \frac{12 \text{ newtons}}{2 \text{ kg}} = 6 \text{ m/sec}^2$$

2.25 LAW OF GRAVITATION

Gravitation is the universal force of attraction that exists between two masses. Isaac Newton showed that the strength of gravitational attraction depends on two factors: the masses of the objects and the distance between them. The greater the mass or the quantity of matter, the greater the attractive force; the greater the distance, the less the attractive force.

The weight of an object is the measure of the planet's gravitational pull on that object. The earth exerts a large pull on an object of large mass and a smaller pull on an object of lower mass. The weight of a body is the size of this pull. If the earth had twice its mass (but the same volume), it would exert twice its gravitational pull and all objects would weigh twice as much as they do now.

On earth, the weight of an object depends on its distance from the center of the earth. Two objects having the same mass and being in the same place on earth will have the same weight. An object that weighs 200 pounds at sea level will weigh less on the top of Pike's Peak. The weight of an object traveling away from the earth changes in inverse proportion to the square of its distance from the center of the earth. An object that doubles its distance reduces its weight to one-fourth, and one that triples the distance has only one-ninth of its original weight. At a sufficiently great distance from the center of the earth, the planet's gravitational attraction becomes almost zero and an object becomes nearly weightless. Thus, although an object's mass remains the same everywhere, its weight will vary according to the gravity of the body with which it is associated. (See Table 2.8.)

The amount of attraction between two objects is directly proportional to the mass of each of the objects involved, but is inversely proportional

Table 2.8. Weight of a 150-pound man on the different planets and the moon

Planet	*Pounds*	*Surface Gravity (Earth = 1)*
Moon (Earth's)	25.0	0.166
Mercury	40.5	0.27
Mars	55.5	0.37
Venus	129.0	0.86
Jupiter	396.0	2.64
Saturn	175.5	1.17
Uranus	138.0	0.92
Neptune	216.0	1.44
Pluto	. . .	unknown

to the square of the distance by which they are separated. This functional relationship can be expressed by:

$$F = \frac{GM_1M_2}{D^2}$$

where F = the mutual force of attraction
M_1 and M_2 = the masses of the objects
D = distance between the objects and
G = a universal constant ($6.670 \times 10^{-8}\ cm^3\ g^{-1}\ s^{-2}$)

Examples and Problems

How is the gravitational force between two objects affected if the distance between them is doubled? tripled? halved?

If the distance is doubled, the force is one-fourth as great; if tripled, one-ninth as great; if halved, four times as great.

2.26 CENTRIFUGAL FORCE

If the centrifugal force exerted on particles in solution greatly exceeds the opposing diffusion, the molecules will sediment down from the surface of the solution leaving behind the pure solvent. The most frequent calculation when using the regular centrifuge, i.e., not an ultracentrifuge, is the conversion of revolutions per minute (rpm) and revolutions per second (rps) to the number of gravitational force units (g force), as it is meaningless to report the speed of a centrifuge as rpm or rps (unless the type of the centrifuge and rotor are specified) because of differences in mass and distance from the center of rotation. The centrifugal force acting on any mass m at a distance r from the center of rotation depends on the speed of rotation. The speed of rotation may be expressed in terms of v (the number of revolutions per second) or ω* (the angular velocity which is the speed of the centrifuge in rps times 2π). Thus the expressions

*The angular velocity, represented by ω (omega), is defined as the ratio of the angular displacement to the elapsed time.

$$\text{Angular velocity} = \frac{\text{angular displacement}}{\text{elapsed time}} = \frac{\theta}{t}$$

Equations involving angular velocity take a simple form when the velocities are expressed in radians/second. However, it is more common to express the angular velocities in revolutions per second (rps) or revolutions per minute (rpm). Since there are 2π radians in one complete revolution, the number of radians/second is equal to 2π times the number of rps, or equal to $2\pi/60$ times the number of rpm.

for the relative centrifugal force (RCF) are

$$\text{RCF} = \omega^2 mr = 4\pi^2 v^2 mr$$

To calculate RCF one must have the following information: the distance in centimeters from the center of the centrifuge tube to the center of the rotor, which is equal to the radius of the centrifuge rotor (r); π^2 which is 9.8697; and the value for v^2, square of the value of revolutions per second $(\text{rps})^2$. Since manufacturers state centrifuge speeds in rpm, one must utilize the relation $(\text{rps})^2 = \dfrac{(\text{rpm})^2}{(60)^2}$ to obtain the above equation in terms of rpm. After such conversion is done, the symbol v' (rpm) in place of v (rps) will be shown in the formula below. Also, one must divide by 980.665 cm/sec^2 (the constant of acceleration due to gravity) since for a given mass the motor of the centrifuge has to be set to exert a given force, and for each given m there exists a constant gravitational pull of 980.665 cm/sec^2.

Therefore

$$\text{RCF in } g \text{ units} = \frac{(4)(9.8697)(r)(v')^2}{(980.665)(3600)} = \frac{39.4788\, r(v')^2}{3{,}530{,}394}$$

$$= 0.00001118\ (r)(v')^2$$

Thus, simply, the relative centrifugal force RCF = 0.00001118 $\times$ the radius of the rotor $\times$ $(\text{rpm})^2$.

Example: A medical technologist who was assigned to separate some blood cells according to some published procedure was stopped at the first step. While the dial on his centrifuge had calibrations in rpm (as the case in all centrifuges), the procedure called for centrifuging at 1000 $\times$ g. What value of rpm should he use?

All this technologist needed to do was to measure the radius of the rotor he had to use. Assume he found it to be 10 cm. If the measurement of the radius was made in inches, multiply by 2.54 to convert the inches to centimeters.

Then simply convert 1000 $\times$ g to rpm by using the simplified equation above in which v' is rpm.

$$\text{RCF in g units} = 0.00001118 \times \text{radius of the rotor} \times (v')^2$$

$$1000 \times \text{g} = 0.00001118 \times 10 \times (v')^2$$

$$(v')^2 = \frac{1000 \times \text{g}}{0.00001118 \times 10}$$

$$v' = \sqrt{\frac{1000 \times \text{g}}{0.00001118 \times 10}} = \sqrt{8944543} = \sim \sqrt{9000000} = 3000 \text{ rpm}$$

If one was faced with the opposite, i.e., was given rpm and had to find the g units, then it is a straightforward multiplication of the $(\text{rpm})^2 \times \text{radius} \times 0.00001118$.

In a certain centrifugation run, a speed of 3000 rpm was given and the rotor's radius was 10 cm. What is the RCF in g units?

$$\text{g units} = (3000)^2 \times 10 \times 0.00001118 = 1006 \text{ (i.e., approximately 1000 g)}$$

Use of Ultracentrifugation in Determining Sedimentation Constants of Solutes

In 1925 T. Svedberg developed an analytical ultracentrifuge which provided centrifugal fields exceeding 250,000 times the force of gravity. Such a high centrifugal field causes particles to sediment from solution opposing the force of diffusion which normally keeps them evenly dispersed in solution. The analytical ultracentrifuge consists of a rotor, rotating about an axis with a high and constant velocity. Placed in this rotor, some distance from the axis of rotation, is a small cell into which the solution to be studied is placed.

After sedimenting away from the meniscus (the curved upper surface of the solution), the molecules form a sharp boundary. The rate of movement of this boundary down the centrifuge cell is observed by optical measurements of the index of refraction at different positions along the cell. The measurements are made photographically at timed intervals during the centrifugation, while the rotor is spinning at some fixed constant speed. A sedimentation velocity pattern employing Schlieren type of optics is shown in Figure 2.34.

The rate of sedimentation is expressed as the *sedimentation coefficient*

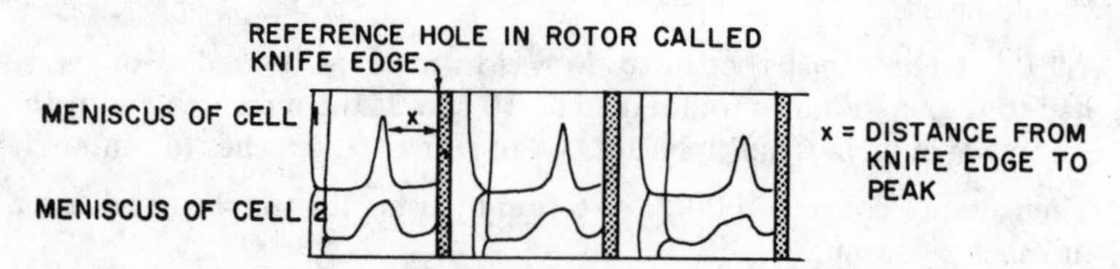

Fig. 2.34. A Schlieren sedimentation velocity pattern. Sedimentation velocity pattern of an enzyme (rabbit phosphorylase *b*) under two different conditions. These pictures were taken in a Model E Analytical Ultracentrifuge at 20° C at 16-min intervals, 40,000 rpm. Top panel in this photograph is an enzyme sample with 5 times more 5′-adenosine monophosphate than the lower panel, otherwise both contained identical concentrations of sodium fluoride and 2-mercaptoethanol. The sedimentation patterns which were automatically photographed in the ultracentrifuge on a film-coated glass plate are sketched here for illustration. Position of the meniscus is not used in the calculation. (Taken from S. A. Assaf and A. A. Yunis, *Biochem. Biophys. Res. Commun.* 42: 865, 1971.)

S and can be found by the following formula:

$$S = \frac{d \log x}{dt} \times \frac{2.303}{60\,\omega^2} = \frac{(\text{slope})(2.303)}{60\,(2\pi \text{rps})^2}$$

where ω = angular velocity in radians/second
t = time elapsed in minutes
x = distance from center of rotation in centimeters after correcting for magnification
rps = revolutions per second
slope = slope of the line obtained from a plot of log x versus time on the abscissa (intervals of the photographs)

A sedimentation coefficient obtained as above which is not corrected for temperature or viscosity of water at 20° is called S_{observed} or S_{obs}. Upon simplification of the above Svedberg equation, one finds that

$$S_{\text{obs}} = \text{a factor} \times \text{slope}$$

$$\text{This factor} = \frac{0.03838}{\omega^2} = \frac{2.303}{60\,(2\pi \text{rps})^2}$$

and hence one can make a table of these factors for each speed of the ultracentrifuge. Such a table was provided by Spinco Instruments,* the manufacturer of the Model E Analytical Ultracentrifuge, and a portion of the table is shown in the Appendix where the factor is equal to a number times 10^{-10}.

A sedimentation coefficient of 1×10^{-13} is called a *Svedberg unit* or simply a Svedberg, abbreviated S. Thus a sedimentation coefficient of 10×10^{-13} would be denoted 10 S. If this same sedimentation coefficient had been corrected for temperature at 20°C, it would be called S_{20} and when it is also corrected for density and viscosity of water at 20°C, it is called $S_{20,w}$.

To convert S_{obs} to $S_{20,w}$ one needs to use the following equation:

$$S_{20,w} = S_{\text{obs}} \left(\frac{\eta_t}{\eta_{20,w}}\right) \left(\frac{\eta}{\eta_0}\right) \left(\frac{1 - \overline{V}\rho_{20,w}}{1 - \overline{V}\,(\rho/\rho_0)\rho_{t,w}}\right)$$

where $\frac{\eta_t}{\eta_{20,w}}$ = ratio of water's viscosity at the temperature of the sedimentation velocity experiment to that at 20°C

$\frac{\eta}{\eta_0}$ = ratio of the viscosity of buffer to that of water at any given temperature

*It should be noted that only one model of an analytical ultracentrifuge is made in the United States. It is called the Model E Ultracentrifuge manufactured by the Spinco Division of Beckman Instruments, Palo Alto, Calif. Methods for analyzing data using the Model E Analytical Ultracentrifuge may be obtained from the above company.

$\overline{V}$ = partial specific volume (0.74 - 0.75 for most proteins)
$\rho_{20,w}$ = density of water at 20°C
ρ/ρ_0 = ratio of the density of the buffer to the density of water at a given temperature
$\rho_{t,w}$ = water's density at the temperature of the sedimentation velocity experiment

Example and Problem

A protein was dissolved in a buffer of known pH, concentration, and density. The solution was placed in the Model E Analytical Ultracentrifuge, preset at a certain temperature, and centrifugation was carried out at 60,000 rpm. Photographs were taken after the peak broke away from the meniscus at 8-minute intervals. What is the S_{obs}, if measurement of x (the distance from the dark knife edge of the plate to the center of the peak) is as follows:

Data:

Time	x (cm)	$\frac{x\text{ (cm)}}{2.1}$	7.35 - corrected x	log X
20 min	1.739	0.8281	6.5219	0.8143
28	1.531	0.7290	6.6210	0.8210
36	1.316	0.6267	6.7233	0.8276
44	1.097	0.5224	6.8276	0.8343

Procedure for reading ultracentrifuge plates:

knife edge - peak reading = x

$$\frac{x}{\text{magnification factor}} = \text{corrected } x$$

7.35 - corrected x = X

log X = the value to be plotted against time to determine the slope

From a plot of log X versus time in minutes, one obtains a slope of 8.4×10^{-4}. Looking in the Appendix, one finds 9.7208 for $F \times 10^{10}$ at 60,000 rpm; hence, the factor = 9.7208×10^{-10}.

Solution: S_{obs} = slope × factor
$= 8.4 \times 10^{-4} \times 9.7208 \times 10^{-10}$
$= 81.65472 \times 10^{-14} = 8.2 \times 10^{-13}$

Since one S, i.e., a Svedberg unit, is equal to 1×10^{-13}, this $S_{obs} = 8.2\ S$.

Temperature Corrections for S_{obs}

To get S_{20} one must correct to the temperature of water at 20°C. This is a large correction if the centrifugation is done at a cold temperature. For example, if the experiment is carried out at 1°C, one multiplies S_{obs}

by 1.723. If the experiments are carried out at 2, 4, and 8° C, one multiplies S_{obs} by 1.665, 1.560, and 1.379, respectively. As the temperature approaches 20° C, the correction for temperature becomes small. Thus at 18° C, one multiplies S_{obs} by 1.0512, at 19° C by 1.0251, and at 20° C by 1.000. At temperatures higher than 20° C, for example, at 22, 24, 26, 28, and 30° C, one multiplies by 0.9525, 0.9082, 0.8671, 0.8291 and 0.7935, respectively. Spinco Instruments provides tables for this purpose.

Correction for $S_{20,w}$ involves multiplication of S_{obs} by the factors indicated in the $S_{20,w}$ formula.

Use of Relative Sedimentation Velocity Values for Approximate Determination of Molecular Weight

For many proteins the sedimentation coefficient is related to the 2/3 power of its molecular weight (MW), i.e., $S \propto MW^{2/3}$. Proteins that are not spheres and contain bound lipids or carbohydrates do not follow this relation.

A simple but a fairly accurate method was developed by Martin and Ames [R. J. Martin and B. N. Ames, *J. Biol. Chem.* 236 (1961): 1372] for the determination of the relative sedimentation velocity of a substance compared to another of known sedimentation velocity or molecular weight. This method is called sucrose density gradient ultracentrifugation. In this technique the sample to be studied is layered carefully on a gradient of sucrose solution, usually 5-20 percent. A material of known sedimentation coefficient is either mixed with the sample to be studied (i.e., internal standard) or layered on another tube containing an identical gradient (external standard). Centrifugation is carried out in a swinging bucket rotor for a period sufficient to cause good separation. A hole is then punched in the bottom of the cellulose acetate centrifuge tube and identical volume of fractions are collected by a fraction collector. Position of the standard and the unknown is then determined by analyzing the concentrations of the standard and the unknown in the fractions. Fraction numbers with peak concentration are used in a calculation involving proportions. Martin and Ames presented a simple equation for such proportions.

$$R = \frac{\text{distance traveled from meniscus by unknown}}{\text{distance traveled from meniscus by standard}}$$

where the meniscus is defined as the last fraction collected.

In terms of sedimentation coefficients,

$$R = \frac{\text{sedimentation coefficient of unknown}}{\text{sedimentation coefficient of standard}}$$

and according to Martin and Ames

$$\frac{\text{sedimentation coefficient of unknown}}{\text{sedimentation coefficient of standard}} = \left(\frac{\text{MW of unknown}}{\text{MW of standard}}\right)^{2/3}$$

Example: A biochemist who carried out a sucrose density gradient ultracentrifugation experiment on an enzyme purified from human blood platelets found this enzyme came in fraction number 16 while that of the standard whose sedimentation coefficient was 11.3 S and MW = 250,000 g/mole came out in tube number 6. A total of 33 identical fractions were collected. Given this data, what is the sedimentation coefficient and molecular weight of the unknown platelet enzyme?

Consider S_1 and MW_1 to be the sedimentation and molecular weight of the unknown and S_2 and MW_2 the values of the standard.

$$R = \frac{33 - 16}{33 - 6} = 0.63$$

$$0.63 = \frac{S_{20,w} \text{ of unknown}}{11.3} = 7.1S$$

From the proportion $\dfrac{S_1}{S_2} = \left(\dfrac{MW_1}{MW_2}\right)^{2/3}$

$$\frac{7.1}{11.3} = \left(\frac{MW_1}{250{,}000}\right)^{2/3}$$

$$(0.63) = \left(\frac{MW_1}{250{,}000}\right)^{2/3}$$

$$(0.63)^{3/2} = \frac{MW_1}{250{,}000}$$

$$(\sqrt{0.63})^3 = \frac{MW_1}{250{,}000}$$

$$0.50 = \frac{MW_1}{250{,}000}$$

$MW_1 = 250{,}000\ (0.50) = 125{,}000$ g/mole, the molecular weight of the unknown platelet enzyme

Use of Sedimentation Equilibrium for Accurate Determination of Molecular Weight

The calculations are based on a simple equation provided by Svedberg in 1959. Using interference optics, the change in concentration is determined by counting the fringes and fractional fringes in a technique described very well in the Spinco Manual of Beckman Instruments. A plot of log concentration in fringes versus x^2 (the symbol r^2 is also used) gives the slope used in the calculation, where x or r is the distance to the center of the rotor.

The molecular weight (MW) equation is

$$\text{MW} = \frac{\text{slope} \times 2 \times 2.303 \times R \times T}{(1 - \overline{V}\rho)\, w^2}$$

where R = gas constant = 8.314×10^7 (ergs/°C)
T = absolute temperature
ρ = density of the buffer and w^2 and $\overline{V}$ as defined above

Example: A crystalline enzyme, which was dissolved in a buffer whose density is 1.005, was run in the analytical ultracentrifuge, using a high-speed sedimentation equilibrium method at a speed of 12,000 rpm, 20°. A plot of log concentration in fringes versus r^2 gave a line whose slope is 0.6357. What is the molecular weight assuming its $\overline{V}$ (partial specific volume) value is 0.75 ml/g?

Data: slope = 0.6357
$R = 8.314 \times 10^7$
$T = 20° + 273° = 293°$
$\overline{V} = 0.75$
$\rho = 1.005$
$w = 2\pi \text{ rps} = 6.2832 \times 200 = 1256.64$
$w^2 = 1.57914 \times 10^6$ (as may also be found in the Appendix)

$$\text{MW} = \frac{0.6357 \times 2 \times 2.303 \times 8.314 \times 10^7 \times 293}{(1 - (0.75 \times 1.005))\,(1.57914 \times 10^6)}$$

$$= \frac{7160.86 \times 10^7}{0.38886 \times 10^6} = 18{,}415.007 \times 10^1 = 184{,}150 \text{ g/mole}$$

2.27 WORK

Whenever a force acts and motion takes place in the direction of the force, then work has been done. This can be expressed by the functional relationship

$$W = FD$$

where W = work done
F = force acting and
D = distance through which the force moves, measured in the same direction as the force.

If the force is measured in pounds and the distance in feet, the amount of work done is expressed in foot-pounds. In the metric system, if the force is measured in newtons and the distance in meters, the work unit is the joule.

Example and Problem

How much work is done when a man carries a 117-pound box up 15 ft to the next floor?

Work = 117 lb × 15 ft = 1755 ft-lb of work against an opposition of gravity.

2.28 ENERGY

Energy is the ability to do work, i.e., the ability to move something against an opposing force. Energy may occur as heat, sound, light, and electric fields. Kinetic energy is the energy of motion. Anything in motion can do work until its motion stops. Potential energy is energy that exists because of the relative position of two objects; it is stored up ready to use. Man cannot create energy, but he can change one form of energy to another, or by destroying a mass one can obtain the energy equivalent of that mass. This is known as the *Law of Conservation of Energy.* It means, in fact, that the total amount of energy in the whole universe has always been the same, and will always be the same. Under certain circumstances, matter can be converted into energy. In an atomic explosion, for example, a minute quantity of matter is converted into a vast amount of energy. This is not a contradiction of the Law of Conservation of Energy, since energy is not being created from nothing but from matter. Matter is just another form of energy.

The conversion between mass and energy can be expressed as $E = mc^2$ where

E = amount of energy involved
m = amount of mass corresponding to the energy E, and
c = velocity of light ($\sim 3 \times 10^8$ meters per sec).

The factor c^2 has a large numerical value (9×10^{16}), thus a very small mass can be equivalent to a very large amount of energy.

2.29 ELECTROMAGNETIC RADIATION

Light rays, radio waves, and X rays are some of the varieties of electromagnetic radiation. All forms of electromagnetic radiation are similar in two respects. First, they travel through space with the same velocity (almost 300,000 km per sec or 186,000 miles per sec in vacuum); this velocity is usually represented by the symbol c. Second, they all have the nature of waves.

The *wavelength* λ of a beam of electromagnetic radiation is the linear distance traversed by one complete wave cycle; i.e., the distance between corresponding points on successive waves. (See Fig. 2.35.) The frequency

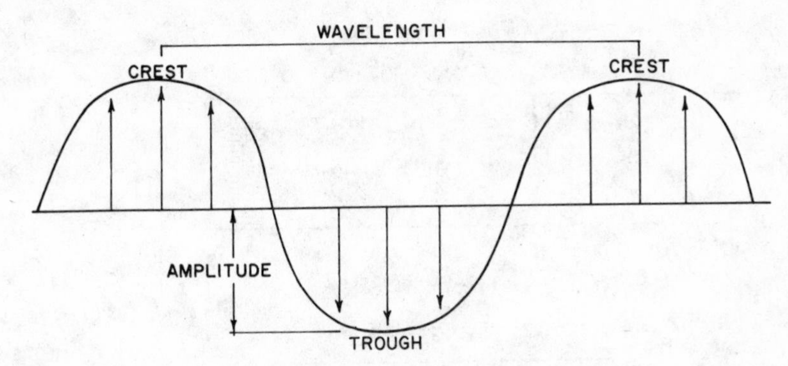

Fig. 2.35. A transverse wave. These are waves whose direction of vibration is perpendicular to the direction of propagation.

v is the number of wave cycles passing a given point in one second. It is readily obtained by dividing the λ into c; thus the velocity of the radiation is $v = \frac{c}{\lambda}$. The velocity varies with the medium through which the radiation is passing, where c in a vacuum is 2.998×10^{10} cm sec^{-1}.

In some of its interaction with matter, radiation can be shown to behave as if it were composed of discrete packets of energy called *photons.* The energy of a photon is variable and depends upon the frequency or wavelength of the radiation. The relationship between the energy E of a photon and frequency v is given by the equation

$$E = hv$$

where h is Planck's constant with a numerical value of 6.63×10^{-27} erg sec. The equivalent expression involving wavelength is

$$E = \frac{hc}{\lambda}$$

Thus it can be seen that light at short wavelengths is more energetic than light at long wavelengths. (See Fig. 2.36).

If the energy absorbed per mole is desired, then one has to multiply the numerator of the equation by 6.023×10^{23}, the number of molecules per mole which is the Avogadro number, N. Thus

$$E = \frac{6.023 \times 10^{23} \times 6.63 \times 10^{-27} \times 2.998 \times 10^{10}}{\lambda}$$

$$= \frac{1.1972 \times 10^{8}}{\lambda} \text{ ergs per mole}$$

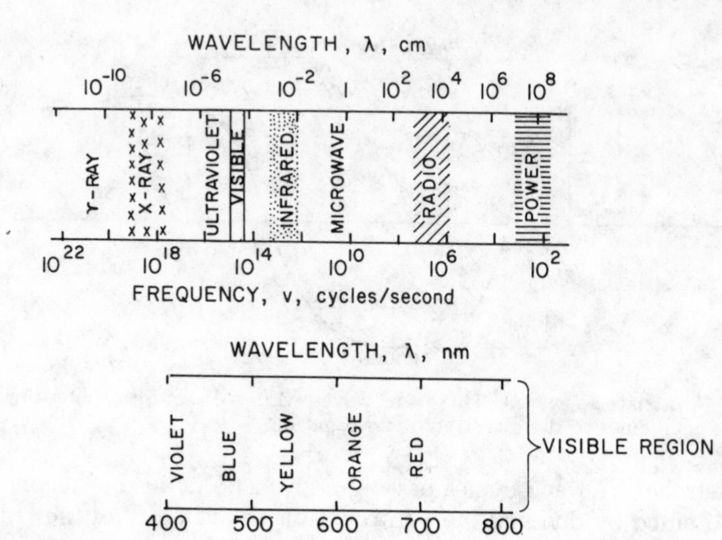

Fig. 2.36. The electromagnetic spectrum. Note that 1 cm = 10^7 nm. Approximate range of some segments of the electromagnetic spectrum: X-rays 0.7–75 nm; UV 75–400 nm; visible 400–950 nm; infrared 950–70,000 nm.

If the wavelength is expressed in Angstrom units, i.e., 10^{-8} cm, as is commonly done, the equation becomes

$$E = \frac{1.196 \times 10^{16}}{\lambda} \text{ ergs per mole}$$

It is more convenient to state the energy in calories or kilocalories, and since 1 cal is equivalent to 4.184×10^7 ergs, it follows that

$$E = \frac{1.196}{\lambda} \times \frac{10^{16}}{4.184 \times 10^7} \text{ cal per mole}$$

$$= \frac{2.859}{\lambda} \times 10^5 \text{ kcal per mole.}$$

Another expression used is *wave number* with the symbol $\bar{\nu}$, which is the number of wavelengths per centimeter and is equal to the reciprocal of the wavelength:

$$\bar{\nu} = \frac{1}{\lambda} \text{ cm}^{-1}$$

and since frequency $v = \dfrac{c}{\lambda}$ then $\bar{\nu} = \dfrac{v}{c}$ which relates the frequency in wave numbers (cm^{-1}) to the true frequency (sec^{-1}).

2.30 HALF-LIFE

Half-life is the time required for half of a quantity of a radioactive isotope to decay or break down into a different isotope. Half-life periods of radioactive isotopes range from a fraction of a second to millions of years. The period is not changed by any variation in pressure or temperature nor by any chemical reaction into which the radioactive substance enters. The half-life period is different for different radioactive isotopes but is the same for different samples of the same radioactive isotope.

The number of disintegrations per unit time in a group of radioactive nuclei of a given sort is proportional to the number of such atoms present:

$$-\frac{dN}{dt} = kN$$

where N is the number of radioactive atoms present, dN is the change in this number in time interval dt, and the constant of proportionality, k, is called the rate of decay or disintegration constant, which is characteristic for each radioactive element. This equation represents an exponential decay type of variation; it states that in a small interval of time, dt, a certain fraction, kdt, of the initial number of atoms decays.

Integrating,

$$N = N_0 e^{-kt}$$

$$kt = \ln \frac{N_0}{N} \quad \text{and} \quad k = \frac{1}{t} \ln \frac{N_0}{N} = \frac{2.303}{t} \log \frac{N_0}{N}$$

where N_0 is the number of atoms at $t = 0$ and
N is the number of atoms after time t.

The half-life period for the decay process is the time after which the concentration of decomposing substance has decreased to half its original concentration, i.e., $N = \frac{N_0}{2}$. Thus the half-life period $t_{1/2}$ is

$$t_{1/2} = \frac{\ln 2}{k} = \frac{2.303 \log 2}{k} = \frac{0.693}{k}$$

The half-life period can also be obtained graphically by plotting the logarithm of the number of radioactive isotopes against the time and reading off the time interval corresponding to a change of 0.301 (i.e., log 2) on the ordinate. (See Fig. 2.37.)

Radioactive dating is the technique of determining the age of an object by measuring the extent to which the radioactive material in it has decayed. Carbon-14 (^{14}C), a radioactive isotope of carbon, is absorbed from the atmosphere by organisms during life but not after death. In liv-

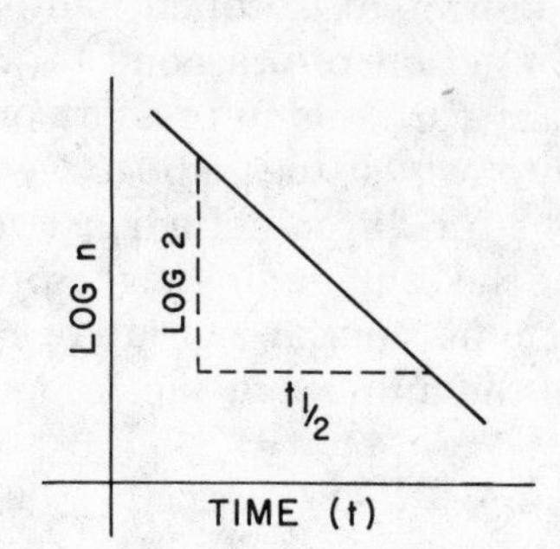

Fig. 2.37. Radioactive decay.

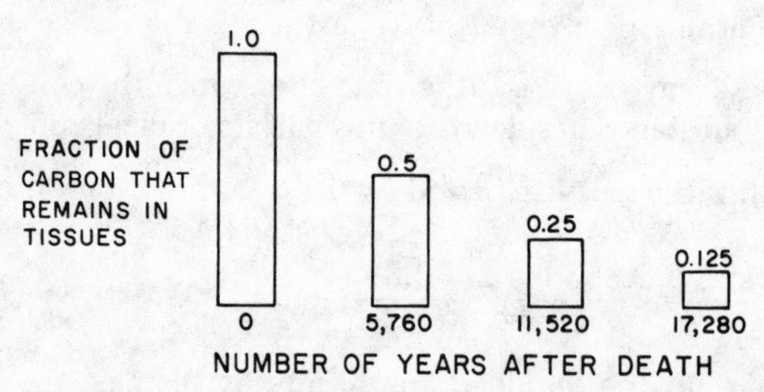

Fig. 2.38. Radioactive decay of carbon-14 (^{14}C).

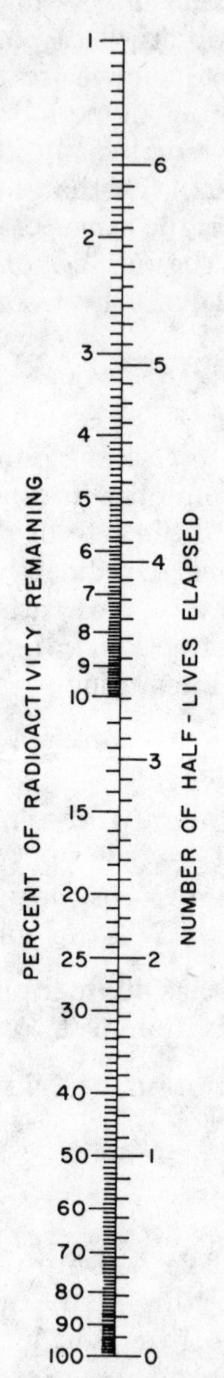

Fig. 2.39. Half-life decay chart.

ing tissues the proportion of carbon-14 to carbon-12 (ordinary carbon which is not radioactive) remains constant. After death of a tissue occurs, no more carbon-14 is absorbed and the carbon-14 already in the tissue breaks down to form nitrogen-14. At the end of 5,760 years (the half-life period of carbon-14), just half of the carbon-14 in the tissue at the time of death has broken down to form nitrogen-14. So, the radioactive level of carbon in a living tree is twice that of a similar tree cut down 5,760 years ago. After another 5,760 years, half of this remainder breaks down, and so on. Thus if one finds that three-fourths of the original carbon-14 has decayed, one knows that the age of the tissues is 11,520 years. (See Figs. 2.38 and 2.39.) By determining the carbon-14 content of specimens (including ancient scrolls made of papyrus) and comparing it with that of the living sources of the raw materials from which they were made, ages may be found.

The *curie* of radioactivity is an amount of material which gives 3.7×10^{10} disintegrations per second. The larger the decay constant, the smaller the amount of an element corresponding to a curie. The millicurie, 10^{-3} curie, and the microcurie, 10^{-6} curie, are terms also in use.

$$\begin{aligned} 1000 \text{ millicuries (mc)} &= 1 \text{ curie (c)} \\ 1000 \text{ microcuries } (\mu c) &= 1 \text{ millicurie (mc)} \\ 250 \text{ microcuries } (\mu c) &= 0.25 \text{ millicurie (mc)} \\ 100 \text{ microcuries } (\mu c) &= 0.10 \text{ millicurie (mc)} \\ 50 \text{ microcuries } (\mu c) &= 0.05 \text{ millicurie (mc)} \\ 1 \text{ microcurie } (\mu c) &= 0.001 \text{ millicurie (mc)} \end{aligned}$$

$$\begin{aligned} 1 \text{ curie} &= 2.22 \times 10^{12} \text{ dpm} = 3.7 \times 10^{10} \text{ dps} \\ 1 \text{ millicurie} &= 2.22 \times 10^{9} \text{ dpm} = 3.7 \times 10^{7} \text{ dps} \\ 1 \text{ microcurie} &= 2.22 \times 10^{6} \text{ dpm} = 3.7 \times 10^{4} \text{ dps} \end{aligned}$$

dpm = disintegrations per minute
dps = disintegrations per second

Time Conversion Factors

$$1 \text{ day} = 1.44 \times 10^{3} \text{ minutes} = 8.64 \times 10^{4} \text{ seconds}$$
$$1 \text{ year} = 5.26 \times 10^{5} \text{ minutes} = 3.16 \times 10^{7} \text{ seconds}$$

Examples and Problems

The half-life of ^{64}Cu is 12.8 hours; calculate k.

$$k = \frac{0.693}{12.8 \text{ hr}}$$

12.8 hr = 4.608×10^{-4} sec
Thus $k = 1.52 \times 10^{-5}$ sec

^{42}K has a half-life of 12.4 hr. What fraction of the initial activity remains after 3.1 hr, 12.4 hr, and approximately $2\frac{1}{2}$ days?

3.1 hr = 85% left (see decay chart)
12.4 hr = 50% left
$2\frac{1}{2}$ days = 60 hr = ~5 half-lives = 1.55% left

What is the half-life of a radionuclide in a sample with 5000 counts per minute and with 3500 counts per minute 5 hr later?

$$k = \frac{2.303}{t} \log \frac{5000}{3500}$$

$$= 0.4606 \log 1.43 = 0.4606 \times 0.1553 = 0.0715 \text{ hr}$$

$$t_{1/2} = \frac{0.693}{0.0715} = 9.7 \text{ hr}$$

Determination of Specific Activity of a Given Isotope of Unknown Concentration

Specific Activity = cpm divided by molar concentration of the isotope.

Example: Determine the specific activity and concentration of an ATP^{32} sample prepared in some laboratory for the radiochemical assay of a certain enzyme system. Data regarding the optical density and cpm of the ATP^{32} solution is described below. The concentration of ATP^{32} is determined as if it were not radioactive, using the molar extinction coefficient of ATP at its maximum absorption ($\epsilon_{259\text{nm}}$ = 15.4 × 10^3 at pH 7.0) from the relation $OD = \epsilon C$. Using a cuvette of 1 centimeter path length, one can calculate concentration.

Assume the solution had to be diluted 1:40 and gave an *OD* of 0.50 at 259 nm. Hence, the number of *OD* units = 0.50 × 40 = 20.

$$\text{Therefore C} = \frac{20}{15 \times 10^3} = 1.33 \times 10^{-3} \text{M.}$$

To be able to read its radioactivity in a scintillation counter, some dilutions had to be made. Assume 1/1000 and then 0.1 ml was taken and mixed with a Dioxane counting fluid. Thus 1 to 1000 final dilution is made. The counts per 0.2 minute were found to be 450,000. Therefore multiplication by 5 yields the cpm which is 2,300,000, i.e., 2.3 × 10^6 . To correct for the dilution of 1 to 1000, multiply cpm by 1000, which gives 2.3 × 10^9 cpm/ml.

$$\text{Specific activity of an isotope} = \frac{\text{cpm}}{\text{molar concentration of the isotope}}$$

$$= \frac{2.3 \times 10^9}{1.33 \times 10^{-3}} = 1.7 \times 10^6 \text{ cpm/mole of ATP}$$

$$= 1.7 \times 10^9 \text{ cpm/m mole of ATP}$$

Every 2.22×10^9 cpm* = 1 millicurie (mCi), so

$$\frac{1.7 \times 10^9}{2.22 \times 10^9} = 0.8 \text{ mCi/m mole.}$$

Determination of cpm/μL of an Isotope Solution of Known Specific Activity and Concentration

Example: If one received from a chemical company after 2 days of delivery time a 0.09 ml bottle of radioactive ATP^{32} solution with this label: specific activity = 19μCi/m mole concentration ATP^{32} = 0.054 μ mole/mCi, determine cpm/μL.

0.054 μ mole/mCi is the same as 0.054×10^{-3} m mole/mCi

Since there are 0.054×10^{-3} m mole ATP^{32} per 0.09 ml bottle, per 1 ml there is how much ATP^{32}?

$$\frac{0.054 \times 10^{-3}}{0.09} = 0.6 \times 10^{-3} \text{ m mole/ml}$$

and since there are 19μCi per 1 m mole of ATP^{32}
i.e., 0.019 mCi in 1 m mole ATP^{32}
then per 0.6 m mole, how many mCi are there?

$$\frac{0.019 \times 0.6 \times 10^{-3}}{1} = 1.14 \times 10^{-5} \text{ mCi}$$

Since we calculated that each 1 ml has 0.6 m mole, then each ml has 1.14×10^{-5} mCi, which is 1.14×10^{-5} mCi/ml.

According to ATP^{32} decay tables (see, for example, *Geigy Pharmentical Documenta: Geigy Scientific Tables*, 6th ed., 1962, p. 278) 90% of the radioactivity remained after the 2 days of shipping time, i.e.,

$$1.14 \times 10^{-5} \times \frac{90}{100} = 1.026 \times 10^{-5} \text{ mCi/ml.}$$

To convert this number to counts per minute (cpm) multiply by

2.22×10^9 which is the number of disintegrations per minute for 1 mCi

So, $1.026 \times 10^{-5} \times 2.22 \times 10^9 = 2.3 \times 10^4$ cpm/ml
$= 230$ cpm/μL

Determination of the μ Moles of an Isotope Incorporated per Milligram of Enzyme per Minute

Example: An enzymologist had an ATP^{32} stock solution which gave 1.63×10^9 cpm/mM. A 0.02 ml aliquot was taken and added to an 0.08 ml enzyme-sub-

*Actually 2.22×10^9 dpm (disintegrations/minute) = 1 mCi and one should use cpm only after correcting for the counting efficiency of the system, although this is frequently neglected.

strate reaction mixture (resulting in a total volume of 0.1 ml). This reaction mixture contained a protein substrate acceptor to which the isotope will bind, depending on the amount of the enzyme added to this reaction mixture. The reaction was stopped after a 20-minute period and a 0.05 ml aliquot of the isotope bound protein gave a count of 30,800 cpm. Determine the μ mole of P^{32} incorporated per milligram of enzyme if the final concentration of the enzyme in the reaction mixture is 0.02 mg/ml.

The isotope dilution factor is $\dfrac{0.02 \text{ ml aliquot}}{0.1 \text{ ml final}} \times 0.05 \text{ ml aliquot} = \dfrac{1}{100}$

The scintillation counter gave a reading of 30,800 cpm for the 0.05 ml sample.

So, per ml one has $30{,}800 \times 100 = 3{,}080{,}000 = 3.08 \times 10^6$ cpm/ml.

We know from the initial stock that 1 mM $ATP^{32} = 1.63 \times 10^9$ cpm and from the experimental data 1 ml of ATP^{32} solution has 3.08×10^6 cpm. Then from proportions:

the concentration of ATP^{32} * $= \dfrac{1 \times 3.08 \times 10^6}{1.63 \times 10^9} = 2.26 \times 10^{-3}$ mM

2.26×10^{-3} mM is incorporated, using 0.02 mg enzyme
x is incorporated using 1 mg enzyme.

$x = \dfrac{2.26 \times 10^{-3} \times 1}{0.02} = 113 \times 10^{-3}$ mM and since the reaction was carried

out for 20 minutes, the amount incorporated per minute =

$\dfrac{113 \times 10^{-3}}{20} = 5.65 \times 10^{-3}$ mM P^{32}/mg/min
$= 5.65\ \mu$ mole P^{32}/mg/min
and this is how an international unit of enzyme activity is expressed, i.e., μ mole product/mg/min under specified conditions of reaction.

2.31 ANGULAR MEASUREMENT

Description of position or apparent distance between bodies on the celestial sphere cannot be given in terms of inches, feet, or miles. All one can do is describe the direction to a body, or the difference between the directions of two objects. Differences of direction are expressed in terms of angles.† Since light travels in straight lines in space, one refers to the

---

*Actually it is P^{32} and not ATP^{32} that is incorporated into a protein acceptor which is phosphorylated in this type of reaction.

†An *angle* may be considered as generated by the rotation of a line, in a fixed plane, about one of its points; the point about which the rotation takes place is called the vertex of the angle, the original position of the line is the initial side, and the final position of the rotating line is the terminal side of the angle.

angles between lines of sight. Three systems of angular measurement are in common use: (1) sexagesimal, (2) radian, and (3) the mil.

Sexagesimal Measure

1 degree (1°) = 1/360 of one revolution
1 minute (1′) = 1/60 of one degree
1 second (1″) = 1/60 of one minute

Radian Measure

A *radian* is defined as the measure of the central angle subtended by an arc of a circle equal in length to the radius of the circle. Since the radius is contained 2π times in the circumference, there are 2π or 6.28 . . . radians in one complete revolution or 360°. In general, if θ represents any arbitrary angle subtended by an arc of length s on the circumference of a circle of radius R, then θ (in radians) is equal to the length of the arc s divided by the radius R. An angle in radians being defined as a ratio of a length to a length is a pure number. (See Fig. 2.40.)

$$1 \text{ radian} = \frac{360°}{2\pi} = 57.3 \text{ degrees*} = 3438 \text{ minutes} = 206{,}265 \text{ seconds}$$

$$1° = 0.01745 \text{ radians}$$

$$360° = 2\pi \text{ radians} = 6.28 \text{ radians}$$
$$180° = \pi \text{ radians} = 3.14 \text{ radians}$$
$$90° = \pi/2 \text{ radians} = 1.57 \text{ radians}$$
$$60° = \pi/3 \text{ radians} = 1.05 \text{ radians}$$

Mil

1 mil = 1/6400 of one revolution = 0.05625° = nearly 1/1000 of a radian.

Angles are commonly measured by a protractor, usually semicircular in form. A straight angle contains 180°, a round angle 360°. The division of a circle into 360° is of very ancient origin, used many centuries ago by the Babylonians in Iraq and Syria. The system probably grew up in connection with their study of the stars. The number 360 is retained to this day because of its convenience and the force of tradition.

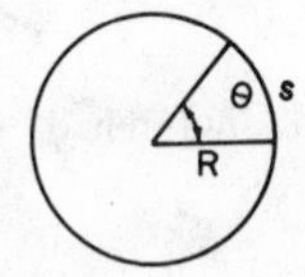

Fig. 2.40. Angle in radians.
$$\theta = \frac{s}{R} \qquad s = \theta R$$

*More accurate value is 57.29578 degrees/radian.

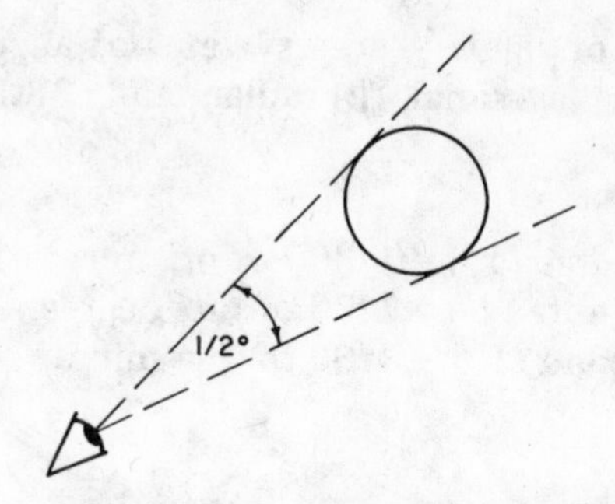

Fig. 2.41. The moon's angular diameter (exaggerated diagram).

Figure 2.41 shows the lines of sight from an observer's eye to the two edges of the moon. At the eye the two lines make an angle of $\frac{1}{2}^\circ$, and thus to the observer the moon's diameter subtends an angle of $\frac{1}{2}^\circ$, i.e., its angular diameter is $\frac{1}{2}^\circ$.

2.32 ANGULAR DIAMETER

To translate the angular diameter of any celestial object into real dimensions, one must know the distance to the object. The formula is as follows:

$$\text{diameter in miles} = \frac{\text{diameter in degrees}}{57.3^\circ} \times \text{distance in miles}$$

$$\text{diameter in miles} = \frac{\text{diameter in minutes}}{3438'} \times \text{distance in miles}$$

$$\text{diameter in miles} = \frac{\text{diameter in seconds}}{206{,}265''} \times \text{distance in miles}$$

Note: 1 radian is equal to $57.3^\circ = 3438' = 206{,}265''$

Astronomical units, parsecs, or any other unit of distance other than miles can be substituted in the above formula.

Example and Problem

Determine the sun's diameter in miles if its angular diameter is 32′ at a distance from the earth of 92,900,000 miles.

$$\text{Diameter in miles} = \frac{32'}{3438'} \times 92{,}900{,}000 = {\sim}864{,}688 \text{ miles}$$

2.33 PARALLAX

Parallax is the apparent change in the position of an object when seen first from one viewpoint and then from another. A star's position against the background of more distant stars appears to change slightly as the earth travels around its orbit. This apparent small angular displacement of position is measured in seconds of arc. The closer the star, the greater its apparent shift, or parallax, with respect to the distant stars.

A base line of known length is laid off and the far-off object is viewed from each end of this base line and the two angles with the base line are noted. Knowing the length of the base line and the number of degrees in the angles at each end, the height of the triangle whose third point is the far-off object can be solved by trigonometry. When a star, for example, is observed from E', star A would be seen at e' on the background of the sky. When observed from E'', A would be seen at e''. A line is then drawn perpendicular to the center of the base line to the star A. One then has two right-angled triangles and the parallax angle (p) at star A. Once the parallax has been determined, the distance SA can be computed by trigonometric calculation. (See Fig. 2.42.)

The solutions of many astronomical problems of distance and dimensions depend on the calculation of the length of one side of a long, slender right triangle, when another side and the smallest angle are known (the parallax angle). For example, the parallax of 1 second corresponds to a distance of about 20 million miles (i.e., 10^{12}) or 3.3 light years or 1 parsec.

Given one side (b) and the parallax angle (p), then the long (c) and

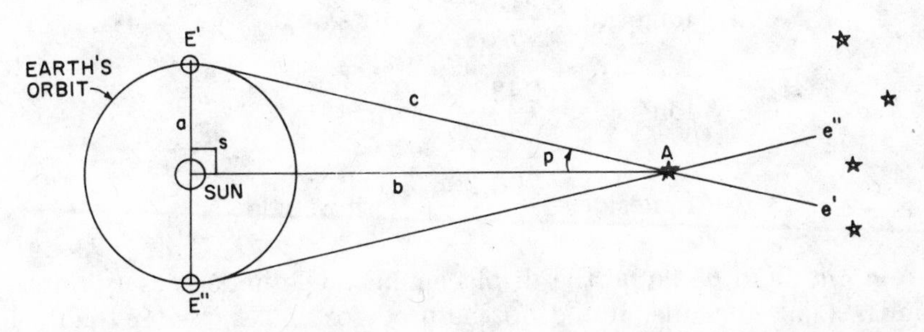

Fig. 2.42. Parallax of a star. Parallax of star A (angle p) is given by the apparent shift of star A against background of more distant stars as the earth moves to opposite side of orbit. Angle p, angle ESA (a right angle), and the distance ES (earth to sun) being known, the distance SA (sun to star) is calculated by trigonometry.

$$c = \frac{a}{\sin p} = \text{earth to star;}$$

$$b = \sqrt{(c + a)(c - a)} = c \cos p = \frac{a}{\tan p} = \text{sun to star } A$$

short (a) sides of right triangles can be found by the following trigonometric relations:

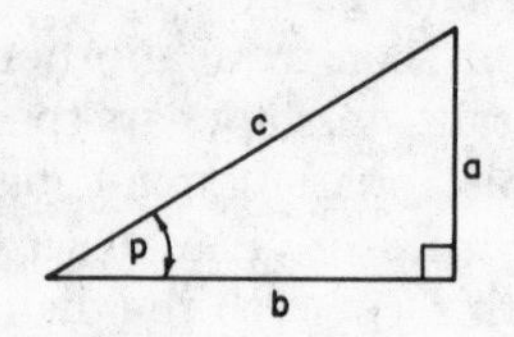

$$\sin p = \frac{a}{c} \qquad \cos p = \frac{b}{c} \qquad \tan p = \frac{a}{b}$$

$$\text{short side } a = b \tan p$$

$$\text{long side } c = \frac{b}{\cos p}$$

Additional identities for the right triangle are

$$a = \sqrt{(c + b)\,(c - b)} = c \sin p$$

$$b = \sqrt{(c + a)\,(c - a)} = c \cos p = \frac{a}{\tan p}$$

$$c = \frac{a}{\sin p} = \sqrt{a^2 + b^2}$$

A simple formula for the calculation of the long side of a long, slender right triangle is

$$\text{long side} = \frac{57.3^\circ}{n^\circ} \times \text{short side}$$

$$\text{long side} = \frac{3438'}{n'} \times \text{short side}$$

$$\text{long side} = \frac{206{,}265''}{n''} \times \text{short side}$$

where n°, n', and n'' equal the displacement angle in degrees, minutes, or seconds. (This formula should not be used for angles greater than a few degrees.)

Examples and Problems

Given a triangle whose parallax angle is 1° and whose short side is 93,000,000 (93×10^6) miles $\left(\frac{1}{2}\text{ the diameter of the orbit of the earth}\right)$ what is the distance (c) to object (0)?

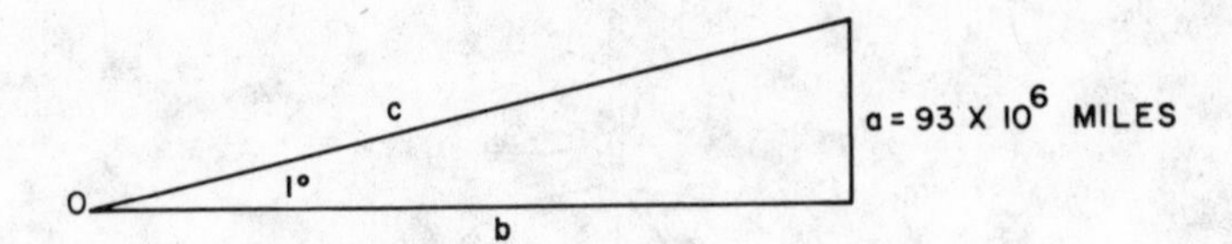

Since the angle measured is in degrees, the formula used is

$$\text{long side } c = \frac{57.3^\circ}{1^\circ} \times (93 \times 10^6 \text{ miles})$$
$$= 57.3 \times (93 \times 10^6 \text{ miles})$$
$$= 5328.9 \times 10^6 \text{ miles}$$
$$\cong 5.3 \times 10^9 \text{ miles}$$

The displacement angle of the moon is 30′ when observed simultaneously from two laboratories 2000 miles apart. What is the distance to the moon?

$$\text{Distance to the moon} = \frac{3438'}{30'} \times 2000 = 114.6 \times 2000 = 229{,}000 \text{ mi}$$

2.34 KEPLER'S LAWS

Kepler's laws are three mathematical laws that describe the motion of planets. They are as follows:

1. Each planet travels in an elliptical orbit with the sun at one focus of the ellipse. The *ellipse* is a curve of such shape that for every point on it, the sum of the distances from two fixed points (the foci) is constant. In Figure 2.43, the sum AF and AF' is the same as the sum BF plus BF' or DF plus DF'. The closer the two foci, the more nearly the form of the ellipse approaches a circle.

An ellipse can be drawn by placing a loop of thread around two pins and while pulling the thread tight with the point of a pencil, draw a curve. (See Fig. 2.44.)

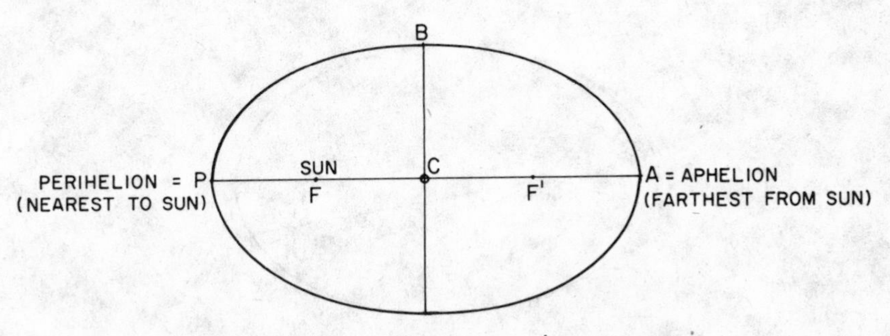

Fig. 2.43. An ellipse. F and F' are the foci.

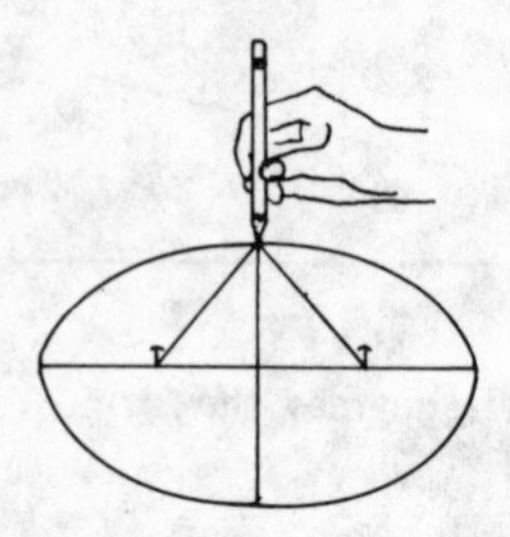

Fig. 2.44. Drawing an ellipse.

2. For equal time intervals the areas swept by a line between the sun and the planet are equal. This means that the speed of a planet in orbit around the sun varies. The planets move most rapidly in the parts of their orbits nearest the sun and most slowly when they are farthest from the sun. (See Fig. 2.45.)

In equal periods of time ($t_1 = t_2$), the radius vectors (r_1 and r_2) sweep equal areas ($A_1 = A_2$). The planet takes the same time to travel from point A to point B as it does from C to D, and all the "triangles" swept out are equal in area.

3. For any two planets, the ratios of the squares of the periods of revolution to the cubes of the *mean distances* of the planets from the sun are the same. The more distant planets have longer periods, not only because they have farther to go but because they move more slowly. The mean distance is the length of the semimajor axis which is one-half the distance from the perihelion (when the planet is nearest the sun) to the aphelion (when the planet is farthest from the sun). (See Fig. 2.46.)

Kepler's third law can be stated as a proportion:

$$\frac{P'^2}{P''^2} = \frac{D'^3}{D''^3}$$

where P' and P'' are the respective periods of revolution of two planets and D' and D'' are the respective mean distances.

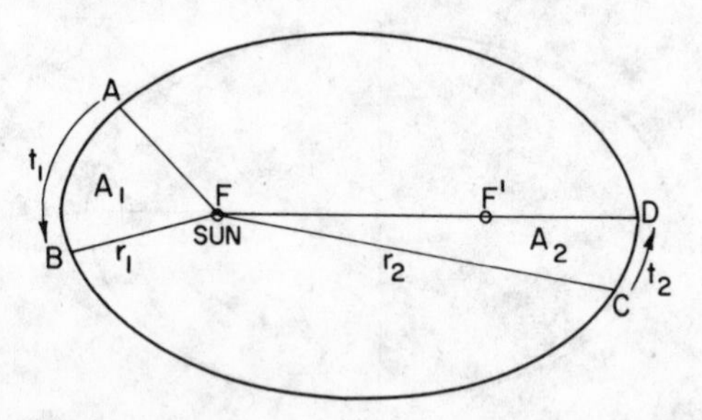

Fig. 2.45. Kepler's law of planetary motion.

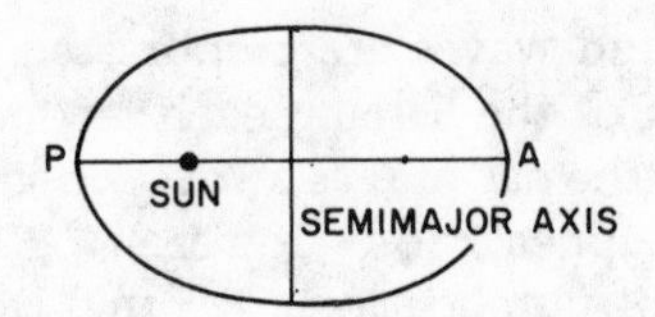

Fig. 2.46. Axes of an ellipse.

Isaac Newton later showed that Kepler's third law was not precisely correct and that the masses of the sun and planets should be considered as follows:

$$\frac{D'^3}{D''^3} = \frac{(M + m')\,P'^2}{(M + m'')\,P''^2}$$

where M is the mass of the sun and m' and m'' are the masses of the two planets. Kepler's simpler form of the law appeared correct because the sun's mass is so large compared with the masses of planets.

Examples and Problems

Saturn has a period of revolution of 29.5 years and is about 9.5 times farther from the sun than the earth. Does Kepler's third law of planetary motion hold true for Saturn?

Yes. $29.5^2 \cong 9.5^3$

If the distance of Venus from the sun is 0.7 times the earth's distance, then what is its period?

$P^2 = D^3$ $\quad P^2 = (0.7)^3$ $\quad P^2 = 0.343$ $\quad P = 0.59$ **earth years (rounded off)**

2.35 DOPPLER'S PRINCIPLE

Doppler's principle is the name given to the physical law that the apparent wavelength of sound or light depends upon the velocities of the observer and the source from which the wavelength proceeds. It is the apparent change of frequency when the distance between source and observer is changing. Light or sound waves are altered if either the source emitting them or the observer is moving.

For example, a man standing in the surf will be struck by a certain number of waves per minute if the waves come in at a regular rate. If he swims into the waves, he will be struck by a greater number per minute, and if he moves toward shore, he will be hit by fewer. Similarly, the frequency (number of waves passing a given point each second) of light and sound waves increases as source and observer draw closer, but decreases when they go apart. The sound of an automobile horn seems to rise in pitch as the vehicle approaches, then drops in pitch after passing. As the car approaches, the sound waves are compressed into a shorter distance; more waves reach the ear of the listener each second and one hears a sound of higher frequency. As the car moves away, the sound waves are stretched out and their frequency appears to be lower.

The most important applications of the Doppler principle are in astronomy in connection with the measurement of the velocities of celestial bodies by observing the displacement that their motion produces in the positions of the lines of their spectra. If a heavenly body is approaching the earth, the lines in the spectrum of that body shift toward the violet end, due to the apparent shortening of each wavelength by motion. If the earth and the heavenly body are traveling away from each other, there is a displacement of spectra lines toward the red end (longer wavelengths). The received wavelength frequency is lower than the emitted frequency when the source and the receiver are moving apart, and higher when they are approaching each other.

The rotation of Saturn's rings has been experimentally demonstrated in this way. Certain stars have been demonstrated to be double as the spectral lines appear to be single when the two components are perpendicular to the line of sight and double when the positions of the component stars are such that one star is approaching the earth while the other is traveling away from the earth.

Astronomers have observed a red shift in the spectra of galaxies and the greater the distance of these galaxies from the Milky Way Galaxy the more pronounced the shift.* It appears that the galaxies of the universe are rushing outward from a common center of origin. Many astronomers take this as evidence for the "big bang" theory.†

Doppler's formula, $\frac{\Delta\lambda}{\lambda_0} = \frac{V}{c}$, is a proportion which can be stated as follows: the change in wavelength is to the normal wavelength of the light source as the velocity of the light source is to the speed of light, where

$\Delta\lambda$ = the change in wavelength
λ_0 = the normal wavelength of the light source
V = velocity of the light source, and
c = the speed of light.

*Some theorists believe that the stars are not moving away from us, and that the red shift is not due to the Doppler effect at all but results from a slowing down of light in the gravitational field of the universe.

†The big bang theory assumes that at one time all matter was concentrated at one point and exploded outward and is presently moving outward.

PART 3

Statistics

3.1 STATISTICAL NOTATION AND VOCABULARY

$$\sum_{i=1}^{i=n} X_i$$

The capital Greek sigma, Σ, is called an operator symbol and simply means the sum of the items indicated. The $i = 1$ means that the items should be summed, starting with the first one and ending with the nth one as indicated by $i = n$ above the Σ. The subscript and superscript are necessary to indicate how many items should be summed. The "$i =$ " in the superscript is usually omitted.

If there were 5 items and only the first 4 were to be summed, the symbol would be written $\sum_{i=1}^{4} X_i$; if all except the first one were to be summed, the symbol would be written $\sum_{i=2}^{5} X_i$.

For nonmathematical readers it is desirable to omit subscripts and superscripts as they distract attention from the respective formula by adding to the apparent complexity $\left(\text{e.g.}, \sum_{i=1}^{i=n} X_i \text{ becomes } \Sigma X\right)$.

Population

A population is the total set of individual observations (actual or potential) about which inferences are to be made.

Sample

A sample is a finite portion or a subset of the population. It is a collection of individual observations selected by a specified procedure.

It is conventional in statistics to use Greek letters for population parameters and Roman letters for sample statistics.

	Mean	Standard Deviation	Variance
Sample	$\overline{X}$	s	s^2
Population	μ	σ	σ^2

Degrees of Freedom

Degrees of freedom refers to the number of items of data that are free to vary independently. If there are N observations of y, each observation represents a separate opportunity for the value of y to vary, and there are therefore N degrees of freedom for variation. For a specified value of the mean, only $(N - 1)$ items are free to vary. The value of the Nth item is then absolutely determined by the values assumed by the other items and by the mean.

In general, one degree of freedom for variation is taken away for each parameter calculated from the observations. If a sampling distribution of n variables is independently variable in $n - p$ of the variables, there are $n - p$ degrees of freedom (where p is the number of linear restraints—e.g., mean, or slope).

Null Hypothesis

A null hypothesis assumes that there are no differences from some particular expected value other than those produced by chance. The interpretation of experimental results is based upon the formation of an appropriate null hypothesis. This hypothesis stems directly from the particular experimental question being investigated. The analysis proceeds by assessing the probability of obtaining by chance experimental findings as extreme as those actually observed. If this probability is small enough, the experimental results are judged to be significant. The conventionally accepted probability leading to a judgment of significance is 0.05 or less. A judgment that the results are highly significant is considered appropriate if the probability is 0.01 or less.

3.2 MEAN, MEDIAN, AND MODE

The *arithmetic mean* or *average* ($\overline{X}$) is computed by adding the values of all the items in a set of data (ΣX) and dividing by the number of items (n). The formula for the mean is

$$\overline{X} = \frac{\Sigma X}{n} \quad \text{or} \quad \overline{X} = \frac{\Sigma fX}{n}$$

where f is frequency.

The *median* is a statistic of location. It is defined as that value of the variable in an ordered array that has an equal number of items on either side of it. The median divides a frequency distribution into two halves. The median is either (1) the middle item in a tabulation with an odd number of items or (2) the average of the two middle items in a tabulation with an even number of items.

The *mode* is the item that appears most frequently in a set of values. When seen on a frequency distribution, it is the value of the variable at the peak of the curve.

Examples and Problems

Given the following numbers, what are the mean, median, and mode?

67, 75, 63, 72, 77, 80, 81, 77, 76

$$\text{Mean} = \frac{\Sigma X}{n} = \frac{668}{9} = 74.22$$

Ordered array of numbers

63, 67, 72, 75, *76*, 77, 77, 80, 81

Median = 76

Mode = 77 (appears twice)

Find the mean, median, and mode.

4.21, 4.25, 4.27, 4.30, 4.33, 4.44

Mean = 4.3

Median = 4.285

Mode = none

3.3 STANDARD DEVIATION

The amount by which a single item in a set of values differs from the mean of those values is the *deviation*. A deviation is considered positive if the item is larger than the mean and vice versa. The standard deviation of the items in a given set of values is a commonly used measure of variation. It is a measure of the precision of the mean and is obtained by (1) squaring the deviations from the mean $(X - \overline{X})^2$, (2) summing the squared deviations and dividing by the number of deviations minus one (the degree of freedom), and (3) finding the square root of the quotient of the division. The formula for the standard deviation of a sample is

$$s = \sqrt{\frac{\Sigma (X - \overline{X})^2}{n - 1}} \quad \text{or} \quad s = \sqrt{\frac{\Sigma f(X - \overline{X})^2}{n - 1}}$$

Examples and Problems

Estimate the standard deviation s for the following series of data:

4.27, 4.21, 4.30, 4.44, 4.25, 4.33

The mean ($\bar{X}$) for this set is 4.3. The individual deviations may then be computed and squared.

X	$(X - \bar{X})$	$(X - \bar{X})^2$
4.21	−0.09	0.0081
4.25	−0.05	0.0025
4.27	−0.03	0.0009
4.30	0	0
4.33	0.03	0.0009
4.44	0.14	0.0196
		0.0320

The denominator $(n - 1)$ is equal to 5. Substituting in the equation

$$s = \sqrt{\frac{0.032}{5}}$$

$$s = \sqrt{0.0064}$$

$$s = 0.08$$

Below are 10,000 blood samples taken from a hypothetical sample, X being the number of abnormal white cells and f being the frequency of occurrence. Find the mean and standard deviation.

X	f	fX	$(X - 40)$	$(X - 40)^2$	$f(X - 40)^2$
33	6	198	−7	49	294
34	35	1,190	−6	36	1,260
35	125	4,375	−5	25	3,125
36	338	12,168	−4	16	5,408
37	740	27,380	−3	9	6,660
38	1303	49,514	−2	4	5,212
39	1810	70,590	−1	1	1,810
40	1940	77,600	0	0	0
41	1640	67,240	1	1	1,640
42	1120	47,040	2	4	4,480
43	600	25,800	3	9	5,400
44	222	9,768	4	16	3,552
45	84	3,780	5	25	2,100
46	30	1,380	6	36	1,080
47	5	235	7	49	245
48	2	96	8	64	128
	10,000	398,354			42,394

$$\overline{X} = \frac{398{,}354}{10{,}000} = 39.8354 = \sim 40 \qquad s = \sqrt{\frac{42{,}394}{9{,}999}} = \sqrt{4.240} = 2.059$$

Note: A large standard deviation (as in the second problem) signifies that the frequency distribution spreads out from the mean, while a small standard deviation (as in the first problem) shows that the observations are closely concentrated around the mean.

3.4 VARIANCE

Variance is the square of the standard deviation. The unbiased estimate of the population variance (σ^2) from sample data is designated s^2. It is given by

$$s^2 = \frac{\Sigma (X - \overline{X})^2}{n - 1}$$

As in standard deviation, take the deviation of each X from the sample mean, $\overline{X}$, and square these individually. The squared deviations are then added to yield $\Sigma (X - \overline{X})^2$ known as the sum of squares (SS).* This is divided by $(n - 1)$ to give s^2.

Examples and Problems

Calculate the variance of the following spectrographic determinations of copper.

X	Y	Z
5.64	5.55	5.31
5.56	5.45	5.46
5.53	5.40	5.29
	5.55	5.59
		5.47
		5.54

Sum of X = 16.73, mean = $5.57\frac{2}{3}$

Sum of Y = 21.95, mean = $5.48\frac{3}{4}$

Sum of Z = 32.66, mean = $5.55\frac{1}{3}$

$(X - \overline{X})$	$(X - \overline{X})^2$	$(Y - \overline{Y})$	$(Y - \overline{Y})^2$	$(Z - \overline{Z})$	$(Z - \overline{Z})^2$
0.06	0.0036	0.06	0.0036	−0.13	0.0169
−0.02	0.0004	−0.04	0.0016	0.02	0.0004
−0.05	0.0025	−0.09	0.0081	−0.15	0.0225
	0.0065	0.06	0.0036	0.15	0.0225
$n - 1 = 2$			0.0169	0.03	0.0009
		$n - 1 = 3$		0.10	0.0100
					0.0732
				$n - 1 = 5$	

Note: The deviations from the mean were rounded off

$$s^2 = \frac{0.0065}{2} + \frac{0.0169}{3} + \frac{0.0732}{5} = 0.00325 + 0.00563 + 0.01464 = \underline{0.02352}$$

*The sum of squares can also be computed by the machine formula $SS = \Sigma X^2 - \frac{(\Sigma X)^2}{n}$.

OR, using the machine formula:

X^2	Y^2	Z^2
31.8096	30.8025	28.1961
30.9136	29.7025	29.8116
30.5809	29.1600	27.9841
93.3041	30.8025	31.2481
	120.4675	29.9209
		30.6916
		177.8524

$$\frac{(\Sigma X)^2}{n} = \frac{279.8929}{3} = 93.2976$$

$$\frac{(\Sigma Y)^2}{n} = \frac{481.8025}{4} = 120.4506$$

$$\frac{(\Sigma Z)^2}{n} = \frac{1066.6756}{6} = 177.7793$$

$$s^2 = \frac{93.3041 - 93.2976}{2} + \frac{120.4675 - 120.4506}{3} + \frac{177.8524 - 177.7793}{5}$$

$$= \frac{0.0065}{2} + \frac{0.0169}{3} + \frac{0.0731}{5} = 0.0035 + 0.00563 + 0.01462 = \underline{0.02375}$$

3.5 NORMAL CURVE

The normal distribution curve is a bell-shaped curve. The random sampling distribution of many statistics is approximately normal in form. As more data are accumulated, the irregularities of a histogram of a random sampling of data tend to smooth out. The distribution approaches more and more closely the shape of a mathematically defined bell-shaped curve, known as the *Gaussian curve of error* or simply the *normal distribution.*

The location and shape of the normal distribution curve are specified completely by two parameters (i.e., numbers that characterize a specific population).

1. The mean (μ) locates the center of the distribution on the axis of measurement (x axis). It is the expected value of a measurement x.

2. The standard deviation (σ) describes the amount of variability or dispersion of the data which in turn determines the breadth of the distribution curve. It is related to the degree with which the values deviate from the mean.

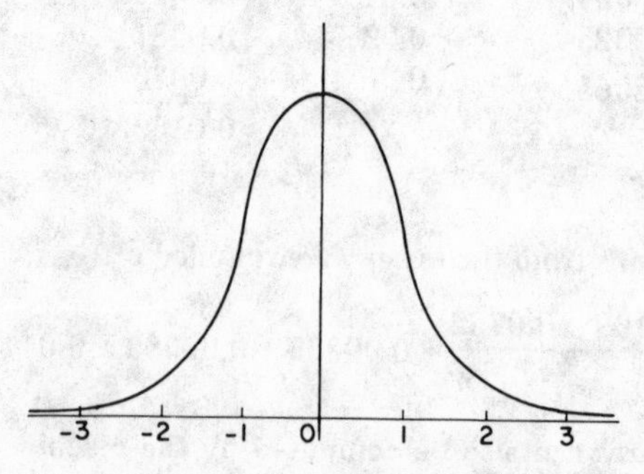

Fig. 3.1. A normal distribution curve.

The normal distribution curve is symmetrical about the mean with points of inflection at plus and minus one standard deviation from the mean. (See Fig. 3.1.) The concept of a normal distribution applies to the shape or form of the distribution, however, and not to the specific mean or standard deviation of the distribution. Thus there is not just one normal distribution; there is an infinity of such curves since the mean and standard deviation can assume an infinity of values. Also, since the curve is symmetrical around the mean, the mean, median, and mode of the normal distribution are all at the same point.

In any normal distribution, the following percentages of the distribution are included in the indicated intervals:

mean ± one standard deviation contains 68.26% of the items
mean ± two standard deviations contains 94.46% of the items
mean ± three standard deviations contains 99.73% of the items

Given any particular normal distribution with a specified mean and standard deviation, the probability of obtaining at random a value greater than or less than any particular assigned number (or between two assigned numbers) can be calculated. The graph of the normal distribution is given by the equation

$$y = \frac{1}{\sigma\sqrt{2\pi}}\, e^{-(X-\mu)^2/2\sigma^2}$$

where y = the height of the curve at any given point along the base line
π = 3.1416 (rounded)
e = 2.7183 (rounded)

The area under this curve is equal to 1.00.

A normal distribution is considered to have a mean of zero and a standard deviation of one. The mean of any distribution can be shifted from a specific value to zero by expressing each value of X as a deviation from the mean (i.e., $x = X - m$), and since the $\Sigma x = 0$, the mean of the distribution on this transformed scale will be 0. Any distribution can also be transformed to a new scale for which the unit of measurement is the standard deviation of the distribution. Divide each value of $X - m$ by the standard deviation of the distribution and on the transformed scale the standard deviation will be 1 $\left(\text{i.e., } z = \dfrac{X - m}{\sigma}\right)$. The deviation from the mean measured in standard deviation units is called a *standard* or *normal deviate*. The normal deviate of a value gives the distance on the base line from X to the mean in standard deviation units. This value can be looked up on a table of the unit normal distribution which shows the probability of obtaining values deviating by various amounts from the mean of 0 for the normal curve. Since the normal curve is symmetrical, the table values are given only for positive values of z, the normal deviate. A unit normal curve table can give values which are the proportion s of the total area between ordinates erected at a given X and the mean, the area in the larger segment of the curve, area in the smaller segment, and the value of y corresponding to the value of z, the normal deviate. (See Appendix Table A.38.)

3.6 PROBABILITY

Probability is the mathematical chance that a particular event will occur. It is expressed as a fraction. The calculation of probability considers only logical relations between events and so sets up simplified models which must be tested against actual conditions. The probability of an event depends on the number of possibilities.

$$\text{Probability} = \frac{\text{number of ways the event occurs}}{\text{number of possible ways the event can occur}}$$

The probability that a tossed coin will land "heads up" is $\frac{1}{2}$. **Assuming** a probability of $\frac{1}{2}$ as in the tossing of a coin, the probability of a couple planning two children to have two girls involves four possibilities:* (1) girl-girl, (2) girl-boy, (3) boy-girl, (4) boy-boy; therefore, the probability is $\frac{1}{4}$. The chance of having three girls is $\left(\frac{1}{2} \times \frac{1}{2} \times \frac{1}{2}\right) = \frac{1}{8}$.

Two or more events are said to be independent if the occurrence or nonoccurrence of any one of them does not affect the probabilities of occurrence of any of the others. Even if a coin has been tossed ten times and has landed heads each time, the eleventh toss may be either heads or tails and is not influenced by the previous tosses. The probability that two or more independent events will take place is equal to the product of their individual probabilities. Thus the probability of getting all girls after ten pregnancies is $\frac{1}{2}$ raised to the 10th power, or one chance in 1,024.

The probability that two independent events will occur simultaneously is equal to the product of the probabilities of the separate events or

$$p = q \times r$$

where p = probability of simultaneous occurrence
q = probability of occurrence of the first event
r = probability of occurrence of the second event

*This assumption disregards biological factors.

For example, the chance of throwing a four is $\frac{1}{6}$, and the chance of throwing a double four with two dice is $\frac{1}{36}$.

The probability that one of two (or more) mutually exclusive events will occur is the sum of the probabilities of their separate occurrence, q and r.

$$p = q + r$$

For example, what is the probability that a die will turn up a 5 or an even number? The chance of a 5 is $\frac{1}{6}$ and the chance of an even number is $\frac{3}{6}$. The events are mutually exclusive, since if it is 5, it cannot be an even number and if an even number turns up, it cannot be a 5.

$$p = \frac{1}{6} + \frac{3}{6} = \frac{2}{3}$$

and there are two chances in three for either a 5 or an even number to turn up.

Examples and Problems

What is the probability of rolling a 3 with a single roll of a die?

answer: $\frac{1}{6}$

What is the probability of getting 2 heads after flipping a coin 2 times?

answer: $\frac{1}{4}$

What is the probability of an odd number appearing on the roll of 1 die?

answer: $p = \frac{3}{6} = \frac{1}{2}$

If one were told to pick any 3-digit whole number, what is the probability that a second person would pick the same number?

answer: $\frac{1}{900}$

Three-digit numbers are from 100 to 999. One through 99 are not 3-digit numbers.

999 minus 99 = 900. Thus there are 900 possibilities of 3-digit numbers.

What is the probability that the sum 5 appears in a single toss of 2 dice?

answer: There are 36 ways 2 dice can be associated with each other. There are 4 ways of obtaining the sum 5 (1,4), (2,3), (3,2) and (4,1). Thus $p = \frac{4}{36} = \frac{1}{9}$.

The probability that experiment A will succeed is $\frac{3}{4}$ and the probability that experiment B will succeed is $\frac{2}{3}$. Find the probability that

(1) both experiments will succeed

$$\left(\frac{3}{4}\right) \times \left(\frac{2}{3}\right) = \frac{6}{12} = \frac{1}{2}$$

(2) only experiment A will succeed

$$\left(\frac{3}{4}\right) - \left(\frac{1}{2}\right) = \left(\frac{3}{4}\right) - \left(\frac{2}{4}\right) = \frac{1}{4}$$

(3) only experiment B will succeed

$$\left(\frac{2}{3}\right) - \left(\frac{1}{2}\right) = \left(\frac{4}{6}\right) - \left(\frac{3}{6}\right) = \frac{1}{6}$$

(4) at least one experiment will succeed

$$\left(\frac{3}{4}\right) + \left(\frac{1}{6}\right) \text{ from part 3} = \frac{18}{24} + \frac{4}{24} = \frac{22}{24} = \frac{11}{12}$$

$$\left(\frac{2}{3}\right) + \left(\frac{1}{4}\right) \text{ from part 2} = \frac{8}{12} + \frac{3}{12} = \frac{11}{12}$$

3.7 PERMUTATIONS AND COMBINATIONS

Permutations are the different orders in which things can be arranged. In general, the basic procedure in making permutations can be summarized as follows:

the first act can be performed in any one of n number of ways
the second act can be performed by $(n - 1)$ different ways
the third in $(n - 2)$ different ways, etc.

then,

the arithmetical procedure is to multiply the number of possibilities for the first position by the number remaining for the second, and so on for as many acts as the situation describes.

Example: There are seven laboratories available in a chemistry department for four professors. In how many different ways can the four professors be assigned to a different laboratory?

$7 \times 6 \times 5 \times 4 = 840$ ways

How many arrangements can be made with the numbers 1, 2, and 3?

$3 \times 2 \times 1 = 6$

To find the number of permutations of n things taken r at a time, the rule is to multiply all the consecutive numbers, beginning with the number of things given, as many times as they are to be taken. Also the formula below can be followed.

$$_nP_r = \frac{n!}{(n-r)!}$$

The number of permutations of 4 digits taken 3 at a time is expressed as $_4P_3$ or P_3^4. The 4 stands for the number of digits, the P for permutation, and the 3 stands for the number taken each time.

Solution: $_4P_3 = 4 \times 3 \times 2 = 24$

or by the formula

$$_4P_3 = \frac{4!}{(4-3)!} = \frac{4 \times 3 \times 2 \times 1}{1} = 24$$

To find the number of permutations of a number of things in which some are the same, the permutations can be formed by dividing the total number of ways by the permutations of the number of repetitions. Thus the permutations of the letters of the word "receded" would be

$$\frac{7!}{3!\,2!} = \frac{7 \times 6 \times 5 \times 4 \times 3 \times 2 \times 1}{3 \times 2 \times 1 \times 2 \times 1} = \frac{5040}{12} = 420$$

since the three e's and the two d's are the same. And, for the word "division," the number of permutations of the letters would be

$$\frac{8!}{3!} = \frac{8 \times 7 \times 6 \times 5 \times 4 \times 3 \times 2 \times 1}{3 \times 2 \times 1} = \frac{40{,}320}{6} = 6{,}720$$

Combinations are the number of different groups consisting of a specific number of units which can be formed from a given total number of units. Combinations differ from permutations in that combinations consider the units in a group without reference to their arrangement. Since in the process of calculating combinations the groups which differ only in arrangement and not in units must be eliminated, the procedure consists of dividing the number of permutations which the situation permits by the number of permutations possible in the group itself. In the case of four digits taken three at a time without reference to their arrangement

$$_4C_3 = \frac{4 \times 3 \times 2 \times 1}{3 \times 2 \times 1} = \frac{24}{6} = 4$$

or by formula

$$_nC_r = \frac{n!}{r!\,(n-r)!} = \frac{4!}{3!\,(4-3)!} = \frac{4 \times 3 \times 2 \times 1}{3 \times 2 \times 1\,(1)} = \frac{24}{6} = 4$$

Another example: Five samples of platelet-rich plasma are to be chosen from 12. In how many ways can they be selected?

$$_{12}C_5 = \frac{12 \times 11 \times 10 \times 9 \times 8}{5 \times 4 \times 3 \times 2 \times 1} = \frac{95{,}040}{120} = 792$$

or by formula

$$_{12}C_5 = \frac{12!}{5!\,(12-5)!} = \frac{95{,}040}{120} = 792$$

Examples and Problems

Evaluate the following:

$_4P_4 = 4 \times 3 \times 2 \times 1 = 24$
$_{64}P_1 = 64$
$_8P_4 = 8 \times 7 \times 6 \times 5 = 1830$
$_6P_3 = 6 \times 5 \times 4 = 120$

There are 7 amino acids in a polypeptide chain. In how many ways could they conceivably be arranged?

answer: $_7P_7 = 7 \times 6 \times 5 \times 4 \times 3 \times 2 \times 1 = 5040$

There are 8 geology professors in a department and 5 elementary geology courses offered. In how many ways can the 8 professors be assigned to the 5 classes?

answer: $$_8P_5 = \frac{8!}{(8-5)!} = \frac{8 \times 7 \times 6 \times 5 \times 4 \times 3 \times 2 \times 1}{3 \times 2 \times 1}$$

$$= 8 \times 7 \times 6 \times 5 \times 4 = 6720$$

In how many ways can 6 samples be divided between 2 research groups so that each group has 3 samples?

answer: $$_6C_3 = \frac{6!}{3!(6-3)!} = \frac{6!}{3!3!} = \frac{6 \times 5 \times 4 \times 3 \times 2 \times 1}{3 \times 2 \times 1 \times 3 \times 2 \times 1}$$

$$= \frac{6 \times 5 \times 4}{3 \times 2 \times 1} = \frac{120}{6} = 20$$

3.8 CHI SQUARE

The *Chi Square test* is a test of compatibility of observed and expected frequencies. Chi Square, χ^2, is defined as

$$\chi^2 = \Sigma \frac{(0 - E)^2}{E}$$

where 0 is the observed frequency and E is a corresponding expected frequency. In the Chi Square table (see Appendix A.39) the column headings are probabilities, the body of the table gives values of χ^2 and the left column the degrees of freedom. If for the required degrees of freedom and at a desired probability on the χ^2 table the χ^2 value is smaller than the calculated value of χ^2, it is then concluded that the χ^2 value is significant. The null hypothesis is rejected and it is concluded that the distribution of frequencies tested deviates significantly from a chance or uniform distribution. A calculated χ^2 value is not significant when it is smaller than a tabular χ^2 value and it is then concluded that the observed distribution does not deviate significantly from the expected distribution; the null hypothesis is not rejected.

Examples and Problems

Example with even expected frequencies:

Data: $N = 795$

Expected frequency $= \frac{1}{3}(795) = 265$

Degrees of freedom = classes − 1 = 2

Observed frequencies of corn blight in three small test fields:

I = 245 II = 200 III = 350

Calculations:

$$\chi^2 = \frac{(245 - 265)^2}{265} + \frac{(200 - 265)^2}{265} + \frac{(350 - 265)^2}{265}$$

$$= \frac{(-20)^2 + (-65)^2 + (85)^2}{265}$$

$$= \frac{400 + 4225 + 7225}{265}$$

$$= \frac{11{,}850}{265} = 44.71698$$

With a calculated $\chi^2 = 44.72$, $DF = 2$, at a probability of 0.01 ($\chi^2_{2,0.01}$ = 9.210). Therefore, one could conclude that the calculated χ^2 is significant (the null hypothesis is rejected) at the 0.01 level of probability and that one or two of the test fields is more prone to corn blight than the other or others.

Example with unequal expected values:

Data: In a genetic experiment a 3:1 ratio is expected.

$N = 200$
Expected frequency = 3:1 = 150 to 50
Degrees of freedom = classes − 1 = 2 − 1 = 1

Observed frequencies in genetic experiment:

Black = 63 Brown = 137

Calculations:

$$\chi^2 = \frac{(63-50)^2}{50} + \frac{(137-150)^2}{150}$$
$$= \frac{(13)^2}{50} + \frac{(-13)^2}{150}$$
$$= \frac{169}{50} + \frac{169}{150}$$
$$= 3.38 + 1.127$$
$$= 4.507$$

With a calculated $\chi^2 = 4.507$, $DF = 1$, at the probability of 0.05 ($\chi^2_{1,0.05} = 3.841$). Therefore, the calculated χ^2 value is significant and the observed results differ significantly from the 3:1 ratio expected.

For cases involving a single degree of freedom where the expected values are small, a correction factor should be used. An arbitrary dividing point between large and small expected frequencies has been set at 20. Thus, if an expected frequency occurs less than 20, simply reduce the magnitude of all deviations by ½ a unit. Chi square is then computed with these reduced deviations in exactly the same manner as before.

Shortcut formula for calculations of χ^2 from a 2 × 2 table:

Table

Row	Column I	Column II	
I	a	b	$a + b$
II	c	d	$c + d$
	$a + c$	$b + d$	$a + b + c + d$ (T)

$$\chi^2 = \frac{(ad - bc)^2 (a + b + c + d)}{(a + c)(b + d)(a + b)(c + d)} \quad \text{(formula from 2 × 2 table)}$$

Note: Putting the formula into words, multiply the two diagonals and square the difference. Multiply this difference by the total (T). This product is then divided by all of the subtotals multiplied by each other.

Example:

	A	B	
X	10	6	16
Y	14	5	19
	24	11	35

$$\chi^2 = \frac{[(10 \times 5) - (6 \times 14)]^2 \ (35)}{24 \times 11 \times 16 \times 19} = \frac{(50 - 84)^2 \ (35)}{80256} = \frac{(34)^2 \ (35)}{80256}$$

$$= \frac{(1156)\ (35)}{80256} = \frac{40{,}460}{80{,}256} = \sim 0.50$$

Calculated $\chi^2 = 0.50$, 1 DF,* does not exceed the χ^2 at the 0.05 level. ($\chi^2_{1,0.05} = 3.841$) Therefore one cannot reject the hypothesis and it is concluded that sample X does not differ significantly from sample Y.

Shortcut formula incorporating the 0.5 correction factor:

$$\chi^2 = \frac{(|ad - bc| - \frac{1}{2}T)^2 \ (T)}{(a + c)\ (b + c)\ (a + b)\ (c + d)}$$

Note: Putting the formula into words, multiply the two diagonals and subtract the smaller product from the larger one. The correction factor is incorporated in the formula by subtracting half the total, squaring the result, and then multiplying the entire product by the total (T). As above, this product is then divided by all the subtotals multiplied by each other.

Examples and Problems

In a particular genetic experiment, theory set a 9:3:3:1 ratio. The data recorded was

Group 1—99 $N = 160$
2—33
3—24
4— 4

Is the data in agreement with the theory?

$$\chi^2 = \frac{(99 - 90)^2}{90} + \frac{(33 - 30)^2}{30} + \frac{(24 - 30)^2}{30} + \frac{(4 - 10)^2}{10}$$

$$= \frac{9^2}{90} + \frac{3^2}{30} + \frac{6^2}{30} + \frac{6^2}{10} = \frac{81}{90} + \frac{9}{30} + \frac{36}{30} + \frac{36}{10}$$

$$= 0.9 + 0.3 + 1.2 + 3.6 = 6.0$$

*To find the degrees of freedom, take one less than the number of items in the row and multiply this number by one less than the number of items in the columns.

The calculated χ^2 of 6.0 is less than the tabulated χ^2 at 0.05 with 3 *DF* ($\chi^2_{3,0.05}$ = 7.815); therefore the null hypothesis is not rejected and we conclude that there is no difference between the experiment and theory.

In a certain city, there was an epidemic of a disease known to have a mortality of 30%. Out of 127 patients, only 23 died. Was the new treatment against this disease effective?

	Die	Not Die
Observed	23.0	104.0
Expected	38.1	88.9

$$\chi^2 = \frac{(23 - 38.1)^2}{38.1} + \frac{(104 - 88.9)^2}{88.9}$$

$$= \frac{(15.1)^2}{38.1} + \frac{(15.1)^2}{88.9} = \frac{228.01}{38.1} + \frac{228.01}{88.9}$$

$$= 5.98 + 2.56 = 8.54$$

Since 8.54 exceeds the tabulated χ^2 value of 6.635 ($\chi^2_{1,0.01}$), it is concluded that the treatment was effective in reducing the mortality rate.

The cure rate in a certain disease is known to be 77%. Twenty patients were found to recover from this disease after being treated by a new therapeutic procedure. Is the new procedure effective?

Observed	20.0	0.0
Expected	15.4	4.6

$$\chi^2 = \frac{(20 - 15.4)^2}{15.4} + \frac{(0 - 4.6)^2}{4.6} = \frac{(4.6)^2}{15.4} + \frac{(4.6)^2}{4.6} = \frac{21.16}{15.4} + \frac{21.16}{4.6}$$

$$= 1.37 + 4.6 = 5.97$$

The calculated χ^2 of 5.97 exceeds the tabulated χ^2 of 3.84 at $P = 0.05$ with 1 *DF*. Therefore the null hypothesis is rejected and it is concluded that the new procedure is effective. (It is not effective at the 0.01 level as $\chi^2 = 6.64$.)

3.9 CORRELATION

Consider a definite group containing a large number of items and measure some attribute *A* of the items and some attribute *B*. For instance, the items might be all stars of a certain magnitude and *A* might represent the parallax of the star while *B* represents its proper motion. These two attributes are correlated, for a star which has a large parallax (therefore, comparatively near to us) is more likely to have a large proper motion than a small proper motion.

Correlation is thus the interdependence between two sets of numbers; it is a measure of the degree of linear association between two variables. A correlation coefficient may be computed whenever observations are paired. Correlation techniques may be used, however, only to measure the strength of a relationship where increases in one variable are associated with constant average increases or decreases of the other variable.

Correlation coefficients (r) may vary from zero (no correlation) to +1 or −1. A positive correlation of +1 indicates a perfect direct relationship between the two variables; a negative correlation of −1 indicates a perfect inverse relationship between two variables (i.e., as X gets larger, Y gets smaller). A correlation of zero indicates no relationship between the two variables. (See Fig. 3.2.)

The Pearson Product Moment coefficient of correlation uses data in raw form.

$$r_{xy} = \frac{\Sigma xy}{\sqrt{(\Sigma x^2)(\Sigma y^2)}} \quad \text{(deviation form)}$$

which is the same as

$$r_{XY} = \frac{\Sigma XY - \dfrac{(\Sigma X)(\Sigma Y)}{N}}{\sqrt{\left(\Sigma X^2 - \dfrac{(\Sigma X)^2}{N}\right)\left(\Sigma Y^2 - \dfrac{(\Sigma Y)^2}{N}\right)}}$$

Note: The formula when stated in deviation form is expressed using the small letters x and y and the computation formula is expressed using the large X and Y (signifying raw data scores).

Example: Correlation between the iron and zinc levels in the blood of nine 2-pound garfish caught in the Everglades.

Null hypothesis: There is no relationship between the iron and zinc levels in the blood of garfish.

Data: X = iron levels, ppm; Y = zinc levels, ppm

Sample Number	X	X^2	Y	Y^2	XY
1	64	4,096	120	14,400	7,680
2	68	4,624	170	28,900	11,560
3	62	3,844	130	16,900	8,060
4	69	4,761	160	25,600	11,040
5	73	5,329	190	36,100	13,870
6	72	5,184	210	44,100	15,120
7	65	4,225	140	19,600	9,100
8	67	4,489	180	32,400	12,060
9	68	4,624	190	36,100	12,920
	608	41,176	1490	254,100	101,410

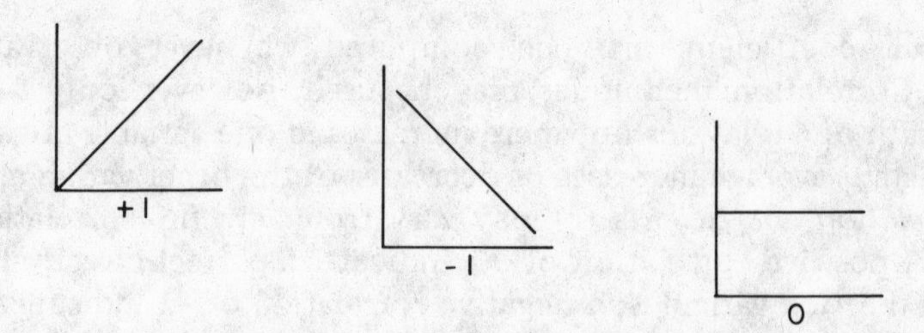

Fig. 3.2. Graphs portraying positive, negative, and zero correlation.

Calculations:

$$\Sigma x^2 = \Sigma X^2 - \frac{(\Sigma X)^2}{N} = 41{,}176 - \frac{(608)^2}{9} = 102.2$$

$$\Sigma y^2 = \Sigma Y^2 - \frac{(\Sigma Y)^2}{N} = 254{,}100 - \frac{(1{,}490)^2}{9} = 7{,}422.2$$

$$\Sigma xy = \Sigma XY - \frac{(\Sigma X)(\Sigma Y)}{N} = 101{,}410 - \frac{(608)(1{,}490)}{9} = 752.2$$

$$\sqrt{(\Sigma x^2)(\Sigma y^2)} = \sqrt{(102.2)(7{,}422.2)} \cong 871.0$$

$$r = \frac{\Sigma xy}{\sqrt{(\Sigma x^2)(\Sigma y^2)}} = \frac{752.2}{871.0} = .864$$

Degrees of freedom are 2 less than the number of pairs. Therefore, with 7 degrees of freedom, the calculated r of 0.864 is highly significant since both the 0.05 (Table $r = 0.6664$) and the 0.01 (Table $r = 0.7977$) level of significance are smaller than the calculated r of 0.864. The null hypothesis is rejected and it is concluded that the two sets of readings are related beyond just the possibility of chance.

The significance of r can be further estimated by

$$t = \sqrt{\frac{r^2(N-2)}{1-r^2}} = 4.54 \text{ (from above data)}$$

where the t table is entered with $N - 2$ degrees of freedom.

The Spearman Rank correlation coefficient pays no attention to raw values. The correlation is between the ranks of variables. The formula is

$$\rho = 1 - \frac{6\Sigma D^2}{N(N^2 - 1)}$$

where D is the difference between the ranks of the paired items.

3.10 LEAST SQUARES

It is sometimes desired to fit a set of experimental data for y as a function of x with the formula giving the most accurate representation of the results. If y is a linear function of x, expressible by a formula $y = a + bx$, it is possible to apply the method of least squares which by its name is designed to reduce the square of the vertical deviations of the experimental points from the calculated values of y to the minimum possible value. Using the standard statistical formulas for a least squares fit, one can calculate the best values for the parameters b and a; i.e., the slope and the intercept of the straight line which best fits a series of equal weighted observations.

The procedure is as follows:

1. The X values for all the points to be taken are added together; this sum is represented by ΣX.

2. The sum of all the Y values is obtained; this is the ΣY.

3. The X value for each point is squared, and the results added together for all the points to give $\Sigma(X^2)$.

4. The X value for each point is multiplied by the Y value and the results are summed over all the points, giving ΣXY.

5. The constants a and b are evaluated by the following equations in which n is the number of experimental points used.

$$b = \frac{\Sigma XY - \dfrac{\Sigma X \Sigma Y}{n}}{\Sigma X^2 - \dfrac{(\Sigma X)^2}{n}}$$

$$a = (\overline{Y} - b\overline{X})$$

The least squares line passes through the general mean $(\overline{X}, \overline{Y})$ and the y-intercept, a. To draw the regression line, simply plot the means and the y-intercept or estimate two convenient points of x from the regression equation and draw a straight line between them. (See examples and figures.)

Examples and Problems

Find the best-fitting line for the following data (obtained from a spectrophotometric reaction in which X is amount added and Y is time of incubation) to fit the equation $y = bx + a$. (See Fig. 3.3.)

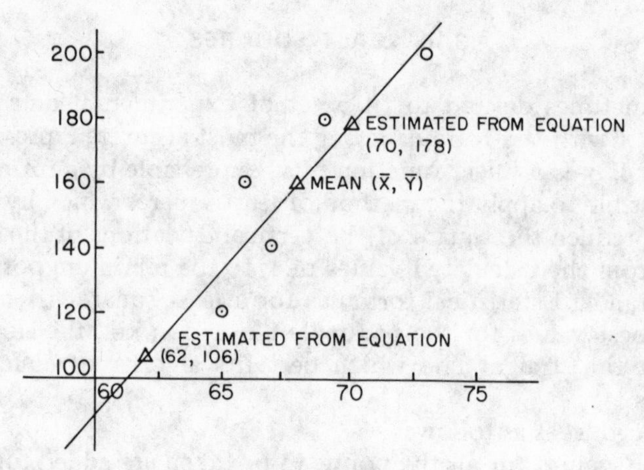

Fig. 3.3. Graph for $y = 9x - 452$

⊙ = points plotted from data

Estimated points:

$\hat{y} = 9(62) - 452$
$\hat{y} = 558 - 452$
$\hat{y} = 106$
Point = (62, 106)

$\hat{y} = 9(70) - 452$
$\hat{y} = 630 - 452$
$\hat{y} = 178$
Point = (70, 178)

Note: In Fig. 3.3, for illustrative purposes, the x and y axis do not cross at zero. If they did intersect at zero, the y-intercept would be -452.

X	X^2	Y	XY
69	4,761	180	12,420
73	5,329	200	14,600
67	4,489	140	9,380
66	4,356	160	10,560
65	4,225	120	7,800
340	23,160	800	54,760

$$\overline{Y} = \frac{800}{5} \qquad \overline{X} = \frac{340}{5}$$

$$\overline{Y} = 160 \qquad \overline{X} = 68$$

$$b = \frac{54{,}760 - \dfrac{(340)(800)}{5}}{23{,}160 - \dfrac{(340)^2}{5}}$$

$$a = 160 - 9(68) = 160 - 612 = -452$$

$$= \frac{54{,}760 - \dfrac{272{,}000}{5}}{23{,}160 - \dfrac{115{,}600}{5}}$$

$$= \frac{54{,}760 - 54{,}400}{23{,}160 - 23{,}120} = \frac{360}{40} = 9$$

Therefore $y = 9x - 452$.

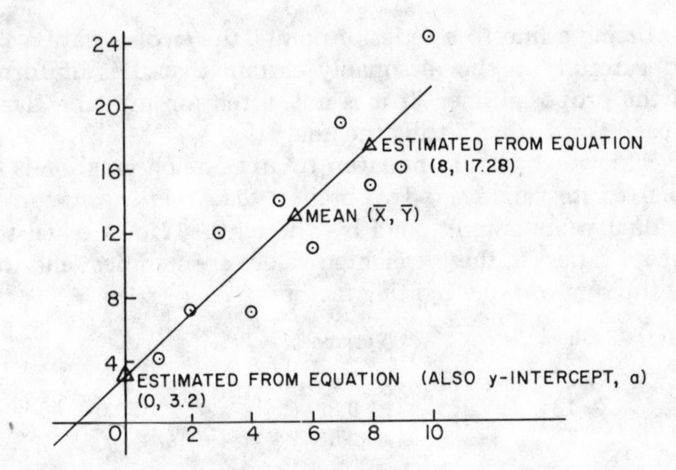

Fig. 3.4. Graph for $y = 1.76x + 3.2$

⊙ = points plotted from data

Estimated points:

$\hat{y} = 1.76(0) + 3.2$
$\hat{y} = 3.2$
Point = (0, 3.2)

$\hat{y} = 1.76(8.0) + 3.2$
$\hat{y} = 14.08 + 3.2$
$\hat{y} = 17.28$
Point = (8, 17.28)

Find the best-fitting line for the following data to fit the equation $y = bx + a$. (See Fig. 3.4.) Assume these points were data for a protein concentration standard curve where X is the amount of protein used in a given case and Y is the instrument reading obtained.

X	X^2	Y	XY
1	1	4	4
2	4	7	14
3	9	12	36
4	16	7	28
5	25	14	70
6	36	11	66
7	49	19	133
8	64	15	120
9	81	16	144
10	100	24	240
55	385	129	855

$$\overline{Y} = \frac{129}{10} \qquad \overline{X} = \frac{55}{10}$$

$$\overline{Y} = 12.9 \qquad \overline{X} = 5.5$$

$$b = \frac{855 - \dfrac{(55)(129)}{10}}{385 - \dfrac{(55)^2}{10}} = \frac{855 - \dfrac{7095}{10}}{385 - \dfrac{3025}{10}}$$

$$= \frac{855 - 709.5}{385 - 302.5} = \frac{145.5}{82.5} = 1.764$$

$$a = 12.9 - (1.764)(5.5)$$
$$= 12.9 - 9.702$$
$$= 3.198$$

Therefore $y = 1.76x + 3.2$.

Before fitting a line to a series of points, the error variation should be considered carefully to be reasonably certain that it is uniform for all regions of the proposed line. If it is not, attention must be given to the weight of each point when fitting the line.

The weight which should be given to any one observation is inversely proportional to its variance. To obtain a *weighted mean*, for example, each individual value is multiplied by a weighting factor equal to the reciprocal of its variance, these weighted values are summed, and the sum is divided by the sum of the weighting factors.

Example:	X values:	$\overline{X}$:	Variances:
	X_1 10	12.67	2.67
	X_2 13		0.33
	X_3 15		2.33

Weighting Factors:

$$X_1 \quad \frac{1}{2.67} = 0.3745$$

$$X_2 \quad \frac{1}{0.33} = 3.0303$$

$$X_3 \quad \frac{1}{2.33} = \frac{0.4292}{3.8340} = \text{sum}$$

Best approximation of $\overline{X}$ is

$$\frac{10(0.3745) + 13(3.0303) + 15(0.4292)}{3.8340} = \frac{49.5769}{3.8340} = 12.93$$

Note: The difference between correlation and regression (least square) should be noted. Correlation portrays the degree of relationship between an x and y variable. Regression portrays the dependence of one variable upon another to obtain its value [i.e., $y = f(x)$].

3.11 ANALYSIS OF VARIANCE (ANOV)

Analysis of variance is the method of comparing two or more sample means; it is a method for deciding whether the variation from a particular source is really significantly greater than variation due to experimental error.

For a two-sample comparison there is a grand mean of the observations in both samples and a mean for each sample. The total variance in any such system is divided into within samples variance (error) and between samples variance. After computing these two variance estimates, an F-ratio is constructed by placing the between sample variance over the within sample variance. The quotient of this ratio is then looked up in an F table by utilizing the two degrees of freedom associated with each variance. If the F ratio quotient exceeds the critical ratio located in the table, it is concluded that the two sample variances could not reasonably be estimates of the population variance.

The calculations of the analysis of variance are usually summarized in a table. The ANOV Table 3.1 containing the sum of squares, degrees of freedom, and variance estimates (also called mean square) presents the initial analysis in a compact form.

The between samples mean square describes the dispersion of the group means around the grand mean. The within sample mean square (also known as the error mean square) gives the average dispersion of the items in each group around the group means. The total mean square is a statistic of dispersion of the total items around the grand mean. It describes the variance in the entire sample due to all causes.

If the calculated F ratio is below the critical table value, the sample means can be compared to determine if they come from a single population or from two populations with the same variance but different means. This is done by comparing the amount of difference between the sample means with an estimate of the standard error and then referring to the t-distribution.

$$t = \frac{(\overline{X}_1 - \overline{X}_2)}{\sqrt{\frac{s^2}{n_1} + \frac{s^2}{n_2}}}$$

The t table is entered with the degrees of freedom contributed by both samples $(n_1 - 1 + n_2 - 1)$ to see if the observed value might have occurred by chance. If the observed t exceeds the tabulated critical value at the

Table 3.1. ANOV Table for Single Classification

Source	*Sum of Squares*	*Degrees of Freedom*	*Mean Square*
Between	$\frac{(X_1)^2}{n_1} + \frac{(X_2)^2}{n_2} + \ldots \frac{(X_t)^2}{n_t} - \frac{(\Sigma X)^2}{N}$	$t - 1$	$\frac{\text{Between } SS}{t - 1}$
Within	Total SS - Between SS	$tn - t$	$\frac{\text{Within } SS}{tn - t}$
Total	$\Sigma X^2 - \frac{(\Sigma X)^2}{N}$	$tn - 1$	$F = \frac{\text{Between } SS}{\text{Within } SS}$

Note: X = individual observations in a group (i.e., group 1 through group t)
t = number of groups (i.e., 1, 2, . . . t)
n = number of individual observations in each group
N = grand total of observations

0.05 level, it is concluded that the samples were not drawn from the same population, i.e., they differ from each other significantly. As the degrees of freedom increase, a smaller value of t is required for significance.

Examples and Problems

The following table contains the results for the determination of protein concentration by three different methods. Using single classification analysis of variance, state a null hypothesis and test it.

Null hypothesis: There is no difference between the means of the results using three different methods for the determination of protein concentration.

Micro-Kjeldahl Method		Folin-Lowry Method		Biuret Method	
X	X^2	X	X^2	X	X^2
7	49	4	16	2	4
10	100	6	36	2	4
10	100	7	49	3	9
11	121	9	81	7	49
12	144	9	81	6	36
50	514	35	263	20	102

$\Sigma X = 105 \quad \Sigma X^2 = 879$

$$\text{Total } SS = 879 - \frac{(105)^2}{15} = 879 - 735 = 144$$

$$\text{Between } SS = \frac{(50)^2 + (35)^2 + (20)^2}{5} - \frac{(105)^2}{15}$$

$$= \frac{2500 + 1225 + 400}{5} - 735$$

$$= \frac{4125}{5} - 735 = 825 - 735 = 90$$

$$\text{Within } SS = 144 - 90 = 54$$

$$\text{Between } MS = \frac{90}{2} = 45 \qquad F = \frac{45}{4.5} = 10$$

$$\text{Within } MS = \frac{54}{12} = 4.5$$

Source	SS	DF	MS	
Between	90	2	45	
Within	54	12	4.5	
Total	144	14		$F = 10$

Calculated F ratio is 10. Tabular F ratio at DF 2 and 12 for 0.05 = 3.88 and at 0.01 = 6.93. Therefore the calculated F ratio is highly significant and the null hypothesis is rejected. There is a difference between the means of the three stated methods.

The following table contains the results for the measurement of enzyme activity by three methods. Is the value of F significant?

Radioactive Method		Spectrophotometric Method		Colorometric Method	
X	X^2	X	X^2	X	X^2
27	729	22	484	37	1,369
45	2,025	24	576	38	1,444
44	1,936	42	1,764	25	625
31	961	41	1,681	47	2,209
38	1,444	31	961	23	529
185	7,095	160	5,466	170	6,176

$\Sigma X = 515 \quad \Sigma X^2 = 18{,}737$

$$\text{Total } SS = 18{,}737 - \frac{(515)^2}{15} = 18{,}737 - 17{,}681.667 = 1{,}055.334$$

$$\text{Between } SS = \frac{(185)^2 + (160)^2 + (170)^2}{5} - \frac{(515)^2}{15}$$

$$= \frac{88{,}725}{5} - 17{,}681.667 = 17{,}745 - 17{,}681.667 = 63.333$$

$$\text{Within } SS = 1{,}055.333 - 63.333 = 992.000$$

$$\text{Between } MS = \frac{63.333}{2} = 31.667$$

$$\text{Within } MS = \frac{992.0}{12} = 82.667$$

Source	SS	DF	MS	
Between	63.334	2	31.667	
Within	992.000	12	82.667	
Total	1055.334	14		$F = \ldots$

Since the mean square for between is smaller than the mean square for within, it is not possible to obtain a significant value of F no matter what the degrees of freedom would be. Therefore there is no significant difference between the means of these methods.

An infectious disease laboratory collected data on two different breeds of dogs which were given the same dosage of the antibiotic tetracycline. The number of days that it took the drug to be fully metabolized in the blood was recorded. Using ANOV, ascertain whether one breed of dog metabolized the drug faster than the other.

Breed X		Breed Y	
X	X^2	Y	Y^2
13	169	20	400
16	256	17	289
19	361	22	484
14	196	24	576
16	256	19	361

$\Sigma X = 78$ $\quad \Sigma Y = 102$

$(\Sigma X)^2 = 6084$ $\quad (\Sigma Y)^2 = 10404$

$N = 10$

Total $N = 180$

$T^2 = 32{,}400$

$\Sigma X^2 = 1238$

$\Sigma Y^2 = 2110$

$\Sigma X^2 + \Sigma Y^2 = 3348$

$$\text{Total } SS = 3348 - \frac{(180)^2}{10} = 3348 - 3240 = 108$$

$$\text{Between } SS = \frac{(78)^2}{5} + \frac{(102)^2}{5} - \frac{(180)^2}{10}$$

$$= \frac{6084}{5} + \frac{10404}{5} - \frac{32{,}400}{10}$$

$$= 1216.8 + 2080.8 - 3{,}240$$

$$= 3297.6 - 3{,}240 = 57.6$$

Within $SS = 108 - 57.6 = 50.4$

Source	SS	DF	MS
Between	57.6	1	57.6
Within	50.4	8	6.3
Total	108.0	9	

$F = 9.14$ with $DF_{1,8}$

The calculated F value exceeds the table F value of $5.32_{0.05}$ so the null hypothesis is rejected and there is a real difference between the time that it takes tetracycline to completely metabolize in the two breeds of dogs. However, the results cannot be considered highly significant at the 0.01 level as the calculated F value of 9.14 does not exceed the tabular value of 11.26.

Sixty-four rabbits were available in a center for hematological research for four investigators. The director of the center assigned the 64 rabbits at random to the four investigators and asked them to determine the arsenic content in the blood of these rabbits. The results are given below. Analyze the data using the analysis of variance.

Number of Rabbits	Investigator I		Investigator II		Investigator III		Investigator IV	
	X	X^2	X	X^2	X	X^2	X	X^2
1	6	36	5	25	4	16	8	64
2	4	16	3	9	7	49	5	25
3	3	9	4	16	3	9	6	36
4	7	49	3	9	7	49	7	49

Number of Rabbits	Investigator I		Investigator II		Investigator III		Investigator IV	
	X	X^2	X	X^2	X	X^2	X	X^2
5	13	169	3	9	4	16	7	49
6	9	81	4	16	8	64	9	81
7	4	16	0	0	7	49	7	49
8	10	100	3	9	4	16	10	100
9	8	64	4	16	8	64	4	16
10	9	81	4	16	4	16	3	9
11	8	64	3	9	11	121	3	9
12	5	25	3	9	13	169	69	487
13	5	25	4	16	9	81		
14	10	100	2	4	89	719		
15	9	81	5	25				
16	15	225	3	9				
17	10	100	3	9				
18	6	36	1	1				
19	4	16	2	4				
20	5	25	59	211				
21	7	49						
	157	1367						

$\Sigma X = 374 \quad \Sigma X^2 = 2784 \quad N = 64$

$$\text{Total } SS = 2784 - \frac{(374)^2}{64} = 2784 - 2185.5625 = 598.438$$

$$\text{Between } SS = \frac{(157)^2}{21} + \frac{(59)^2}{19} + \frac{(89)^2}{13} + \frac{(69)^2}{11} - \frac{(374)^2}{64}$$

$$= \frac{24{,}649}{21} + \frac{3481}{19} + \frac{7921}{13} + \frac{4761}{11} - \frac{139{,}876}{64}$$

$$= 1173.76 + 183.21 + 609.308 + 432.82 - 2185.562$$

$$= 2399.098 - 2185.5625 = 213.5355$$

$$\text{Within } SS = 598.438 - 213.536 = 384.902$$

$$\text{Between } MS = \frac{213.536}{3} = 71.18$$

$$\text{Within } MS = \frac{384.902}{60} = 6.42 \qquad F = \frac{71.18}{6.42} = 11.09$$

Source	SS	DF	MS	
Between	213.536	3	71.18	
Within	384.902	60	6.42	
Total	598.438	63		$F = 11.09$

Calculated F ratio is 11.09. Tabular F ratio at DF 3 and 60 for 0.05 = 8.57 and at 0.01 = 26.3. Therefore the calculated F ratio is significant at the 0.05 level and not significant at the 0.01 level. The null hypothesis is rejected and it is concluded that there is a significant difference between the means of these groups at the 0.05 level of significance.

Note: The only special feature in handling unequal sample sizes is that the Between *SS* is separately calculated as T^2/N for each sample, dividing each total by the respective number in that group, adding these together, and then subtracting the correction factor.

3.12 CONFIDENCE LIMITS

Confidence limits are a range of values which are believed (with a preassigned degree of confidence) to include the particular value of some parameter or statistic being estimated.

A confidence interval can be estimated for every sample from a population. Some of the intervals will contain the parameter or statistic being estimated, some will not. The *predetermined percentage* (called *confidence coefficient*) of the intervals that will contain the unknown is most often chosen to be 95% or 99%. If a 95% confidence interval is determined for each of 1,000 samples of a fixed size, about 950 of these 1,000 confidence intervals will contain the unknown parameter or statistic. Thus for a given sample with a 95% confidence interval for a parameter or statistic, it is said that one is 95% confident that in repeated sampling such an inference concerning the parameter or statistic would be correct 95 times in 100; it is only hoped that the sample is not one of the "bad" 5%.

The *t distribution* is used when setting confidence limits for the means of samples from normal frequency distributions whose population standard deviation is unknown. The number of degrees of freedom pertinent to a t distribution are the same as the number of degrees of freedom of the sample standard deviation in the ratio $X - \overline{X}/\ s/\sqrt{n}$. Although confidence limits are a useful measure of the reliability of a sample statistic, they are not commonly given in scientific publications, the statistic ± its standard error being cited instead.

To look up the *critical values* of t for a given number of degrees of freedom, look up the degrees of freedom in the left column of Table A.42 and read off the desired values of t in that row. The percentages given usually represent the area beyond the critical values of $\pm\ t$ in both tails of the distribution. For example, for 25 degrees of freedom, $t_{0.05} = 2.06$ and $t_{0.01} = 2.787$. The last value indicates that 1% of the area of the t distribution (with 25 degrees of freedom) is beyond $t = \pm\ 2.787$ with $\frac{1}{2}$% in each tail. If a one-tailed test* is desired, the probabilities at the head of this type of table must be halved.

---

*One-tailed test—When there is a prior hypothesis about the direction of the difference between two proportions. A one-tailed test is directional.

Two-tailed test—A two-tailed test is nondirectional, i.e., both the possibilities of a proportion being smaller or larger than the other proportion are being considered.

Steps in computing sample confidence limits for the mean:

1. Compute raw scores

$$\Sigma X = 1597 \qquad \Sigma X^2 = 141{,}823 \qquad n = 20 \qquad \overline{X} = 79.85$$

2. Compute sample standard deviation

$$s^2 = \frac{\Sigma X^2 - \frac{(\Sigma X)^2}{n}}{n - 1} = \frac{141{,}823 - \frac{(1597)^2}{20}}{20 - 1}$$

$$= \frac{141{,}823 - \frac{2550409}{20}}{19}$$

$$= \frac{141{,}823 - 127{,}520}{19} = \frac{14{,}303}{19}$$

$$= 752.79$$

$$s = \sqrt{752.79} = 27.44$$

3. Compute the sample standard error of the mean

$$s_{\overline{x}} = \frac{s}{\sqrt{n}} = \frac{27.44}{\sqrt{20}} = \frac{27.44}{4.47} = 6.14$$

4. Compute confidence limits

$\overline{X} = 79.85$ $\quad s_{\overline{x}} = 6.14$ $\quad n = 20$ $\quad DF = 19$ $\quad t = 2.093$ for 19 DF at 95% confidence level

95%:
- 79.85 + (2.093)(6.14) upper limit
- 79.85 − (2.093)(6.14) lower limit

95%:
- 92.70 upper limit
- 67.00 lower limit

This same technique can be used for setting confidence limits to any given statistic as long as it follows normal distribution.

Appendix

A.1 DETERMINATION OF pK USING THE HENDERSON-HASSELBALCH EQUATION

Each buffer has a pK which is the pH at which a solution of the buffer is 50% ionized; i.e., for a buffer made up of an acid component and a basic component there will be a pH where one can obtain in solution an equal mixture of acidic ions and basic ions. A buffer is best employed near its pK as it will have an effective resistance to changes in pH, i.e., have a good buffering capacity.

Henderson-Hasselbalch Equation

$$pH = pK + \log \frac{salt}{acid}$$

Example using this equation:

Make 100 ml of 1 M, pH 7.5 potassium phosphate buffer. Since phosphate has pK values around 3, 7, and 10 and the pH of the buffer required is 7.5, one should choose the pK around 7—which is 6.75 (found in tables).

Substituting in the Henderson-Hasselbalch equation above for

$$pH = 7.5$$
$$pK = 6.75$$
$$\text{and } \frac{salt}{acid} = \frac{K_2HPO_4}{KH_2PO_4}$$

Rearranging the Henderson-Hasselbalch one obtains:

$$\log \frac{salt}{acid} = pH - pK$$
$$= 7.5 - 6.75 = 0.75$$

Antilog of 0.75 = 5.623

Let salt be S and acid be A for convenience. So,

$$\frac{S}{A} = 5.623$$

$$\therefore S = (5.623)\ (A)$$

Concentration of $S + A = 1$ M (as stated in this problem)

Substituting for S

$$5.623\,A + A = 1\ \text{M}$$
$$6.623\,A = 1\ \text{M}$$
$$A = \frac{1}{6.623} = 0.151\ \text{M}$$

Since $S + A = 1$ M

$$\therefore S = 1\ \text{M} - 0.151\ \text{M}$$
$$= 0.849\ \text{M}$$

Thus, for the salt K_2HPO_4 (formula weight 174.183) one would need

$$0.849 \times 174 = 147.7\ \text{g}$$

Since one needs to make only 100 ml in this problem and not 1 liter

$$147.7 \times \frac{100}{1000} = 14.77\ \text{g are required of the salt } K_2HPO_4$$

Similarly for the acid component KH_2PO_4 (formula weight 136.091)

$$0.151 \times 136.09 \times \frac{100}{1000} = 2.055\ \text{g } KH_2PO_4$$

So if one dissolves 14.77 g K_2HPO_4 and 2.055 g KH_2PO_4 into 100 ml H_2O, one would obtain a 1 M phosphate buffer with a pH of 7.5.

To find a pK experimentally, one mixes equal volumes of equal concentrations of a salt and its acid and reads the pH in a pH meter. This reading is approximately the pK value of the solute in question. Alternatively, as above, the pK can be determined using the Henderson-Hasselbalch equation if the pH and the concentrations of the salt and the acid are known.

Calculation of Ionic Strength

To calculate the ionic strength of the solution in the preceding problem: Note that the solution KH_2PO_4 ionizes into K^+ and $H_2PO_4^-$ (0.151 M K^+ and 0.151 M H_2PO_4, each ion having a charge of one) and the

K_2HPO_4 ionizes in solution into K_2^+ and HPO_4^{--} (0.849 M K_2^+, 2 ions of charge one, and 0.849 M HPO_4^{--}, one ion of charge two).

$$\text{Ionic strength} = \frac{1}{2}\,(M_1Z_1^2 + M_2Z_2^2 \cdots M_nZ_n^2)$$

where M = molarity and Z = charge.
Ionic strength of 1 M phosphate

$$= \frac{1}{2}\left((0.151)\,(1)^2 + (0.151)\,(1)^2 + (0.849)\,(2)\,(1)^2 + (0.849)\,(2)^2\right)$$

$$= \frac{1}{2}\,(0.151 + 0.151 + 1.698 + 3.396)$$

$= \frac{5.396}{2} = 2.698$ is the ionic strength of the 1 M solution of K_2HPO_4 and KH_2PO_4, pH 7.5, which is equivalent in ionic strength to 2.698 M of any monovalent salt, as for example sodium chloride.

Examples and Problems

A solution of 0.05 M acid, pK = 5.0, is mixed with an equal volume of a 0.1 M solution of its sodium salt. What is the pH of the final mixture?

The final concentration of the acid is 0.025 M and the salt is 0.05 M because equal volumes of the two solutions were mixed.

The molar ratio of the concentrations of salt to acid is

$$\frac{[\text{salt}]}{[\text{acid}]} = \frac{0.05}{0.025} = 2.0$$

Knowing this ratio and the pK value,

$$\text{pH} = \text{pK} + \log \frac{[\text{salt}]}{[\text{acid}]}$$

$= 5.0 + \log 2.0 = 5.0 + 0.30 = 5.3$ pH of the final mixture.

If 20 ml of 0.18 M acid, when mixed with 40 ml of 0.30 M potassium salt of that acid, gave a solution with a pH of 6.35, what is the pK for the acid?

Total volume is 60 ml.

$$\text{Final concentration of acid} = \frac{20}{60} \times \frac{0.18\text{ M}}{1} = 0.06\text{ M}$$

$$\text{Final concentration of salt} = \frac{40}{60} \times \frac{0.30\text{ M}}{1} = 0.20\text{ M}$$

$$\frac{[\text{salt}]}{[\text{acid}]} = \frac{0.20\ \text{M}}{0.06\ \text{M}} = 3.33$$

$$6.35 = \text{pK} + \log 3.33$$
$$= \text{pK} + 0.52$$
$$\text{pK} = 6.35 - 0.52 = 5.83$$

What is the ionic strength of 3 M NaCl (this salt ionizes into Na^+ and Cl^-) using the equation for ionic strength?

$$\text{Ionic strength} = \frac{1}{2}(M_1 Z_1^2 + M_2 Z_2^2 \ldots + MnZ_n^2$$

Since M = 3 for both Na^+ and Cl^- and the charge Z of these two ions is 1

∴ by substitution

$$\text{Ionic strength} = \frac{1}{2}((3)(1)^2 + (3)(1)^2)$$
$$= \frac{1}{2}(6)$$
$$= 3\ \text{M}$$

Thus a solution of NaCl which is 3 M in its concentration is also 3 M in its ionic strength.

A.2 GRAPHIC REPRESENTATION OF ORDERS OF REACTION

Order	Equation	Graphical Representation
Zero	$x = kt$ where x is amount reacted k is the rate constant t is time k units = concentration × time^{-1}	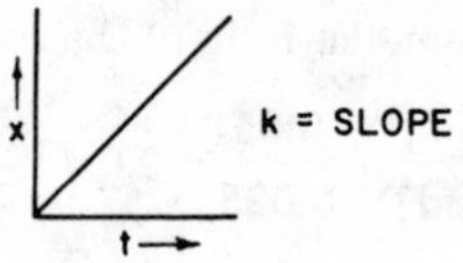 Plot of x vs. t →; k = SLOPE
1st	$\log \frac{a}{(a-x)} = \frac{kt}{2.303}$ where a is the concentration of the reactant at zero time x is the amount which reacted at a given time k units = time^{-1}. (See also note on page 227.)	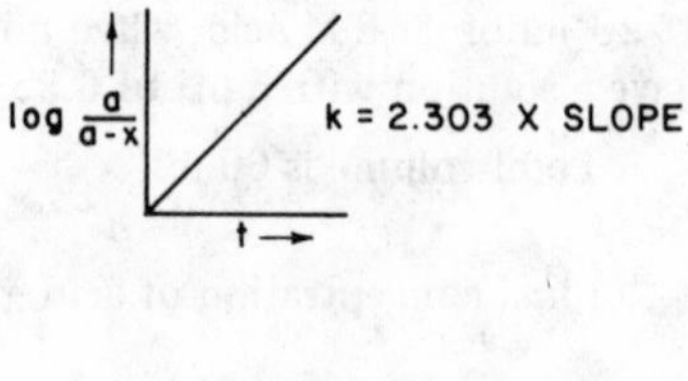 Plot of $\log \frac{a}{a-x}$ vs. t →; k = 2.303 X SLOPE

A.2 *(continued)*

Order	Equation	Graphical Representation
2nd	$\log \frac{b(a-x)}{a(b-x)} = \frac{k(a-b)t}{2.303}$ where a is one reactant and b is the other reactant. k units = concentration^{-1} $\times$ time^{-1}	$\log \frac{b(a-x)}{a(b-x)}$ vs t; $k = \frac{2.303}{(a-b)} \times$ SLOPE
n	For reactions of higher order than 2nd, plot $1/C^{n-1}$ *vs* time where C is the concentration of the reactant, t is time, and n is the order of the reaction. This plot should be linear; if it is not, then that order is not the order of that reaction. The rate constant (k) for a reaction of order n has the dimensions of concentration^{n-1} $\times$ time^{-1}.	

Note for first order reactions:

The half-life or half-life period after which the concentration of a reacting substance has decreased to half its original concentration corresponds to one-half the time ($t_{1/2}$) where

$$t_{1/2} = \frac{\ln 2}{k} = \frac{2.303 \log 2}{k} = \frac{0.693}{k}$$

This $t_{1/2}$ can be obtained from the above first order plot by reading off the time interval which makes 90° with the number 0.301 (i.e., log 2) on the ordinate.

A.3 KINETIC PARAMETERS IN SIMPLE ENZYME KINETICS

The reaction of an enzyme with its substrate* to yield products is a complex phenomenon represented briefly as follows:

$$E + S \underset{k_2}{\overset{k_1}{\rightleftharpoons}} ES \underset{k_4}{\overset{k_3}{\rightleftharpoons}} E + P$$

where E = enzyme
S = substrate
P = product
k = rate of reaction

*Specific substance or compound which the enzyme binds and transforms into a product or products.

The initial rate or velocity in an enzyme catalyzed reaction is measured by following the disappearance of a reactant (substrate) with time or the formation of the product with time at a given condition of pH, ionic strength, and temperature. Michaelis-Menten gave the following equation to express the relation between an enzyme's initial velocity, its maximal velocity, substrate molar concentration, and "disaffinity."

$$v = \frac{[S]\ V_{max}}{[S] + K_m}$$

where V_{max} = maximal velocity in $\frac{\mu \text{ moles product or reactant formed/min}}{\text{mg of enzyme used}}$
$[S]$ = substrate molar concentration and
K_m = disaffinity

The latter is called the Michaelis-Menten constant for substrate, or simply the Michaelis constant. We have called it substrate "disaffinity" for simplification because the greater this constant is, the poorer is the enzyme's affinity for such substrate.

Using this equation, if one plots velocity (v) *versus* substrate concentration $[S]$ for a simple enzyme reaction with its substrate, one obtains a hyperbola. (See Figure A.4.) Such a graph just gives approximate values of the parameters V_{max} and K_m, therefore, graphical forms of this equation to obtain straight lines has been derived to evaluate these parameters. Most popular is the Lineweaver-Burk plot presented in 1938 which puts the Michaelis-Menten equation in the form of $y = mx + b$. (See Figure A.5.)

Equation:

$$\frac{1}{v} = \frac{K_m}{V_{max}} \frac{1}{[S]} + \frac{1}{V_{max}}$$

$$\downarrow \quad \downarrow \quad \downarrow \quad \downarrow$$

$$y = m \quad x + b$$

Plot $\frac{1}{\text{velocity}}$ *versus* $\frac{1}{\text{substrate concentration}}$ (See Figure A.6.)

Note:

$$V_{max} = \frac{1}{\left(\frac{1}{V_{max}}\right)} \text{ or } \frac{1}{y \text{ intercept value}}$$

$$K_m = \frac{1}{\left(\frac{1}{K_m}\right)} \text{ or } \frac{1}{\text{positive value of } x \text{ intercept}}$$

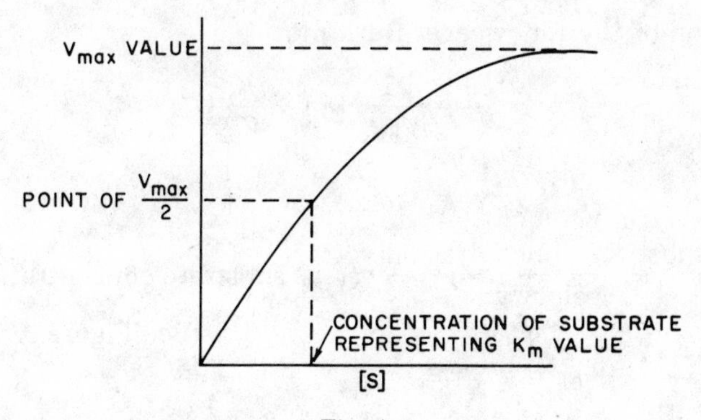

Fig. A.4.

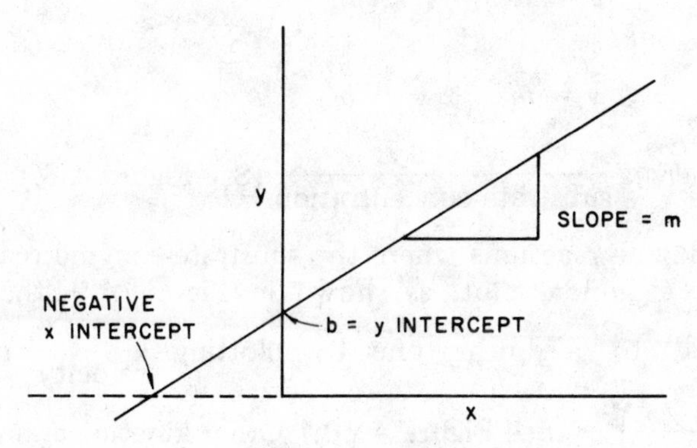

Fig. A.5.

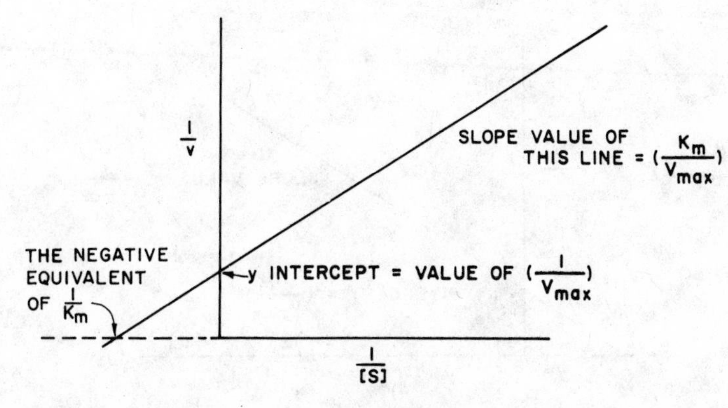

Fig. A.6.

A modification of the Lineweaver-Burk plot is

$$\frac{[S]}{v} = \frac{1}{V_{max}}[S] + \frac{K_m}{V_{max}}$$
$$\downarrow \quad \downarrow \quad \quad \downarrow$$
$$y = m \quad x + \quad b$$

Plot $\frac{\text{substrate concentration}}{\text{velocity}}$ *versus* substrate concentration

(See Figure A.7.)

The Michaelis-Menten equation may be also rearranged in another form: $y = mx + b$.

$$v = -K_m \frac{v}{[S]} + V_{max}$$
$$\downarrow \quad \downarrow \quad \downarrow \quad \downarrow$$
$$y = m \quad x \quad b$$

Plot velocity *versus* $\frac{\text{velocity}}{\text{substrate concentration}}$ (See Figure A.8.)

In some complex enzyme reactions where the substrate-enzyme reaction gives an S curve or sigmoidal plot, as shown in Figure A.9, the plot can be transformed to a linear one by plotting $\frac{1}{\text{velocity}}$ *versus* $\left(\frac{1}{\text{substrate concentration}}\right)^2$. (See Figure A.10.) Other kinetic parameters for a given enzyme may be obtained from kinetic data using other plots. These may be found in books devoted entirely to enzyme kinetics.

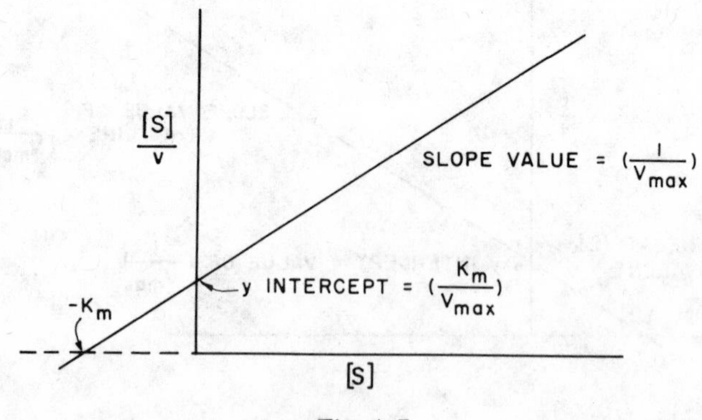

Fig. A.7.

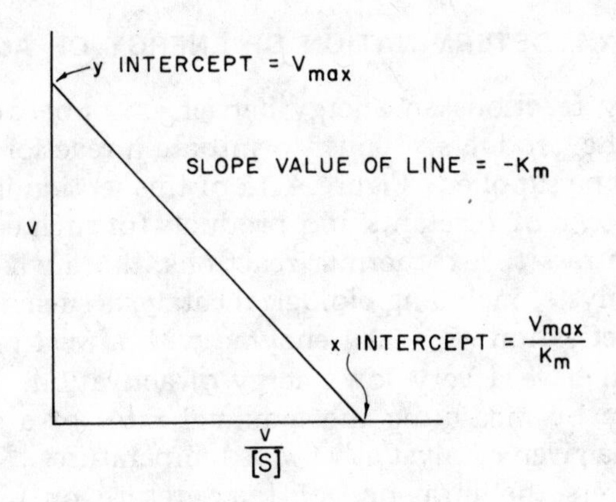

Fig. A.8.

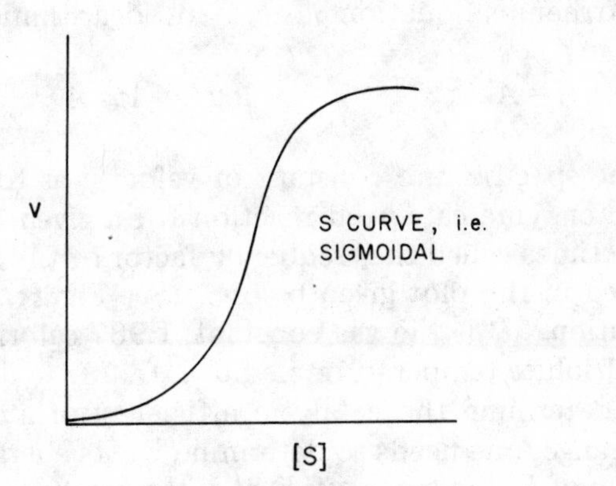

Fig. A.9.

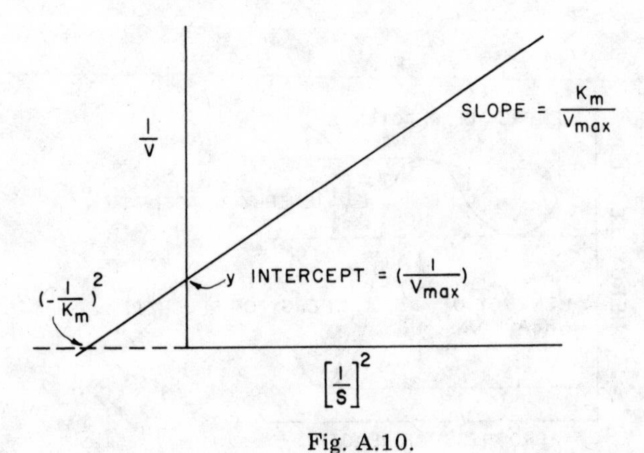

Fig. A.10.

A.4 DETERMINATION OF ENERGY OF ACTIVATION

In many reactions an energy barrier exists between that of the reactants and the products. Thus, to initiate a reaction in such cases, some energy must be supplied. Figure A.11 of this section illustrates the relation between energy of reactants and products for such endothermic reactions (i.e., in contrast to exothermic reactions, those which require energy to start). Catalysts, including biological catalysts such as enzymes, lower the energy of activation (E_a). An enzyme with a relatively great catalytic efficiency will have a very low energy of activation. E_a is determined experimentally by measuring the maximal rates of a given reaction in the presence of a given catalyst at various temperatures.

To express the influence of temperature on reaction velocity, one utilizes the equation given by S. Arrhenius in 1889.

Arrhenius Equation	In Logarithmic Form
$k = Ae^{-Ea/RT}$	$\log k = \log A - \dfrac{E_a}{2.303\,RT}$

where k is the specific rate constant or velocity as for example maximal velocity in an enzyme catalyzed reaction for a given temperature. A is a constant sometimes called the frequency factor of the reaction in question and is not used in the plot given below. (See Figure A.12.) E_a is the energy of activation. R is the gas constant, 1.987 calories degree^{-1} mole^{-1}, and T is the absolute temperature, i.e., 0°C + 273.

Thus, to determine the catalytic efficiency of a catalyst in a certain temperature range, one needs to determine E_a, its energy of activation; the lower the value of E_a the more efficient is the catalyst.

E_a can be determined graphically from a plot of log k versus $\frac{1}{T}$ where

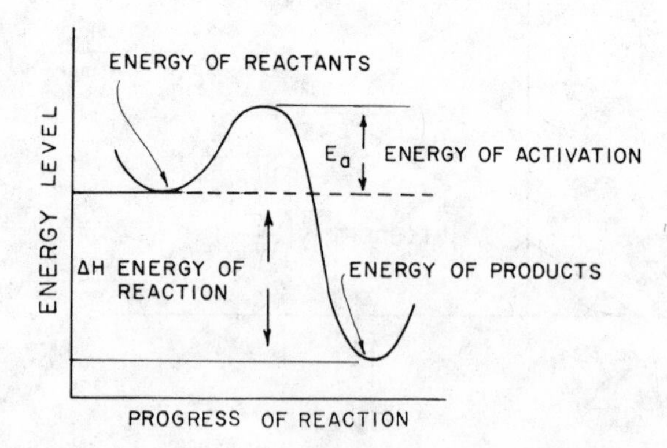

Fig. A.11.

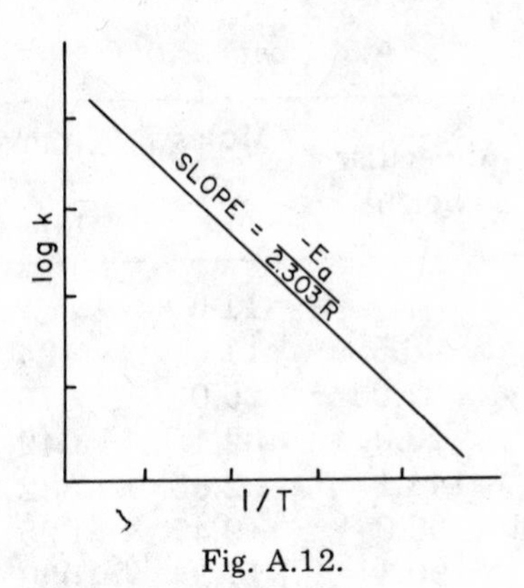

Fig. A.12.

the slope is equal to $\frac{-Ea}{2.303R}$. Since the slope is always negative, the slope becomes a positive value equal to $\frac{Ea}{2.303R}$.

E_a in calories = (slope) (2.303) (1.987). E_a is always positive, meaning that energy must be supplied.

Alternatively, if log k is known at two absolute temperatures, one can calculate E_a from the following relationship:

$$\log \frac{k_2}{k_1} = \frac{E_a}{2.303R} \left(\frac{T_2 - T_1}{T_1 T_2} \right)$$

Consequently, if the values of k_2 and k_1 at the temperatures T_2 and T_1 are known, the value of E_a can be readily calculated. The dimension of E_a is expressed in calories or kilocalories.

A.5 CONCENTRATION OF ACIDS
(Common Commercial Strengths)

Acid	Molecular weight	Moles per liter	Grams per liter	Percent by weight	Specific gravity
Acetic acid	60.05	17.4 N = 17.4	1045	99.5	1.05
Butyric acid	88.1	10.3	912	95.0	0.96
Formic acid	46.02	23.4	1080	90.0	1.20
Hydriodic acid	127.9	7.57	969	57.0	1.70
Hydrobromic acid	80.92	8.89	720	48.0	1.50

A.5 *(continued)*

Acid	Molecular weight	Moles per liter	Grams per liter	Percent by weight	Specific gravity
		N = 11.6			
Hydrochloric acid	36.5	11.6	424	36.0	1.18
Hydrocyanic acid	27.03	25.0	676	97.0	0.697
Hydrofluoric acid	20.01	32.1	642	55.0	1.167
Hydrofluosilicic acid	144.1	2.65	382	30.0	1.27
Hypophosphorous acid	66.0	9.47	625	50.0	1.25
Lactic acid	90.1	11.3	1020	85.0	1.2
Nitric acid	63.02	15.99	1008	71.0	1.42
		N = 16.0			
Perchloric acid	100.5	11.65	1172	70.0	1.67
Phosphoric acid	98.0	14.7	1445	85.0	1.70
Sulfuric acid	98.1	18.0	1766	96.0	1.84
		N = 36.0			
Sulfurous acid	82.1	0.74	61.2	6.0	1.02

A.6 AMINO ACIDS

Name	Formula	Molecular Weight
alanine	$C_3H_7NO_2$	89.1
arginine	$C_6H_{14}N_2O_2$	174.2
asparagine	$C_4H_8N_2O_3$	132.1
aspartic acid	$C_4H_7NO_4$	133.1
cysteine	$C_3H_7NO_2S$	121.2
cystine	$C_6H_{12}N_2O_4S_2$	240.3
glutamic acid	$C_5H_9NO_4$	147.1
glutamine	$C_5H_{10}N_2O_3$	146.1
glycine	$C_2H_5NO_2$	75.1
histidine	$C_6H_9N_3O_2$	155.2
isoleucine	$C_6H_{13}NO_2$	131.2
leucine	$C_6H_{13}NO_2$	131.2
lysine	$C_6H_{14}N_2O_2$	146.2
methionine	$C_5H_{11}NO_2S$	149.2
phenylalanine	$C_9H_{11}NO_2$	165.2

A.6 *(continued)*

Name	Formula	Molecular Weight
proline	$C_5H_9NO_2$	115.1
serine	$C_3H_7NO_3$	105.1
threonine	$C_4H_9NO_3$	119.1
tryptophan	$C_{11}H_{12}N_2O_2$	204.2
tyrosine	$C_9H_{11}NO_3$	181.2
valine	$C_5H_{11}NO_2$	117.1

A.7 pH OF SOME COMMON SOLUTIONS AT 0.1 M CONCENTRATION

Solution	pH
Ammonium phosphate primary	4.0
Ammonium sulfate	5.5
Ammonia water	11.3
Boric acid	5.3
Citric acid	2.1
Hydrochloric acid	1.1
Potassium acetate	9.7
Potassium bicarbonate	8.2
Potassium carbonate	11.5
Sodium benzoate	8.0
Sodium hydroxide	12.9
Tartaric acid	2.0

A.8 VAPOR PRESSURE OF WATER

Temperature		Pressure in mm of mercury	Temperature		Pressure in mm of mercury
°F	°C		°F	°C	
32.0	0	4.6	77.0	25	23.6
41.0	5	6.5	78.8	26	25.1
46.4	8	8.0	80.6	27	26.5
48.2	9	8.6	82.4	28	28.1
50.0	10	9.2	84.2	29	29.8
51.8	11	9.8	86.0	30	31.5
53.6	12	10.5	87.8	31	33.4

A.8 (*continued*)

Temperature °F	Temperature °C	Pressure in mm of mercury	Temperature °F	Temperature °C	Pressure in mm of mercury
55.4	13	11.2	89.6	32	35.4
57.2	14	11.9	91.4	33	37.4
59.0	15	12.7	93.2	34	39.6
60.8	16	13.5	95.0	35	41.8
62.6	17	14.4	104.0	40	54.9
64.4	18	15.4	122.0	50	92.5
66.2	19	16.3	140.0	60	148.9
68.0	20	17.5	158.0	70	233.3
69.8	21	18.5	176.0	80	354.9
71.6	22	19.7	194.0	90	525.5
73.4	23	20.9	212.0	100	760.0
75.2	24	22.2			

A.9 SYMBOLS

$+$	plus; positive
$-$	minus; negative
$\pm$	plus or minus; positive or negative
$=$	equals
$\neq$	does not equal
$\doteq$	almost
ab, a·b, a × b	*a* times *b*; *a* multiplied by *b*
a/b, $a \div b$, $a:b$	*a* divided by *b*; the ratio of *a* to *b*
$a/b = c/d$	a proportion; *a* is to *b* as *c* is to *d*
$\cong$ or ≌	approximately equal; congruent
$\sim$ or $\backsim$	equivalent; similar; approximately equal; about
$>$	greater than
$<$	less than
$\geqslant$ or $\geqslant$	greater than or equal to
$\leqslant$ or $\leqslant$	less than or equal to
a^n	$a \cdot a \cdot a \ldots$ to *n* factors
$\sqrt{a}$, $a^{1/2}$	the positive square root of *a*; for positive *a*
$\sqrt[n]{a}$, $a^{1/n}$	n^{th} root of *a*; usually means the principal n^{th} root
a^{-n}	the reciprocal of a^n ; $1/a^n$
e	base of the system of natural log 2.71828 . . .
$\log a$ or $\log_{10} a$	common logarithms of *a* to the base 10
$\ln a$ or $\log_e a$	natural logarithms of *a* to the base *e*

$n!$ (or $\lfloor n$)	n factorial; $1 \cdot 2 \cdot 3 \ldots n$
a°	a degrees (angle)
a'	a minutes (angle)
a''	a seconds (angle)
$\angle$, $\measuredangle$s	angle(s)
$\perp$	perpendicular
$\parallel$	parallel
$\therefore$	therefore; hence
π	the ratio of the circumference of a circle to the diameter; the Greek letter pi; equal to 3.1415926536 . . .
∞	infinity
(x, y)	rectangular coordinates of a point in a plane
$P(x, y)$	point P with coordinates x and y in the plane
m	slope
$f(x)$, $F(x)$	function of x
Σ	sum of certain terms, the terms being indicated by the context or by added notation
$\sum_{1}^{n}$ or $\sum_{i=1}^{n}$	sum of n terms, one for each positive integer from 1 to n
Δy	an increment of y
$\int$	integral of
$\int_a^b f(x)dx$	the definite integral of $f(x)$ between the limits a and b
dy	differential of y
$\frac{dy}{dx}$ or $f'(x)$	first derivation of $y = f(x)$ with respect to x
$\frac{d^2y}{dx^2}$ or $f''(x)$	second derivation of $y = f(x)$ with respect to x
χ^2	Chi Square
df or DF	degrees of freedom
F	F ratio
r	correlation coefficient (Pearson Product Moment)
s	standard deviation from a sample
σ	standard deviation of a population
t	t distribution; confidence level factor
$\overline{X}$	arithmetic mean from a sample
μ	arithmetic mean from a population
I or μ	ionic strength
k	rate constant
n	refractive index

η	viscosity
λ	wavelength
ρ	density
EMF	electromotive force
m	molality
M	molarity
N	normality
N or L	Avogadro number
R	molar gas constant
ϵ	extinction coefficient
h	Planck's constant
G	gravitation constant
g	acceleration of free fall
w	angular velocity
$T_{1/2}$	half-life
b.p.	boiling point
m.p.	melting point
Ci	Curie
$^{\circ}K$	degrees Kelvin
$^{\circ}$C	degrees Celcius
dil	dilute
STP	standard temperature and pressure
avdp.	avoirdupois weight
c	concentration or velocity of light
e	natural log base
E	energy in general
E_a	energy of activation
eq	equivalent

A.10 STEMS, PREFIXES, AND SUFFIXES USED IN THE FORMATION OF BIOMEDICAL WORDS

Stem, Prefix, or Suffix	Meaning
a	without
hyper	greater
hypo	lesser
iso	same
macro	large
mega	large
meniscus	crescent
meta	change
micro	small
mono	one

A.10 (*continued*)

Stem, Prefix, or Suffix	Meaning
poly	many
post	after
pro	before

A.11 GREEK ALPHABET

Name of letter	Capital	Lower case	Transliteration
alpha	Α	α	a
beta	Β	ϐ or β	b
gamma	Γ	γ	g
delta	Δ	δ	d
epsilon	Ε	ϵ	e short
zeta	Ζ	ζ	z
eta	Η	η	e long
theta	Θ	θ	th
iota	Ι	ι	i
kappa	Κ	κ	k, c
lambda	Λ	λ	l
mu	Μ	μ	m
nu	Ν	ν	n
xi	Ξ	ξ	x
omicron	Ο	ο	o short
pi	Π	π	p
rho	Ρ	ρ	r
sigma	Σ	σ or ς	s
tau	Τ	τ	t
upsilon	ϒ	υ	y
phi	Φ	ϕ or φ	f
chi	Χ	χ	ch as in German echt
psi	Ψ	ψ	ps
omega	Ω	ω	o long

Source: The Merck Index, Eighth Edition, copyright 1968 by Merck & Co., Inc., Rahway, New Jersey, U.S.A., by permission of the publishers.

A.12 RUSSIAN ALPHABET

Cyrillic print	Transliteration	Pronunciation
А а	a	*a* in far
Б б	b	*b*
В в	v	*v*
Г г	g (h)	*g* in gay
Д д	d	*d*
Е е	e	*e* in fell; also *ye* in yell
Ж ж	zh	*z* in azure
З з	z	*z* in zeal
И и	i	*i* in meet
Й й	ĭ	*y* in boy
К к	k	*k*
Л л	l	*l*
М м	m	*m*
Н н	n	*n*
О о	o	*o* in or
П п	p	*p*
Р р	r	*r*
С с	s	*s* in say
Т т	t	*t*
У у	u	*oo* in boot
Ф ф	f	*f*
Х х	kh	like German *ch*
Ц ц	t͡s	*ts* in hoots
Ч ч	ch	*ch* in church
Ш ш	sh	*sh*
Щ щ	shch	*shch*, as in fre*sh ch*eese
Ъ ъ	″	mute*
Ы ы	y	*y* in rhythm (hard)
Ь ь	′	mute (softens preceding consonant)
Э э	ė	*e* in met
Ю ю	i͡u	*u* in union
Я я	i͡a	*ya* in yard

* Hard sign; used to separate a consonant from a soft vowel especially in foreign words; frequently replaced by an apostrophe.

Source: The Merck Index, Eighth Edition, copyright 1968 by Merck & Co., Inc., Rahway, New Jersey, U.S.A., by permission of the publishers.

A.13 ROMAN NUMERALS

I	II	III	IV	V	VI	VII	VIII	IX	X
1	2	3	4	5	6	7	8	9	10
XX	XXX	XL	L	LX	LXX	LXXX	XC	IC	C
20	30	40	50	60	70	80	90	99	100
CC	CCC	CD	D	DC	DCC	DCCC	CM	XM	M
200	300	400	500	600	700	800	900	990	1000

Source: *The Merck Index*, Eighth Edition, copyright 1968 by Merck & Co., Inc., Rahway, New Jersey, U.S.A., by permission of the publishers.

A.14 NUMERICAL CONSTANTS

$\pi = 3.14159$ $\quad e = 2.71828$ $\quad \sqrt{2} = 1.41421$

$1/\pi = 0.31830$ $\quad 1/e = 0.36787$ $\quad \sqrt[3]{2} = 1.25992$

$\pi^2 = 9.86960$ $\quad e^2 = 7.38905$ $\quad \sqrt{3} = 1.73205$

$\sqrt[3]{3} = 1.44224$

A.15 ASTRONOMICAL MEASUREMENT

One astronomical unit = 93,000,000 miles (approximately)—the mean distance between the sun and the earth

One light year = 59×10^{11} miles—the distance traveled by light in one year

Parsec = about 3.3 light years—the distance if a star at which the angle subtended by the radius of the earth's orbit is 1″

A.16 PHYSICAL CONSTANTS

Avogadro number = $6.023 \times 10^{+23}$ molecules/g mole

Planck constant = 6.626×10^{-27} erg sec molecule^{-1}

Density of mercury at 0° C = 13.5955 g ml

1 atmosphere = 760 mm mercury or 1,013,250 dynes cm^{-2}

Density of water at 3.98° C = 1.0 g/ml

Gas constant = 1.987 cal mole^{-1} degree^{-1}
0.082 liter atm mole^{-1} degree^{-1}
8.314×10^7 ergs/degree mole

Absolute zero = −273.16° C

Gas molar volume at STP (at 0° C and 760 mmHg)	= 22.4 liters = 22,413.8 cm^3 /mole
Acceleration of gravity	= 980.665 cm sec^{-2} or 32 ft/sec^{-2}
Constant of gravitation	= 6.673×10^{-8} cm^3 $g^{-1} \times sec^{-2}$
Velocity of light	= 2.9978×10^{10} cm/sec
Equatorial radius of the earth	= 6378.388 kilometers or 3963.32 miles
1° of latitude at 40°	= 69 miles or 111 kilometers
Mean density of the earth	= 5.522 g/cm^3 or 344.7 lb/ft^3

A.17 AMERICAN SYSTEM OF WEIGHTS AND MEASURES

Length

12 inches = 1 foot
36 inches = 3 feet = 1 yard
16 1/2 feet = 5 1/2 yards = 1 rod
220 yards (1/8 mile) = 1 furlong
1760 yards = 320 rods = 5280 feet = 1 land mile
6076 feet = 1 nautical mile
3 nautical miles = 1 league

Area

144 square inches = 1 square foot
9 square feet = 1 square yard
30 1/4 square yards = 1 square rod
160 square rods = 1 acre
640 acres = 1 square mile

Volume

1728 cubic inches = 1 cubic foot
27 cubic feet = 1 cubic yard

Capacity

Dry Measure	2 pints = 1 quart
	8 quarts = 1 peck
	4 pecks = 1 bushel
Liquid Measure	16 fluid ounces = 1 pint
	32 fluid ounces = 2 pints = 1 quart
	4 quarts = 1 gallon
	8 1/3 pounds = 231 cubic inches = 1 gallon

Avoirdupois Weight (used for all ordinary purposes)

1 dram = 27 1/3 grains
1 ounce = 16 drams = 437.5 grains
16 ounces = 1 pound = 7000 grains
2000 pounds = 1 ton
2240 pounds = 1 long ton

Apothecaries Weight (used by doctors and druggists)

20 grains = 1 scruple
3 scruples = 1 dram
8 drams = 1 ounce
12 ounces = 1 pound
5760 grains = 1 pound

Troy Weight (used for precious metals and stones and coins)

24 grains = 1 pennyweight
20 pennyweights = 1 ounce
12 ounces = 1 pound
5760 grains = 1 pound

A.18 METRIC SYSTEM OF WEIGHTS AND MEASURES (most used)

Length

10 millimeters	= 1 centimeter	= 0.3937 inch
100 centimeters	= 1 meter	= 39.37 inches
1000 meters	= 1 kilometer	= 0.6214 mile

Capacity

1000 cubic milliliters	= 1 liter	= 1.1 quarts
1000 cubic liters	= 1 kiloliter	= 264.2 gallons

Weight

1000 micrograms	= 1 milligram	= 0.035×10^{-3} ounces
1000 milligrams	= 1 gram	= 0.035 ounce
1000 grams	= 1 kilogram	= 2.2 pounds

A.19 EQUIVALENTS OF THE METRIC AND U.S. SYSTEMS OF MEASURES AND WEIGHTS

Length

1 millimeter	= 0.03937 inch	1 inch	= 2.540 centimeters
1 centimeter	= 0.3937 inch	1 foot	= 30.480 centimeters
1 meter	= 39.37 inches	1 foot	= 0.3048 meter
1 meter	= 3.2808 feet	1 yard	= 91.440 centimeters
1 meter	= 1.09361 yards	1 yard	= 0.9144 meter
1 meter	= 0.1988 rod	1 rod	= 5.0292 meters
1 kilometer	= 0.6214 mile	1 mile	= 1.6093 kilometers

Conversion Table for Length

Inches	= Millimeters	Millimeters	= Inches
1/32	0.794	1	0.03937
1/16	1.588	2	0.07874
1/8	3.175	3	0.11811
1/4	6.350	4	0.15748
1/2	12.700	5	0.19685
3/4	19.050	6	0.23622
3/8	9.525	7	0.27559
		8	0.31496
		9	0.35433

Inches	= Centimeters	Centimeters	= Inches
1	2.54	1	0.39370
2	5.08	2	0.78740
3	7.62	3	1.18110
4	10.16	4	1.57480
5	12.70	5	1.96851
6	15.24	6	2.36220
7	17.78	7	2.75591
8	20.32	8	3.14961
9	22.86	9	3.54331

Feet	= Meters	Meters	= Feet
1	0.3048	1	3.28084
2	0.6096	2	6.56168
3	0.9144	3	9.84252
4	1.2192	4	13.12336
5	1.5240	5	16.40420
6	1.8288	6	19.68504
7	2.1336	7	22.96588
8	2.4383	8	26.24672
9	2.7432	9	29.52726

Yards	=	Meters	Meters	=	Yards
1		0.9144	1		1.09361
2		1.8288	2		2.18723
3		2.7432	3		3.28984
4		3.6576	4		4.37445
5		4.5720	5		5.46807
6		5.4864	6		6.56168
7		6.4008	7		7.65529
8		7.1352	8		8.74891
9		8.2296	9		9.84252

Miles (land)	=	Kilometers	Kilometers	=	Miles
1		1.60934	1		0.62137
2		3.21869	2		1.24274
3		4.82803	3		1.86411
4		6.43738	4		2.48585
5		8.04672	5		3.10686
6		9.65606	6		3.78227
7		11.26541	7		4.34960
8		12.87475	8		4.97097
9		14.48410	9		5.59234

Area

1 square millimeter = 0.00155 square inch
1 square centimeter = 0.155 square inch
1 square meter = 1550.0 square inches
1 square meter = 10.764 square feet
1 square meter = 1.196 square yards
1 square kilometer = 0.3861 square mile
1 square inch = 6.4516 square centimeters
1 square foot = 929.0341 square centimeters
1 square foot = 0.0929 square meter
1 square yard = 0.8361 square meter
1 square mile = 2.589 square kilometers

Volume

1 cubic centimeter = 0.06102 cubic inch
1 cubic decimeter = 61.0234 cubic inches
1 cubic meter = 35.3145 cubic feet
1 cubic meter = 1.3079 cubic yards
1 cubic inch = 16.3972 cubic centimeters
1 cubic foot = 28.317 cubic decimeters
1 cubic foot = 0.02832 cubic meter
1 cubic yard = 0.7646 cubic meter

Capacity

1 milliliter	= 0.0338 fluid ounce	1 fluid ounce	= 29.573 milliliters
1 liter	= 33.8148 fluid ounces	1 pint	= 473.166 milliliters
1 liter	= 2.1143 pints	1 quart	= 946.332 milliliters
1 liter	= 1.0567 quarts	1 gallon	= 3.785 liters
1 liter	= 0.2642 gallon	1 cubic inch	= 16.387 milliliters
		1 cubic foot	= 28.316 liters

Weight

1 milligram	= 0.015432 grain
1 gram	= 15.432 grains
1 gram	= 0.03527 avoirdupois ounce
1 kilogram	= 35.274 avoirdupois ounces
1 kilogram	= 2.2046 avoirdupois pounds
1 metric ton	= 2204.62 avoirdupois pounds
1 grain	= 64.7989 milligrams
1 grain	= 0.0648 gram
1 avoirdupois ounce	= 28.3495 grams
1 avoirdupois pound	= 453.5924 grams

Conversion Table for Weight

Avoirdupois weight	= Grams	Avoirdupois weight	= Grams
1/16 ounce	1.772	7 ounces	198.447
1/8 ounce	3.544	8 ounces	226.796
1/4 ounce	7.087	9 ounces	255.146
1/2 ounce	14.175	10 ounces	283.495
1 ounce	28.350	11 ounces	311.845
2 ounces	56.699	12 ounces	340.194
3 ounces	85.049	13 ounces	368.544
4 ounces	113.398	14 ounces	396.893
5 ounces	141.748	15 ounces	425.243
6 ounces	170.097		

Grams	= Grain
0.010	0.1543
0.020	0.3086
0.030	0.4629
0.040	0.6173
0.050	0.7716
0.060	0.9259
0.070	1.0803
0.080	1.2346
0.090	1.3889
0.100	1.5432

Avoirdupois weight	= Kilograms	Kilograms	= Avoirdupois weight
1	0.45359	1	2.20462
2	0.90718	2	4.40924
3	1.36078	3	6.61387
4	1.81437	4	8.81849
5	2.26796	5	11.02311
6	2.72155	6	13.22773
7	3.17515	7	15.43236
8	3.62873	8	17.63698
9	4.08233	9	19.84160

A.20 TABLE OF CONVERSION FACTORS

To convert from	to	Multiply by
Centimeters	inches	0.3937
Centimeters	yards	0.01093611
Cubic centimeters	cubic inches	0.061023
Cubic feet	cubic inches	1728
Cubic feet	cubic meters	0.02832
Cubic feet	cubic yards	0.037037
Cubic inches	cubic centimeters	16.387162
Cubic inches	cubic feet	5.78704×10^{-4}
Cubic inches	gallons	0.004329
Cubic inches	liters	0.0163868
Cubic millimeters	cubic inches	0.000061
Cubic yards	cubic feet	27
Cubic yards	cubic meters	0.76455945
Degrees	minutes	60
Degrees	radians	0.0174533
Dynes	pounds	2.2481×10^{-6}
Ergs	foot-pounds	7.3756×10^{-8}
Ergs/second	horsepower	1.3410×10^{-10}
Ergs/second	kilowatts	1×10^{-10}
Fathoms	feet	6
Feet	centimeters	30.4801
Feet	inches	12
Feet	meters	0.3048
Furlongs	miles	0.125
Quarts	liters	0.94636
Gallons	cubic inches	231
Gallons	liters	3.78533
Gallons	quarts	4
Grains	grams	0.0648
Grains	milligrams	64.798918

A.20 (*continued*)

To convert from	to	Multiply by
Grains	ounces (avdp.)	0.0022857
Grams	grains	15.4324
Inches	centimeters	2.540005
Inches	feet	1/12
Kilograms	grams	1000
Kilograms	pounds	2.20462
Kilograms	tons (metric)	0.001
Kilometers	meters	1000
Kilometers	miles (nautical)	0.539593
Kilometers	miles (land)	0.62137
Knots/hour	feet/hour	6080.20
Knots/hour	miles/hour	1.15155
Liters	cubic inches	61.025
Liters	gallons	0.264178
Liters	quarts (liquid)	1.05668
Meters	feet	3.2808333
Meters	inches	39.37
Meters	yards	1.093611
Microns	meters	1×10^{-6}
Miles (nautical)	feet	6080.20
Miles (nautical)	kilometers	1.85325
Miles (nautical)	miles (land)	1.1516
Miles (land)	feet	5280
Miles (land)	kilometers	1.609347
Miles/hour	feet/minute	88
Miles/hour	knots/hour	0.8684
Milligrams	grains	0.01543236
Milligrams	grams	0.001
Millimeters	inches	0.0394
Millimeters	meters	0.001
Mils	inches	0.001
Minutes	radians	2.90888×10^{-4}
Minutes (angle or time)	seconds	60
Milliliters	ounces	0.03381
Ounces (avdp.)	grams	28.349527
Ounces (avdp.)	pounds (avdp.)	1/16
Ounces (fluid)	cubic centimeters	29.5737
Ounces (fluid)	liters	0.295729
Ounces (fluid)	pints (liquid)	1/16
Ounces (fluid)	milliliters	29.574
Pints (dry)	cubic centimeters	550.61

A.20 (*continued*)

To convert from	to	Multiply by
Pints (dry)	quarts (dry)	0.5
Pints (liquid)	cubic centimeters	473.179
Pints (liquid)	gallons	0.125
Pints (liquid)	quarts (liquid)	0.5
Pounds	grams	453.59
Pounds (avdp.)	kilograms	0.45359
Pounds (avdp.)	ounces (avdp.)	16
Quarts (dry)	cubic centimeters	1101.23
Quarts (dry)	cubic inches	67.2006
Quarts (dry)	pints (dry)	2
Quarts (liquid)	cubic centimeters	946.358
Quarts (liquid)	cubic inches	57.749
Quarts (liquid)	gallons	0.25
Quarts (liquid)	liters	0.946333
Radians	degrees	57.29578
Rods (surveyor's)	yards	5.5
Seconds	minutes	1/60
Square centimeters	square inches	0.155
Square feet	square inches	144
Square feet	square inches	0.0929
Square inches	square centimeters	6.4516
Square inches	square feet	1/144
Square meters	square inches	1550.0
Square meters	square yards	1.195985
Square miles	acres	640
Square millimeters	square inches	0.00155
Square yards	square feet	9
Square yards	square meters	0.83613
Yards	feet	3
Yards	meters	0.9144

A.21 ENERGY CONVERSION FACTORS

	Calorie	Erg	Joule	Liter Atm	Einstein
Calorie	1	4.186×10^{7}	4.186	4.131×10^{-2}	$3.50 \times 10^{-9} \times \lambda$
Erg	2.39×10^{-8}	1	1×10^{-7}	9.87×10^{-10}	$8.33 \times 10^{-17} \times \lambda$
Joule	2.39×10^{-1}	1×10^{-7}	1	9.87×10^{-3}	$8.33 \times 10^{-10} \times \lambda$
Liter Atm	2.421×10^{1}	1.013×10^{9}	1.013×10^{2}	1	$8.47 \times 10^{-8} \times \lambda$
Einstein (in A°)	2.86×10^{3}	1.20×10^{16}	1.20×10^{9}	1.18×10^{7}	1

A.22 TEMPERATURE CONVERSION TABLE

Use of Conversion Table: First find in the F or C column the value which you wish to convert. If the value is Fahrenheit, you will find to its left the Centigrade equivalent; if Centigrade, you will find to its right the Fahrenheit equivalent. Example: If the value is 100° and Fahrenheit, its C equivalent in the column to the left is 37.78° C. If 100° is C, its F equivalent in the column to the right is 212° F.

To C	←F or C→	To F	To C	←F or C→	To F	To C	←F or C→	To F
−40	**−40**	−40	−12.22	**10**	50	15.56	**60**	140
−39.44	**−39**	−38.2	−11.67	**11**	51.8	16.11	**61**	141.8
−38.89	**−38**	−36.4	−11.11	**12**	53.6	16.67	**62**	143.6
−38.33	**−37**	−34.6	−10.56	**13**	55.4	17.22	**63**	145.4
−37.78	**−36**	−32.8	−10	**14**	57.2	17.78	**64**	147.2
−37.22	**−35**	−31	−9.44	**15**	59	18.33	**65**	149
−36.67	**−34**	−29.2	−8.89	**16**	60.8	18.89	**66**	150.8
−36.11	**−33**	−27.4	−8.33	**17**	62.6	19.44	**67**	152.6
−35.56	**−32**	−25.6	−7.78	**18**	64.4	20	**68**	154.4
−35	**−31**	−23.8	−7.22	**19**	66.2	20.56	**69**	156.2
−34.44	**−30**	−22	−6.67	**20**	68	21.11	**70**	158
−33.89	**−29**	−20.2	−6.11	**21**	69.8	21.67	**71**	159.8
−33.33	**−28**	−18.4	−5.56	**22**	71.6	22.22	**72**	161.6
−32.78	**−27**	−16.6	−5	**23**	73.4	22.78	**73**	163.4
−32.22	**−26**	−14.8	−4.44	**24**	75.2	23.33	**74**	165.2
−31.67	**−25**	−13	−3.89	**25**	77	23.89	**75**	167
−31.11	**−24**	−11.2	−3.33	**26**	78.8	24.44	**76**	168.8
−30.56	**−23**	−9.4	−2.78	**27**	80.6	25	**77**	170.6
−30	**−22**	−7.6	−2.22	**28**	82.4	25.56	**78**	172.4
−29.44	**−21**	−5.8	−1.67	**29**	84.2	26.11	**79**	174.2
−28.89	**−20**	−4	−1.11	**30**	86	26.67	**80**	176
−28.33	**−19**	−2.2	−0.56	**31**	87.8	27.22	**81**	177.8
−27.78	**−18**	−0.4	0	**32**	89.6	27.78	**82**	179.6
−27.22	**−17**	1.4	.56	**33**	91.4	28.33	**83**	181.4
−26.67	**−16**	3.2	1.11	**34**	93.2	28.89	**84**	183.2
−26.11	**−15**	5	1.67	**35**	95	29.44	**85**	185
−25.56	**−14**	6.8	2.22	**36**	96.8	30	**86**	186.8
−25	**−13**	8.6	2.78	**37**	98.6	30.56	**87**	188.6
−24.44	**−12**	10.4	3.33	**38**	100.4	31.11	**88**	190.4
−23.89	**−11**	12.2	3.89	**39**	102.2	31.67	**89**	192.2
−23.33	**−10**	14	4.44	**40**	104	32.22	**90**	194
−22.78	**−9**	15.8	5	**41**	105.8	32.78	**91**	195.8
−22.22	**−8**	17.6	5.56	**42**	107.6	33.33	**92**	197.6
−21.67	**−7**	19.4	6.11	**43**	109.4	33.89	**93**	199.4
−21.11	**−6**	21.2	6.67	**44**	111.2	34.44	**94**	201.2
−20.56	**−5**	23	7.22	**45**	113	35	**95**	203
−20	**−4**	24.8	7.78	**46**	114.8	35.56	**96**	204.8
−19.44	**−3**	26.6	8.33	**47**	116.6	36.11	**97**	206.6
−18.89	**−2**	28.4	8.89	**48**	118.4	36.67	**98**	208.4
−18.33	**−1**	30.2	9.44	**49**	120.2	37.22	**99**	210.2
−17.78	**0**	32	10	**50**	122	37.78	**100**	212
−17.22	**1**	33.8	10.56	**51**	123.8	38.33	**101**	213.8
−16.67	**2**	35.6	11.11	**52**	125.6	38.89	**102**	215.6
−16.11	**3**	37.4	11.67	**53**	127.4	39.44	**103**	217.4
−15.56	**4**	39.2	12.22	**54**	129.2	40	**104**	219.2
−15	**5**	41	12.78	**55**	131	40.56	**105**	221
−14.44	**6**	42.8	13.33	**56**	132.8	41.11	**106**	222.8
−13.89	**7**	44.6	13.89	**57**	134.6	41.67	**107**	224.6
−13.33	**8**	46.4	14.44	**58**	136.4	42.22	**108**	226.4
−12.78	**9**	48.2	15	**59**	138.2	42.78	**109**	228.2

A.22 (*continued*)

To C	←F or C→	To F	To C	←F or C→	To F	To C	←F or C→	To F
43.33	**110**	230	62.78	**145**	293	82.22	**180**	356
43.89	**111**	231.8	63.33	**146**	294.8	82.78	**181**	357.8
44.44	**112**	233.6	63.89	**147**	296.6	83.33	**182**	359.6
45	**113**	235.4	64.44	**148**	298.4	83.89	**183**	361.4
45.56	**114**	237.2	65	**149**	300.2	84.44	**184**	363.2
46.11	**115**	239	65.56	**150**	302	85	**185**	365
46.67	**116**	240.8	66.11	**151**	303.8	85.56	**186**	366.8
47.22	**117**	242.6	66.67	**152**	305.6	86.11	**187**	368.6
47.78	**118**	244.4	67.22	**153**	307.4	86.67	**188**	370.4
48.33	**119**	246.2	67.78	**154**	309.2	87.22	**189**	372.2
48.89	**120**	248	68.33	**155**	311	87.78	**190**	374
49.44	**121**	249.8	68.89	**156**	312.8	88.33	**191**	375.8
50	**122**	251.6	69.44	**157**	314.6	88.89	**192**	377.6
50.56	**123**	253.4	70	**158**	316.4	89.44	**193**	379.4
51.11	**124**	255.2	70.56	**159**	318.2	90	**194**	381.2
51.67	**125**	257	71.11	**160**	320	90.56	**195**	383
52.22	**126**	258.8	71.67	**161**	321.8	91.11	**196**	384.8
52.78	**127**	260.6	72.22	**162**	323.6	91.67	**197**	386.6
53.33	**128**	262.4	72.78	**163**	325.4	92.22	**198**	388.4
53.89	**129**	264.2	73.33	**164**	327.2	92.78	**199**	390.2
54.44	**130**	266	73.89	**165**	329	93.33	**200**	392
55	**131**	267.8	74.44	**166**	330.8	93.89	**201**	393.8
55.56	**132**	269.6	75	**167**	332.6	94.44	**202**	395.6
56.11	**133**	271.4	75.56	**168**	334.4	95	**203**	397.4
56.67	**134**	273.2	76.11	**169**	336.2	95.56	**204**	399.2
57.22	**135**	275	76.67	**170**	338	96.11	**205**	401
57.78	**136**	276.8	77.22	**171**	339.8	96.67	**206**	402.8
58.33	**137**	278.6	77.78	**172**	341.6	97.22	**207**	404.6
58.89	**138**	280.4	78.33	**173**	343.4	97.78	**208**	406.4
59.44	**139**	282.2	78.89	**174**	345.2	98.33	**209**	408.2
60	**140**	284	79.44	**175**	347	98.89	**210**	410
60.56	**141**	285.8	80	**176**	348.8	99.44	**211**	411.8
61.11	**142**	287.6	80.56	**177**	350.6	100	**212**	413.6
61.67	**143**	289.4	81.11	**178**	352.4			
62.22	**144**	291.2	81.67	**179**	354.2			

Source: The Merck Index, Eighth Edition, copyright 1968 by Merck & Co., Inc., Rahway, New Jersey, U.S.A., by permission of the publishers.

A.23 DEFINITIONS OF POLYGONS

1. *Polygon*—a closed broken line; a plane rectilinear figure.
2. *Similar Polygons*—the respective angles equal and the corresponding sides proportional.
3. *Regular Polygons*—all sides equal and all angles equal.
4. *Triangle*—polygon of three sides
   a. *isosceles triangle*—two sides and their opposite angles equal.
   b. *equilateral triangle*—all three sides and all three angles equal.
5. *Quadrilateral*—polygon of four sides
   a. *parallelogram*—both pairs of opposite sides parallel and equal.
   b. *rhomboid*—no right angles and, in general, adjacent sides not equal.
   c. *rhombus*—no right angles, but all sides equal.
   d. *rectangle*—oniy right angles and in general adjacent sides not equal.
   e. *square*—only right angles and all sides equal.

6. *Trapezium*—no two sides parallel.
7. *Trapezoid*—two opposite sides parallel.
   a. *isosceles trapezoid*—nonparallel sides equal.
8. *Other Polygons*
   Pentagon—5 sides
   Hexagon—6 sides
   Heptagon—7 sides
   Octagon—8 sides
   Enneagon or nonagon—9 sides
   Decagon—10 sides
   Dodecagon—12 sides

A.24 FORMULAS

Notation: a, b, c = lines; α, γ = angles; h = altitude; s = sides; d, d_1 = diagonals; p = perimeter; A = area; r = radius; C = circumference; T = total surface; S = area of curved surface.

Any Triangle: $A = \frac{bh}{2}$; $p = a + b + c$

Square: $A = s^2 = \frac{d^2}{2}$; $p = 4s$; $s = \frac{d}{\sqrt{2}}$; $d = s\sqrt{2}$

Rectangle: $A = ab$; $p = 2(a + b)$; $d = \sqrt{a^2 + b^2}$

Parallelogram: $A = bh = ba \sin \gamma$; $p = 2(a + b)$

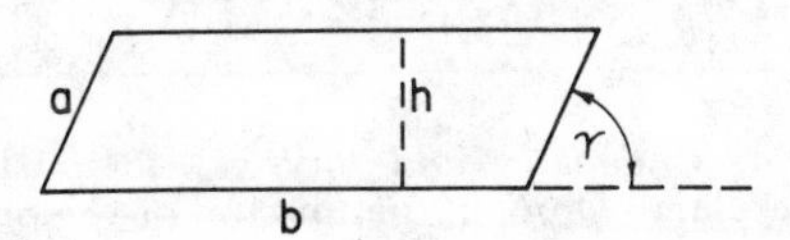

Trapezoid: $A = \frac{h(a + b)}{2}$; $p = a + b + c + d$

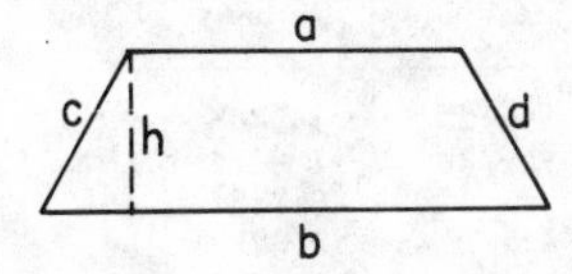

Circle: $A = \pi r^2$ or $0.7854\, d^2$; $C = 2\pi r$ or πd

Ellipse: $A = \pi ab$

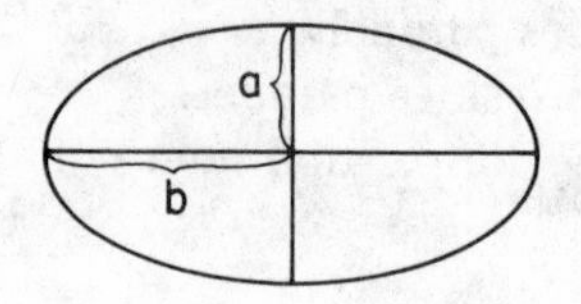

Sphere: $T = 4\pi r^2 = \pi d^2$; $V = \dfrac{4}{3}\pi r^3$ or $\dfrac{\pi d^3}{6}$

Cube: $V = s^3$; $T = 6s^2$

Rectangular Box: $V = lwh$; $T = ph + 2A$

Cylinder: $V = \pi r^2 h$; $T = 2\pi rh + 2\pi r^2$; $S = 2\pi rh$ or πdh

Cone: $V = \dfrac{\pi r^2 h}{3}$; $T = \pi rh + \pi r^2$; $S = \pi rh$

A.25 DECIMAL EQUIVALENTS OF COMMON FRACTIONS

		1/64 =	0.015 625		11/32	22/64 =	0.343 75			43/64 =	0.671 875
	1/32	2/64 =	.031 25			23/64 =	.359 375	11/16	22/32	44/64 =	.687 5
		3/64 =	.046 875	3/8	12/32	24/64 =	.375			45/64 =	.703 125
1/16	2/32	4/64 =	.062 5			25/64 =	.390 625		23/32	46/64 =	.718 75
		5/64 =	.078 125		13/32	26/64 =	.406 25			47/64 =	.734 375
	3/32	6/64 =	.093 75			27/64 =	.421 875	3/4	24/32	48/64 =	.75
		7/64 =	.109 375	7/16	14/32	28/64 =	.437 5			49/64 =	.765 625
1/8	4/32	8/64 =	.125			29/64 =	.453 125		25/32	50/64 =	.781 25
		9/64 =	.140 625		15/32	30/64 =	.468 75			51/64 =	.796 875
	5/32	10/64 =	.156 25			31/64 =	.484 375	13/16	26/32	52/64 =	.812 5
		11/64 =	.171 875	1/2	16/32	32/64 =	.50			53/64 =	.828 125
3/16	6/32	12/64 =	.187 5			33/64 =	.515 625		27/32	54/64 =	.843 75
		13/64 =	.203 125		17/32	34/64 =	.531 25			55/64 =	.859 375
	7/32	14/64 =	.218 75			35/64 =	.546 875	7/8	28/32	56/64 =	.875
		15/64 =	.234 375	9/16	18/32	36/64 =	.562 5			57/64 =	.890 625
1/4	8/32	16/64 =	.25			37/64 =	.578 125		29/32	58/64 =	.906 25
		17/64 =	.265 625		19/32	38/64 =	.593 75			59/64 =	.921 875
	9/32	18/64 =	.281 25			39/64 =	.609 375	15/16	30/32	60/64 =	.937 5
		19/64 =	.296 875	5/8	20/32	40/64 =	.625			61/64 =	.953 125
5/16	10/32	20/64 =	.312 5			41/64 =	.640 625		31/32	62/64 =	.968 75
		21/64 =	.328 125		21/32	42/64 =	.656 25			63/64 =	.984 375

Source: Handbook of Chemistry and Physics, 48th edition, 1967–1968, The Chemical Rubber Co., Cleveland, Ohio, by permission of the publishers.

A.26 RADIAN MEASURE

Multiples and Fractions of π Radians in Degrees

Radians	Radians	Deg.	Radians	Radians	Deg.	Radians	Radians	Deg.
π	3.1416	180	π/2	1.5708	90	2π/3	2.0944	120
2π	6.2832	360	π/3	1.0472	60	3π/4	2.3562	135
3π	9.4248	540	π/4	0.7854	45	5π/6	2.6180	150
4π	12.5664	720	π/5	0.6283	36	7π/6	3.6652	210
5π	15.7080	900	π/6	0.5236	30	5π/4	3.9270	225
6π	18.8496	1080	π/7	0.4488	25.714	4π/3	4.1888	240
7π	21.9911	1260	π/8	0.3927	22.5	3π/2	4.7124	270
8π	25.1327	1440	π/9	0.3491	20	5π/3	5.2360	300
9π	28.2743	1620	π/10	0.3142	18	7π/4	5.4978	315
10π	31.4159	1800	π/12	0.2618	15	11π/6	5.7596	330

A.26 *(continued)*

The table below facilitates conversion of an angle expressed in degrees and decimal fractions into radians. To convert 25.78 into radians, find the equivalents, successively, of 20°, 5°, 0°.7, 0°.08 and add.

Deg.	Radians	Deg.	Radians	Deg.	Radians	Deg.	Radians	Deg.	Radians
10	0.174533	1	0.017453	0.1	0.001745	0.01	0.000175	0.001	0.000017
20	0.349066	2	.034907	.2	.003491	.02	.000349	.002	.000035
30	0.523599	3	.052360	.3	.005236	.03	.000524	.003	.000052
40	0.698132	4	.069813	.4	.006981	.04	.000698	.004	.000070
50	0.872665	5	.087266	.5	.008727	.05	.000873	.005	.000087
60	1.047198	6	.104720	.6	.010472	.06	.001047	.006	.000105
70	1.221730	7	.122173	.7	.012217	.07	.001222	.007	.000122
80	1.396263	8	.139626	.8	.013963	.08	.001396	.008	.000140
90	1.570796	9	.157080	.9	.015708	.09	.001571	.009	.000157

Radians	Degrees	Radians	Degrees	Radians	Degrees	Radians	Degrees
1	57.2958	0.1	5.7296	0.01	0.5730	0.001	0.0573
2	114.5916	.2	11.4592	.02	1.1459	.002	.1146
3	171.8873	.3	17.1887	.03	1.7189	.003	.1719
4	229.1831	.4	22.9183	.04	2.2918	.004	.2292
5	286.4789	.5	28.6479	.05	2.8648	.005	.2865
6	343.7747	.6	34.3775	.06	3.4377	.006	.3438
7	401.0705	.7	40.1070	.07	4.0107	.007	.4011
8	458.3662	.8	45.8366	.08	4.5837	.008	.4584
9	515.6620	.9	51.5662	.09	5.1566	.009	.5157
10	572.9578	1.0	57.2958	.10	5.7296	.010	.5730

Source: Handbook of Chemistry and Physics, 48th edition, 1967–1968, The Chemical Rubber Co., Cleveland, Ohio, by permission of the publishers.

A.27 GRAPHS FOR REFERENCE

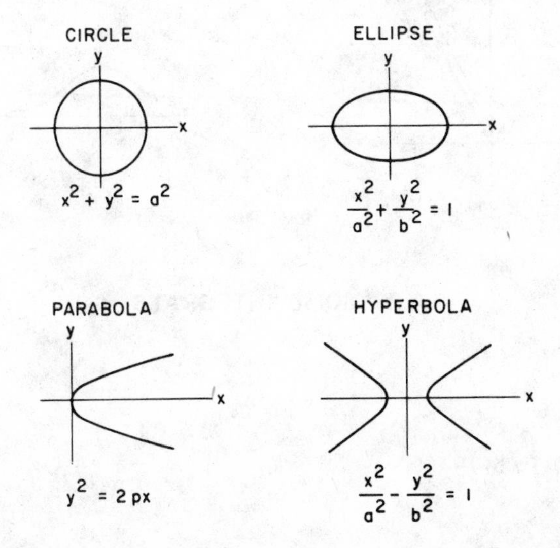

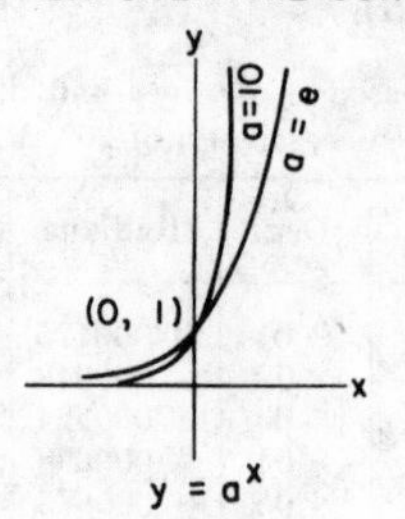

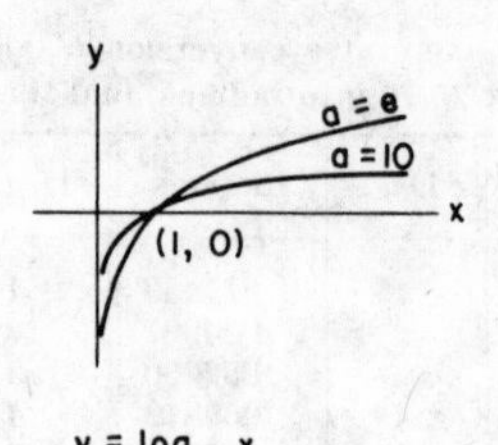

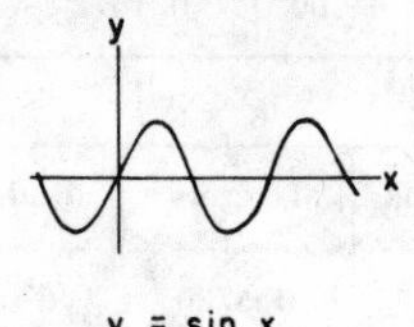

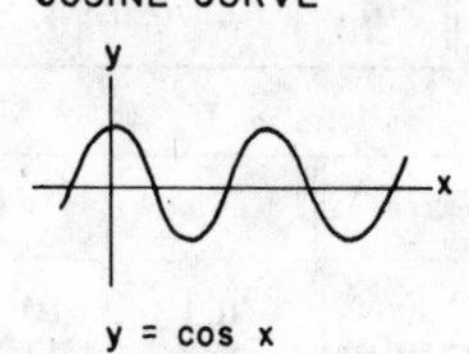

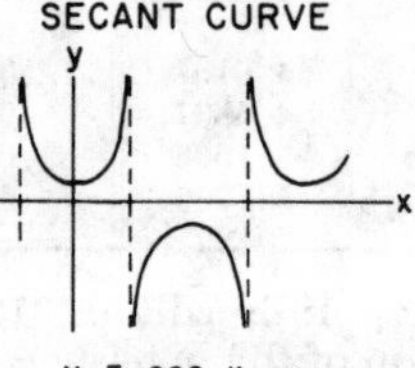

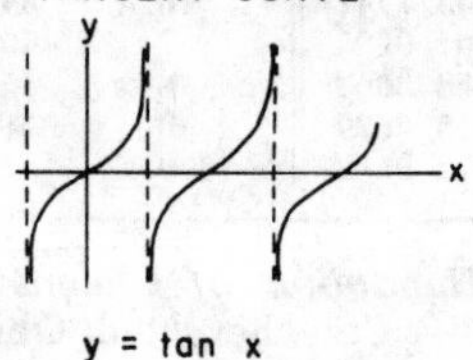

y

$y = 2x + 1$

$y = 2x$

x

y = 2x and y = 2x + 1

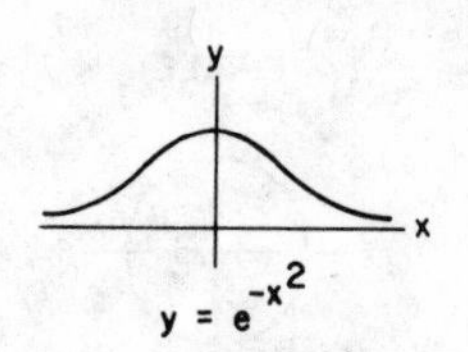

A.28 SOME INTEGRALS

$$\int a\, dx = ax$$

$$\int a \cdot f(x) dx = a \int f(x) dx$$

A.28 *(continued)*

$$\int \phi(y)\,dx = \int \frac{\phi(y)}{y'}\,dy, \qquad \text{where } y' = \frac{dy}{dx}$$

$$\int (u+v)\,dx = \int u\,dx + \int v\,dx, \text{ where } u \text{ and } v \text{ are any functions of } x$$

$$\int u\,dv = u\int dv - \int v\,du = uv - \int v\,du$$

$$\int u\frac{dv}{dx}\,dx = uv - \int v\frac{du}{dx}\,dx$$

$$\int x^n\,dx = \frac{x^{n+1}}{n+1}, \qquad \text{except } n = -1$$

$$\int \frac{f'(x)\,dx}{f(x)} = \log f(x), \qquad (df(x) = f'(x)\,dx)$$

$$\int \frac{dx}{x} = \log x$$

$$\int \frac{f'(x)\,dx}{2\sqrt{f(x)}} = \sqrt{f(x)}, \qquad (df(x) = f'(x)\,dx)$$

$$\int e^x\,dx = e^x$$

$$\int e^{ax}\,dx = e^{ax}/a$$

$$\int b^{ax}\,dx = \frac{b^{ax}}{a\log b}, (b > 0)$$

$$\int \log x\,dx = x\log x - x$$

$$\int a^x \log a\,dx = a^x, (a > 0)$$

$$\int \frac{dx}{a^2 + x^2} = \frac{1}{a}\tan^{-1}\frac{x}{a}$$

$$\int \log x\,dx = x\log x - x$$

A.28 (*continued*)

$$\int x \log x \, dx = \frac{x^2}{2} \log x - \frac{x^2}{4}$$

$$\int x^2 \log x \, dx = \frac{x^3}{3} \log x - \frac{x^3}{9}$$

$$\int \sin x \, dx = -\cos x$$

$$\int \cos x \, dx = \sin x$$

$$\int \tan x \, dx = -\log \cos x = \log \sec x$$

$$\int \cot x \, dx = \log \sin x = -\log \csc x$$

$$\int \sec x \, dx = \log (\sec x + \tan x) = \log \tan \left(\frac{\pi}{4} + \frac{x}{2}\right)$$

$$\int \csc x \, dx = \log (\csc x - \cot x) = \log \tan \frac{x}{2}$$

$$\int \sin^2 x \, dx = -\frac{1}{2} \cos x \sin x + \frac{1}{2} x = \frac{1}{2} x - \frac{1}{4} \sin 2x$$

$$\int \sin^3 x \, dx = -\frac{1}{3} \cos x (\sin^2 x + 2)$$

$$\int \sin^n x \, dx = -\frac{\sin^{n-1} x \cos x}{n} + \frac{n-1}{n} \int \sin^{n-2} x \, dx$$

$$\int \cos^2 x \, dx = \frac{1}{2} \sin x \cos x + \frac{1}{2} x = \frac{1}{2} x + \frac{1}{4} \sin 2x$$

$$\int \cos^3 x \, dx = \frac{1}{3} \sin x (\cos^2 x + 2)$$

$$\int \cos^n x \, dx = \frac{1}{n} \cos^{n-1} x \sin x + \frac{n-1}{n} \int \cos^{n-2} x \, dx$$

$$\int \sin \frac{x}{a} \, dx = -a \cos \frac{x}{a}$$

A.28 *(continued)*

$$\int \cos \frac{x}{a}\, dx = a \sin \frac{x}{a}$$

Source: Handbook of Chemistry and Physics, 48th edition, 1967–1968, The Chemical Rubber Co., Cleveland, Ohio, by permission of the publishers.

A.29 SOME TRIGONOMETRIC IDENTITIES

Where A, B, C denote the side and a, b, c the corresponding opposite angles:

Addition Identities

$$\cos (a + b) = \cos a \cos b - \sin a \sin b$$
$$\cos (a - b) = \cos a \cos b + \sin a \sin b$$
$$\sin (a + b) = \sin a \cos b + \cos a \sin b$$
$$\sin (a - b) = \sin a \cos b - \cos a \sin b$$

$$\tan (a + b) = \frac{\tan a + \tan b}{1 - \tan a \tan b} \qquad \tan (a - b) = \frac{\tan a - \tan b}{1 + \tan a \tan b}$$

Sum and Product Identities

$$2 \sin a \cos b = \sin (a + b) + \sin (a - b)$$
$$2 \cos a \cos b = \cos (a + b) + \cos (a - b)$$
$$-2 \sin a \sin b = \cos (a + b) - \cos (a - b)$$

$$\sin A + \sin B = 2 \sin \frac{A + B}{2} \cos \frac{A - B}{2}$$

$$\sin A - \sin B = 2 \cos \frac{A + B}{2} \sin \frac{A - B}{2}$$

$$\cos A + \cos B = 2 \cos \frac{A + B}{2} \cos \frac{A - B}{2}$$

$$\cos A - \cos B = -2 \sin \frac{A + B}{2} \sin \frac{A - B}{2}$$

Relations Between Sides and Angles

$$a + b + c = 180^\circ$$

Law of sines $\dfrac{A}{\sin a} = \dfrac{B}{\sin b} = \dfrac{C}{\sin c}$

Law of cosines $A^2 = B^2 + C^2 - 2\,BC \cos a \qquad A = B \cos c + C \cos b$

Law of tangents $\dfrac{A + B}{A - B} = \dfrac{\tan 1/2\,(a + b)}{\tan 1/2\,(a - b)}$

A.30 VALUES OF TRIGONOMETRIC FUNCTIONS

Angle	*Sin*	*Cos*	*Tan*	*Cot*	*Sec*	*Csc*	
0° 00′	.0000	1.0000	.0000	- - - -	1.000	- - - -	**90° 00′**
10′	.0029	1.0000	.0029	343.8	1.000	343.8	**50′**
20′	.0058	1.0000	.0058	171.9	1.000	171.9	**40′**
30′	.0087	1.0000	.0087	114.6	1.000	114.6	**30′**
40′	.0116	.9999	.0116	85.94	1.000	85.95	**20′**
50′	.0145	.9999	.0145	68.75	1.000	68.76	**10′**
1° 00′	.0175	.9998	.0175	57.29	1.000	57.30	**89° 00′**
10′	.0204	.9998	.0204	49.10	1.000	49.11	**50′**
20′	.0233	.9997	.0233	42.96	1.000	42.98	**40′**
30′	.0262	.9997	.0262	38.19	1.000	38.20	**30′**
40′	.0291	.9996	.0291	34.37	1.000	34.38	**20′**
50′	.0320	.9995	.0320	31.24	1.001	31.26	**10′**
2° 00′	.0349	.9994	.0349	28.64	1.001	28.65	**88° 00′**
10′	.0378	.9993	.0378	26.43	1.001	26.45	**50′**
20′	.0407	.9992	.0407	24.54	1.001	24.56	**40′**
30′	.0436	.9990	.0437	22.90	1.001	22.93	**30′**
40′	.0465	.9989	.0466	21.47	1.001	21.49	**20′**
50′	.0494	.9988	.0495	20.21	1.001	20.23	**10′**
3° 00′	.0523	.9986	.0524	19.08	1.001	19.11	**87° 00′**
10′	.0552	.9985	.0553	18.07	1.002	18.10	**50′**
20′	.0581	.9983	.0582	17.17	1.002	17.20	**40′**
30′	.0610	.9981	.0612	16.35	1.002	16.38	**30′**
40′	.0640	.9980	.0641	15.60	1.002	15.64	**20′**
50′	.0669	.9978	.0670	14.92	1.002	14.96	**10′**
4° 00′	.0698	.9976	.0699	14.30	1.002	14.34	**86° 00′**
10′	.0727	.9974	.0729	13.73	1.003	13.76	**50′**
20′	.0756	.9971	.0758	13.20	1.003	13.23	**40′**
30′	.0785	.9969	.0787	12.71	1.003	12.75	**30′**
40′	.0814	.9967	.0816	12.25	1.003	12.29	**20′**
50′	.0843	.9964	.0846	11.83	1.004	11.87	**10′**
5° 00′	.0872	.9962	.0875	11.43	1.004	11.47	**85° 00′**
10′	.0901	.9959	.0904	11.06	1.004	11.10	**50′**
20′	.0929	.9957	.0934	10.71	1.004	10.76	**40′**
30′	.0958	.9954	.0963	10.39	1.005	10.43	**30′**
40′	.0987	.9951	.0992	10.08	1.005	10.13	**20′**
50′	.1016	.9948	.1022	9.788	1.005	9.839	**10′**
6° 00′	.1045	.9945	.1051	9.514	1.006	9.567	**84° 00′**
10′	.1074	.9942	.1080	9.255	1.006	9.309	**50′**
20′	.1103	.9939	.1110	9.010	1.006	9.065	**40′**
30′	.1132	.9936	.1139	8.777	1.006	8.834	**30′**
40′	.1161	.9932	.1169	8.556	1.007	8.614	**20′**
50′	.1190	.9929	.1198	8.345	1.007	8.405	**10′**
7° 00′	.1219	.9925	.1228	8.144	1.008	8.206	**83° 00′**
10′	.1248	.9922	.1257	7.953	1.008	8.016	**50′**
20′	.1276	.9918	.1287	7.770	1.008	7.834	**40′**
30′	.1305	.9914	.1317	7.596	1.009	7.661	**30′**
40′	.1334	.9911	.1346	7.429	1.009	7.496	**20′**
50′	.1363	.9907	.1376	7.269	1.009	7.337	**10′**
8° 00′	.1392	.9903	.1405	7.115	1.010	7.185	**82° 00′**
10′	.1421	.9899	.1435	6.968	1.010	7.040	**50′**
20′	.1449	.9894	.1465	6.827	1.011	6.900	**40′**
30′	.1478	.9890	.1495	6.691	1.011	6.765	**30′**
40′	.1507	.9886	.1524	6.561	1.012	6.636	**20′**
50′	.1536	.9881	.1554	6.435	1.012	6.512	**10′**
9° 00′	.1564	.9877	.1584	6.314	1.012	6.392	**81° 00′**
	Cos	*Sin*	*Cot*	*Tan*	*Csc*	*Sec*	*Angle*

A.30 (*continued*)

Angle	Sin	Cos	Tan	Cot	Sec	Csc	
9° 00′	.1564	.9877	.1584	6.314	1.012	6.392	**81° 00′**
10′	.1593	.9872	.1614	6.197	1.013	6.277	**50′**
20′	.1622	.9868	.1644	6.084	1.013	6.166	**40′**
30′	.1650	.9863	.1673	5.976	1.014	6.059	**30′**
40′	.1679	.9858	.1703	5.871	1.014	5.955	**20′**
50′	.1708	.9853	.1733	5.769	1.015	5.855	**10′**
10° 00′	.1736	.9848	.1763	5.671	1.015	5.759	**80° 00′**
10′	.1765	.9843	.1793	5.576	1.016	5.665	**50′**
20′	.1794	.9838	.1823	5.485	1.016	5.575	**40′**
30′	.1822	.9833	.1853	5.396	1.017	5.487	**30′**
40′	.1851	.9827	.1883	5.309	1.018	5.403	**20′**
50′	.1880	.9822	.1914	5.226	1.018	5.320	**10′**
11° 00′	.1908	.9816	.1944	5.145	1.019	5.241	**79° 00′**
10′	.1937	.9811	.1974	5.066	1.019	5.164	**50′**
20′	.1965	.9805	.2004	4.989	1.020	5.089	**40′**
30′	.1994	.9799	.2035	4.915	1.020	5.016	**30′**
40′	.2022	.9793	.2065	4.843	1.021	4.945	**20′**
50′	.2051	.9787	.2095	4.773	1.022	4.876	**10′**
12° 00′	.2079	.9781	.2126	4.705	1.022	4.810	**78° 00′**
10′	.2108	.9775	.2156	4.638	1.023	4.745	**50′**
20′	.2136	.9769	.2186	4.574	1.024	4.682	**40′**
30′	.2164	.9763	.2217	4.511	1.024	4.620	**30′**
40′	.2193	.9757	.2247	4.449	1.025	4.560	**20′**
50′	.2221	.9750	.2278	4.390	1.026	4.502	**10′**
13° 00′	.2250	.9744	.2309	4.331	1.026	4.445	**77° 00′**
10′	.2278	.9737	.2339	4.275	1.027	4.390	**50′**
20′	.2306	.9730	.2370	4.219	1.028	4.336	**40′**
30′	.2334	.9724	.2401	4.165	1.028	4.284	**30′**
40′	.2363	.9717	.2432	4.113	1.029	4.232	**20′**
50′	.2391	.9710	.2462	4.061	1.030	4.182	**10′**
14° 00′	.2419	.9703	.2493	4.011	1.031	4.134	**76° 00′**
10′	.2447	.9696	.2524	3.962	1.031	4.086	**50′**
20′	.2476	.9689	.2555	3.914	1.032	4.039	**40′**
30′	.2504	.9681	.2586	3.867	1.033	3.994	**30′**
40′	.2532	.9674	.2617	3.821	1.034	3.950	**20′**
50′	.2560	.9667	.2648	3.776	1.034	3.906	**10′**
15° 00′	.2588	.9659	.2679	3.732	1.035	3.864	**75° 00′**
10′	.2616	.9652	.2711	3.689	1.036	3.822	**50′**
20′	.2644	.9644	.2742	3.647	1.037	3.782	**40′**
30′	.2672	.9636	.2773	3.606	1.038	3.742	**30′**
40′	.2700	.9628	.2805	3.566	1.039	3.703	**20′**
50′	.2728	.9621	.2836	3.526	1.039	3.665	**10′**
16° 00′	.2756	.9613	.2867	3.487	1.040	3.628	**74° 00′**
10′	.2784	.9605	.2899	3.450	1.041	3.592	**50′**
20′	.2812	.9596	.2931	3.412	1.042	3.556	**40′**
30′	.2840	.9588	.2962	3.376	1.043	3.521	**30′**
40′	.2868	.9580	.2994	3.340	1.044	3.487	**20′**
50′	.2896	.9572	.3026	3.305	1.045	3.453	**10′**
17° 00′	.2924	.9563	.3057	3.271	1.046	3.420	**73° 00′**
10′	.2952	.9555	.3089	3.237	1.047	3.388	**50′**
20′	.2979	.9546	.3121	3.204	1.048	3.356	**40′**
30′	.3007	.9537	.3153	3.172	1.049	3.326	**30′**
40′	.3035	.9528	.3185	3.140	1.049	3.295	**20′**
50′	.3062	.9520	.3217	3.108	1.050	3.265	**10′**
18° 00′	.3090	.9511	.3249	3.078	1.051	3.236	**72° 00′**
	Cos	Sin	Cot	Tan	Csc	Sec	Angle

A.30 (*continued*)

Angle	*Sin*	*Cos*	*Tan*	*Cot*	*Sec*	*Csc*	
18° 00′	.3090	.9511	.3249	3.078	1.051	3.236	**72° 00′**
10′	.3118	.9502	.3281	3.047	1.052	3.207	**50′**
20′	.3145	.9492	.3314	3.018	1.053	3.179	**40′**
30′	.3173	.9483	.3346	2.989	1.054	3.152	**30′**
40′	.3201	.9474	.3378	2.960	1.056	3.124	**20′**
50′	.3228	.9465	.3411	2.932	1.057	3.098	**10′**
19° 00′	.3256	.9455	.3443	2.904	1.058	3.072	**71° 00′**
10′	.3283	.9446	.3476	2.877	1.059	3.046	**50′**
20′	.3311	.9436	.3508	2.850	1.060	3.021	**40′**
30′	.3338	.9426	.3541	2.824	1.061	2.996	**30′**
40′	.3365	.9417	.3574	2.798	1.062	2.971	**20′**
50′	.3393	.9407	.3607	2.773	1.063	2.947	**10′**
20° 00′	.3420	.9397	.3640	2.747	1.064	2.924	**70° 00′**
10′	.3448	.9387	.3673	2.723	1.065	2.901	**50′**
20′	.3475	.9377	.3706	2.699	1.066	2.878	**40′**
30′	.3502	.9367	.3739	2.675	1.068	2.855	**30′**
40′	.3529	.9356	.3772	2.651	1.069	2.833	**20′**
50′	.3557	.9346	.3805	2.628	1.070	2.812	**10′**
21° 00′	.3584	.9336	.3839	2.605	1.071	2.790	**69° 00′**
10′	.3611	.9325	.3872	2.583	1.072	2.769	**50′**
20′	.3638	.9315	.3906	2.560	1.074	2.749	**40′**
30′	.3665	.9304	.3939	2.539	1.075	2.729	**30′**
40′	.3692	.9293	.3973	2.517	1.076	2.709	**20′**
50′	.3719	.9283	.4006	2.496	1.077	2.689	**10′**
22° 00′	.3746	.9272	.4040	2.475	1.079	2.669	**68° 00′**
10′	.3773	.9261	.4074	2.455	1.080	2.650	**50′**
20′	.3800	.9250	.4108	2.434	1.081	2.632	**40′**
30′	.3827	.9239	.4142	2.414	1.082	2.613	**30′**
40′	.3854	.9228	.4176	2.394	1.084	2.595	**20′**
50′	.3881	.9216	.4210	2.375	1.085	2.577	**10′**
23° 00′	.3907	.9205	.4245	2.356	1.086	2.559	**67° 00′**
10′	.3934	.9194	.4279	2.337	1.088	2.542	**50′**
20′	.3961	.9182	.4314	2.318	1.089	.2525	**40′**
30′	.3987	.9171	.4348	2.300	1.090	.2508	**30′**
40′	.4014	.9159	.4383	2.282	1.092	.2491	**20′**
50′	.4041	.9147	.4417	2.264	1.093	.2475	**10′**
24° 00′	.4067	.9135	.4452	2.246	1.095	2.459	**66° 00′**
10′	.4094	.9124	.4487	2.229	1.096	2.443	**50′**
20′	.4120	.9112	.4522	2.211	1.097	2.427	**40′**
30′	.4147	.9100	.4557	2.194	1.099	2.411	**30′**
40′	.4173	.9088	.4592	2.177	1.100	2.396	**20′**
50′	.4200	.9075	.4628	2.161	1.102	2.381	**10′**
25° 00′	.4226	.9063	.4663	2.145	1.103	2.366	**65° 00′**
10′	.4253	.9051	.4699	2.128	1.105	2.352	**50′**
20′	.4279	.9038	.4734	2.112	1.106	2.337	**40′**
30′	.4305	.9026	.4770	2.097	1.108	2.323	**30′**
40′	.4331	.9013	.4806	2.081	1.109	2.309	**20′**
50′	.4358	.9001	.4841	2.066	1.111	2.295	**10′**
26° 00′	.4384	.8988	.4877	2.050	1.113	2.281	**64° 00′**
10′	.4410	.8975	.4913	2.035	1.114	2.268	**50′**
20′	.4436	.8962	.4950	2.020	1.116	2.254	**40′**
30′	.4462	.8949	.4986	2.006	1.117	2.241	**30′**
40′	.4488	.8936	.5022	1.991	1.119	2.228	**20′**
50′	.4514	.8923	.5059	1.977	1.121	2.215	**10′**
27° 00′	.4540	.8910	.5095	1.963	1.122	2.203	**63° 00′**
	Cos	*Sin*	*Cot*	*Tan*	*Csc*	*Sec*	*Angle*

A.30 (*continued*)

Angle	Sin	Cos	Tan	Cot	Sec	Csc	
27° 00′	.4540	.8910	.5095	1.963	1.122	2.203	**63° 00′**
10′	.4566	.8897	.5132	1.949	1.124	2.190	**50′**
20′	.4592	.8884	.5169	1.935	1.126	2.178	**40′**
30′	.4617	.8870	.5206	1.921	1.127	2.166	**30′**
40′	.4643	.8857	.5243	1.907	1.129	2.154	**20′**
50′	.4669	.8843	.5280	1.894	1.131	2.142	**10′**
28° 00′	.4695	.8829	.5317	1.881	1.133	2.130	**62° 00′**
10′	.4720	.8816	.5354	1.868	1.134	2.118	**50′**
20′	.4746	.8802	.5392	1.855	1.136	2.107	**40′**
30′	.4772	.8788	.5430	1.842	1.138	2.096	**30′**
40′	.4797	.8774	.5467	1.829	1.140	2.085	**20′**
50′	.4823	.8760	.5505	1.816	1.142	2.074	**10′**
29° 00′	.4848	.8746	.5543	1.804	1.143	2.063	**61° 00′**
10′	.4874	.8732	.5581	1.792	1.145	2.052	**50′**
20′	.4899	.8718	.5619	1.780	1.147	2.041	**40′**
30′	.4924	.8704	.5658	1.767	1.149	2.031	**30′**
40′	.4950	.8689	.5696	1.756	1.151	2.020	**20′**
50′	.4975	.8675	.5735	1.744	1.153	2.010	**10′**
30° 00′	.5000	.8660	.5774	1.732	1.155	2.000	**60° 00′**
10′	.5025	.8646	.5812	1.720	1.157	1.990	**50′**
20′	.5050	.8631	.5851	1.709	1.159	1.980	**40′**
30′	.5075	.8616	.5890	1.698	1.161	1.970	**30′**
40′	.5100	.8601	.5930	1.686	1.163	1.961	**20′**
50′	.5125	.8587	.5969	1.675	1.165	1.951	**10′**
31° 00′	.5150	.8572	.6009	1.664	1.167	1.942	**59° 00′**
10′	.5175	.8557	.6048	1.653	1.169	1.932	**50′**
20′	.5200	.8542	.6088	1.643	1.171	1.923	**40′**
30′	.5225	.8526	.6128	1.632	1.173	1.914	**30′**
40′	.5250	.8511	.6168	1.621	1.175	1.905	**20′**
50′	.5275	.8496	.6208	1.611	1.177	1.896	**10′**
32° 00′	.5299	.8480	.6249	1.600	1.179	1.887	**58° 00′**
10′	.5324	.8465	.6289	1.590	1.181	1.878	**50′**
20′	.5348	.8450	.6330	1.580	1.184	1.870	**40′**
30′	.5373	.8434	.6371	1.570	1.186	1.861	**30′**
40′	.5398	.8418	.6412	1.560	1.188	1.853	**20′**
50′	.5422	.8403	.6453	1.550	1.190	1.844	**10′**
33° 00′	.5446	.8387	.6494	1.540	1.192	1.836	**57° 00′**
10′	.5471	.8371	.6536	1.530	1.195	1.828	**50′**
20′	.5495	.8355	.6577	1.520	1.197	1.820	**40′**
30′	.5519	.8339	.6619	1.511	1.199	1.812	**30′**
40′	.5544	.8323	.6661	1.501	1.202	1.804	**20′**
50′	.5568	.8307	.6703	1.492	1.204	1.796	**10′**
34° 00′	.5592	.8290	.6745	1.483	1.206	1.788	**56° 00′**
10′	.5616	.8274	.6787	1.473	1.209	1.781	**50′**
20′	.5640	.8258	.6830	1.464	1.211	1.773	**40′**
30′	.5664	.8241	.6873	1.455	1.213	1.766	**30′**
40′	.5688	.8225	.6916	1.446	1.216	1.758	**20′**
50′	.5712	.8208	.6959	1.437	1.218	1.751	**10′**
35° 00′	.5736	.8192	.7002	1.428	1.221	1.743	**55° 00′**
10′	.5760	.8175	.7046	1.419	1.223	1.736	**50′**
20′	.5783	.8158	.7089	1.411	1.226	1.729	**40′**
30′	.5807	.8141	.7133	1.402	1.228	1.722	**30′**
40′	.5831	.8124	.7177	1.393	1.231	1.715	**20′**
50′	.5854	.8107	.7221	1.385	1.233	1.708	**10′**
36° 00′	.5878	.8090	.7265	1.376	1.236	1.701	**54° 00′**
	Cos	Sin	Cot	Tan	Csc	Sec	Angle

A.30 *(continued)*

Angle	Sin	Cos	Tan	Cot	Sec	Csc	
36° 00′	.5878	.8090	.7265	1.376	1.236	1.701	**54° 00′**
10′	.5901	.8073	.7310	1.368	1.239	1.695	**50′**
20′	.5925	.8056	.7355	1.360	1.241	1.688	**40′**
30′	.5948	.8039	.7400	1.351	1.244	1.681	**30′**
40′	.5972	.8021	.7445	1.343	1.247	1.675	**20′**
50′	.5995	.8004	.7490	1.335	1.249	1.668	**10′**
37° 00′	.6018	.7986	.7536	1.327	1.252	1.662	**53° 00′**
10′	.6041	.7969	.7581	1.319	1.255	1.655	**50′**
20′	.6065	.7951	.7627	1.311	1.258	1.649	**40′**
30′	.6088	.7934	.7673	1.303	1.260	1.643	**30′**
40′	.6111	.7916	.7720	1.295	1.263	1.636	**20′**
50′	.6134	.7898	.7766	1.288	1.266	1.630	**10′**
38° 00′	.6157	.7880	.7813	1.280	1.269	1.624	**52° 00′**
10′	.6180	.7862	.7860	1.272	1.272	1.618	**50′**
20′	.6202	.7844	.7907	1.265	1.275	1.612	**40′**
30′	.6225	.7826	.7954	1.257	1.278	1.606	**30′**
40′	.6248	.7808	.8002	1.250	1.281	1.601	**20′**
50′	.6271	.7790	.8050	1.242	1.284	1.595	**10′**
39° 00′	.6293	.7771	.8098	1.235	1.287	1.589	**51° 00′**
10′	.6316	.7753	.8146	1.228	1.290	1.583	**50′**
20′	.6338	.7735	.8195	1.220	1.293	1.578	**40′**
30′	.6361	.7716	.8243	1.213	1.296	1.572	**30′**
40′	.6383	.7698	.8292	1.206	1.299	1.567	**20′**
50′	.6406	.7679	.8342	1.199	1.302	1.561	**10′**
40° 00′	.6428	.7660	.8391	1.192	1.305	1.556	**50° 00′**
10′	.6450	.7642	.8441	1.185	1.309	1.550	**50′**
20′	.6472	.7623	.8491	1.178	1.312	1.545	**40′**
30′	.6494	.7604	.8541	1.171	1.315	1.540	**30′**
40′	.6517	.7585	.8591	1.164	1.318	1.535	**20′**
50′	.6539	.7566	.8642	1.157	1.322	1.529	**10′**
41° 00′	.6561	.7547	.8693	1.150	1.325	1.524	**49° 00′**
10′	.6583	.7528	.8744	1.144	1.328	1.519	**50′**
20′	.6604	.7509	.8796	1.137	1.332	1.514	**40′**
30′	.6626	.7490	.8847	1.130	1.335	1.509	**30′**
40′	.6648	.7470	.8899	1.124	1.339	1.504	**20′**
50′	.6670	.7451	.8952	1.117	1.342	1.499	**10′**
42° 00′	.6691	.7431	.9004	1.111	1.346	1.494	**48° 00′**
10′	.6713	.7412	.9057	1.104	1.349	1.490	**50′**
20′	.6734	.7392	.9110	1.098	1.353	1.485	**40′**
30′	.6756	.7373	.9163	1.091	1.356	1.480	**30′**
40′	.6777	.7353	.9217	1.085	1.360	1.476	**20′**
50′	.6799	.7333	.9271	1.079	1.364	1.471	**10′**
43° 00′	.6820	.7314	.9325	1.072	1.367	1.466	**47° 00′**
10′	.6841	.7294	.9380	1.066	1.371	1.462	**50′**
20′	.6862	.7274	.9435	1.060	1.375	1.457	**40′**
30′	.6884	.7254	.9490	1.054	1.379	1.453	**30′**
40′	.6905	.7234	.9545	1.048	1.382	1.448	**20′**
50′	.6926	.7214	.9601	1.042	1.386	1.444	**10′**
44° 00′	.6947	.7193	.9657	1.036	1.390	1.440	**46° 00′**
10′	.6967	.7173	.9713	1.030	1.394	1.435	**50′**
20′	.6988	.7153	.9770	1.024	1.398	1.431	**40′**
30′	.7009	.7133	.9827	1.018	1.402	1.427	**30′**
40′	.7030	.7112	.9884	1.012	1.406	1.423	**20′**
50′	.7050	.7092	.9942	1.006	1.410	1.418	**10′**
45° 00′	.7071	.7071	1.000	1.000	1.414	1.414	**45° 00′**
	Cos	*Sin*	*Cot*	*Tan*	*Csc*	*Sec*	*Angle*

Source: From *Modern Algebra and Trigonometry, Structure and Method*, Book 2, by Dolciani, Berman, and Wooton, copyright 1965 by Houghton Mifflin Company. Reprinted by permission.

A.31 ATOMIC WEIGHTS

For the sake of completeness all known elements are included in the list. Several of those more recently discovered are represented only by the unstable isotopes. The value in parenthesis in the atomic weight column is, in each case, the mass number of the most stable isotope.**

Name	Symbol	At. No.	International atomic weight 1966	International atomic weight 1959	Valence
Actinium	Ac	89	....	(227)	....
Aluminum	Al	13	26.9815	26.98	3
Americium	Am	95	....	(243)	3, 4, 5, 6
Antimony, stibium	Sb	51	121.75	121.76	3, 5
Argon	Ar	18	39.948	39.944	0
Arsenic	As	33	74.9216	74.92	3, 5
Astatine	At	85	....	(210)	1, 3, 5, 7
Barium	Ba	56	137.34	137.36	2
Berkelium	Bk	97	....	(247)	3, 4
Beryllium	Be	4	9.0122	9.013	2
Bismuth	Bi	83	208.980	208.99	3, 5
Boron	B	5	10.811	10.82	3
Bromine	Br	35	79.904[1]	79.916	1, 3, 5, 7
Cadmium	Cd	48	112.40	112.41	2
Calcium	Ca	20	40.08	40.08	2
Californium	Cf	98	....	(251)	....
Carbon	C	6	12.01115	12.011	2, 4
Cerium	Ce	58	140.12	140.13	3, 4
Cesium	Cs	55	132.905	132.91	1
Chlorine	Cl	17	35.453	35.457	1, 3, 5, 7
Chromium	Cr	24	51.996	52.01	2, 3, 6
Cobalt	Co	27	58.9332	58.94	2, 3
Columbium, see *Niobium*					
Copper	Cu	29	63.546[2]	63.54	1, 2
Curium	Cm	96	....	(247)	3
Dysprosium	Dy	66	162.50	162.51	3
Einsteinium	Es	99	....	(254)	....
Erbium	Er	68	167.26	167.27	3
Europium	Eu	63	151.96	152.0	2, 3
Fermium	Fm	100	....	(257)	....
Fluorine	F	9	18.9984	19.00	1
Francium	Fr	87	....	(223)	1
Gadolinium	Gd	64	157.25	157.26	3
Gallium	Ga	31	69.72	69.72	2, 3
Germanium	Ge	32	72.59	72.60	4
Gold, aurum	Au	79	196.967	197.0	1, 3
Hafnium	Hf	72	178.49	178.50	4
Helium	He	2	4.0026	4.003	0
Holmium	Ho	67	164.930	164.94	3
Hydrogen	H	1	1.00797	1.0080	1
Indium	In	49	114.82	114.82	3
Iodine	I	53	126.9044	126.91	1, 3, 5, 7
Iridium	Ir	77	192.2	192.2	3, 4
Iron, ferrum	Fe	26	55.847	55.85	2, 3
Krypton	Kr	36	83.80	83.80	0
Lanthanum	La	57	138.91	138.92	3
Lead, plumbum	Pb	82	207.19	207.21	2, 4
Lithium	Li	3	6.939	6.940	1
Lutetium	Lu	71	174.97	174.99	3
Magnesium	Mg	12	24.312	24.32	2
Manganese	Mn	25	54.9380	54.94	2, 3, 4, 6, 7
Mendelevium	Md	101	....	(256)	....
Mercury, hydrargyrum	Hg	80	200.59	200.61	1, 2
Molybdenum	Mo	42	95.94	95.95	3, 4, 6
Neodymium	Nd	60	144.24	144.27	3
Neon	Ne	10	20.183	20.183	0
Neptunium	Np	93	....	(237)	4, 5, 6
Nickel	Ni	28	58.71	58.71	2, 3
Niobium (columbium)	Nb	41	92.906	92.91	3, 5
Nitrogen	N	7	14.0067	14.008	3, 5
Nobelium	No	102	....	(254)	....
Osmium	Os	76	190.2	190.2	2, 3, 4, 8
Oxygen	O	8	15.9994	16.000	2
Palladium	Pd	46	106.4	106.4	2, 4, 6
Phosphorus	P	15	30.9738	30.975	3, 5
Platinum	Pt	78	195.09	195.09	2, 4
Plutonium	Pu	94	....	(244)	3, 4, 5, 6
Polonium	Po	84	....	(209)	....
Potassium, kalium	K	19	39.102	39.100	1
Praseodymium	Pr	59	140.907	140.92	3
Promethium	Pm	61	....	(145)	3
Protactinium	Pa	91	....	(231)	....
Radium	Ra	88	....	(226)	2
Radon	Rn	86	....	(222)	0
Rhenium	Re	75	186.2	186.22	....
Rhodium	Rh	45	102.905	102.91	3
Rubidium	Rb	37	85.47	85.48	1
Ruthenium	Ru	44	101.07	101.1	3, 4, 6, 8
Samarium	Sm	62	150.35	150.35	2, 3
Scandium	Sc	21	44.956	44.96	3
Selenium	Se	34	78.96	78.96	2, 4, 6
Silicon	Si	14	28.086	28.09	4
Silver, argentum	Ag	47	107.868[3]	107.873	1
Sodium, natrium	Na	11	22.9898	22.991	1
Strontium	Sr	38	87.62	87.63	2
Sulfur	S	16	32.064	32.066*	2, 4, 6
Tantalum	Ta	73	180.948	180.95	5
Technetium	Tc	43	....	(97)	6, 7
Tellurium	Te	52	127.60	127.61	2, 4, 6
Terbium	Tb	65	158.924	158.93	3
Thallium	Tl	81	204.37	204.39	1, 3
Thorium	Th	90	232.038	(232)	4
Thulium	Tm	69	168.934	168.94	3
Tin, stannum	Sn	50	118.69	118.70	2, 4
Titanium	Ti	22	47.90	47.90	3, 4
Tungsten (wolfram)	W	74	183.85	183.86	6
Uranium	U	92	238.03	238.07	4, 6
Vanadium	V	23	50.942	50.95	3, 5
Xenon	Xe	54	131.30	131.30	0
Ytterbium	Yb	70	173.04	173.04	2, 3
Yttrium	Y	39	88.905	88.91	3
Zinc	Zn	30	65.37	65.38	2
Zirconium	Zr	40	91.22	91.22	4

1. ±0.002
2. ±0.001
3. ±0.001

Source: Handbook of Chemistry and Physics, 48th edition, 1967–1968, The Chemical Rubber Co., Cleveland, Ohio, by permission of the publishers.

*Because of natural variations in the relative abundances of the isotopes of sulfur, the atomic weight of this element has a range of ±0.003.

**The 1959 atomic weights are based on 0 = 16.000 (all other atomic weights are multiples of these same weights) whereas those of 1966 are based on the isotope C^{12}, i.e., based on the value 12 for the relative atomic mass of the carbon isotope C^{12}.

A.32 PERIODIC TABLE OF THE ELEMENTS

KEY TO CHART: Atomic Number → 50 ; Symbol → Sn ; Atomic Weight → 118.69 ; +2 +4 ← Oxidation States ; -18-18-4 ← Electron Configuration

1a	2a	3b	4b	5b	6b	7b	8	8	8	1b	2b	3a	4a	5a	6a	7a	0	Orbit
1 H +1 −1 1.00797 1																	2 He 0 4.0026 2	K
3 Li +1 6.939 2-1	4 Be +2 9.0122 2-2											5 B +3 10.811 2-3	6 C +2 +4 −4 12.01115 2-4	7 N +1 +2 +3 +4 +5 −1 −2 −3 14.0067 2-5	8 O −2 15.9994 2-6	9 F −1 18.9984 2-7	10 Ne 0 20.183 2-8	K-L
11 Na +1 22.9898 2-8-1	12 Mg +2 24.312 2-8-2	Transition Elements					Group 8			Transition Elements		13 Al +3 26.9815 2-8-3	14 Si +2 +4 −4 28.086 2-8-4	15 P +3 +5 −3 30.9738 2-8-5	16 S +4 +6 −2 32.064 2-8-6	17 Cl +1 +5 +7 −1 35.453 2-8-7	18 Ar 0 39.948 2-8-8	K-L-M
19 K +1 39.102 -8-8-1	20 Ca +2 40.08 -8-8-2	21 Sc +3 44.956 -8-9-2	22 Ti +2 +3 +4 47.90 -8-10-2	23 V +2 +3 +4 +5 50.942 -8-11-2	24 Cr +2 +3 +6 51.996 -8-13-1	25 Mn +2 +3 +4 +7 54.9380 -8-13-2	26 Fe +2 +3 55.847 -8-14-2	27 Co +2 +3 58.9332 -8-15-2	28 Ni +2 +3 58.71 -8-16-2	29 Cu +1 +2 63.546 -8-18-1	30 Zn +2 65.37 -8-18-2	31 Ga +3 69.72 -8-18-3	32 Ge +2 +4 72.59 -8-18-4	33 As +3 +5 −3 74.9216 -8-18-5	34 Se +4 +6 −2 78.96 -8-18-6	35 Br +1 +5 −1 79.904 -8-18-7	36 Kr 0 83.80 -8-18-8	-L-M-N
37 Rb +1 85.47 -18-8-1	38 Sr +2 87.62 -18-8-2	39 Y +3 88.905 -18-9-2	40 Zr +4 91.22 -18-10-2	41 Nb +3 +5 92.906 -18-12-1	42 Mo +6 95.94 -18-13-1	43 Tc +4 +6 +7 (97) -18-13-2	44 Ru +3 101.07 -18-15-1	45 Rh +3 102.905 -18-16-1	46 Pd +2 +4 106.4 -18-18-0	47 Ag +1 107.868 -18-18-1	48 Cd +2 112.40 -18-18-2	49 In +3 114.82 -18-18-3	50 Sn +2 +4 118.69 -18-18-4	51 Sb +3 +5 −3 121.75 -18-18-5	52 Te +4 +6 −2 127.60 -18-18-6	53 I +1 +5 +7 −1 126.9044 -18-18-7	54 Xe 0 131.30 -18-18-8	-M-N-O
55 Cs +1 132.905 -18-8-1	56 Ba +2 137.34 -18-8-2	57* La +3 138.91 -18-9-2	72 Hf +4 178.49 -32-10-2	73 Ta +5 180.948 -32-11-2	74 W +6 183.85 -32-12-2	75 Re +4 +6 +7 186.2 -32-13-2	76 Os +3 +4 190.2 -32-14-2	77 Ir +3 +4 192.2 -32-15-2	78 Pt +2 +4 195.09 -32-16-2	79 Au +1 +3 196.967 -32-18-1	80 Hg +1 +2 200.59 -32-18-2	81 Tl +1 +3 204.37 -32-18-3	82 Pb +2 +4 207.19 -32-18-4	83 Bi +3 +5 208.980 -32-18-5	84 Po +2 +4 (210) -32-18-6	85 At (209) -32-18-7	86 Rn 0 (222) -32-18-8	-N-O-P
87 Fr +1 (223) -18-8-1	88 Ra +2 (226) -18-8-2	89** Ac +3 (227) -18-9-2																-O-P-Q

															Orbit
*Lanthanides	58 Ce +3 +4 140.12 -19-9-2	59 Pr +3 140.907 -20-9-2	60 Nd +3 144.24 -22-8-2	61 Pm +3 (145) -23-8-2	62 Sm +2 +3 150.35 -24-8-2	63 Eu +2 +3 151.96 -25-8-2	64 Gd +3 157.25 -25-9-2	65 Tb +3 158.924 -26-9-2	66 Dy +3 162.50 -28-8-2	67 Ho +3 164.930 -29-8-2	68 Er +3 167.26 -30-8-2	69 Tm +3 168.934 -31-8-2	70 Yb +2 +3 173.04 -32-8-2	71 Lu +3 174.97 -32-9-2	-N-O-P
**Actinides	90 Th +4 232.038 -19-9-2	91 Pa +5 +4 (231) -20-9-2	92 U +3 +4 +5 +6 238.03 -21-9-2	93 Np +3 +4 +5 +6 (237) -22-9-2	94 Pu +3 +4 +5 +6 (244) -23-9-2	95 Am +3 +4 +5 +6 (243) -24-9-2	96 Cm +3 (247) 25-9-2	97 Bk +3 +4 (247) -26-9-2	98 Cf +3 (251) 28-8 2	99 Es (254) -29-8-2	100 Fm (257) -30-8-2	101 Md (256) -31-8-2	102 (254) -32-8-2	103 Lw	-O-P-Q

Numbers in parentheses are mass numbers of most stable isotope of that element.

Source: Handbook of Chemistry and Physics, 48th edition, 1967–1968, The Chemical Rubber Co., Cleveland, Ohio, by permission of the publishers.

Note: The Periodic Law states that the properties of the elements are a periodic function of their atomic number. This means that certain elements exhibit similar properties, and that these similarities occur periodically. Thus, Li (atomic number 3) is very similar to Na (atomic number 11) which is similar to K (atomic number 19), etc.

A.33 SQUARES AND SQUARE ROOTS

N	N^2	$\sqrt{N}$	$\sqrt{10N}$	N	N^2	$\sqrt{N}$	$\sqrt{10N}$
1.0	1.00	1.000	3.162	**5.5**	30.25	2.345	7.416
1.1	1.21	1.049	3.317	**5.6**	31.36	2.366	7.483
1.2	1.44	1.095	3.464	**5.7**	32.49	2.387	7.550
1.3	1.69	1.140	3.606	**5.8**	33.64	2.408	7.616
1.4	1.96	1.183	3.742	**5.9**	34.81	2.429	7.681
1.5	2.25	1.225	3.873	**6.0**	36.00	2.449	7.746
1.6	2.56	1.265	4.000	**6.1**	37.21	2.470	7.810
1.7	2.89	1.304	4.123	**6.2**	38.44	2.490	7.874
1.8	3.24	1.342	4.243	**6.3**	39.69	2.510	7.937
1.9	3.61	1.378	4.359	**6.4**	40.96	2.530	8.000
2.0	4.00	1.414	4.472	**6.5**	42.25	2.550	8.062
2.1	4.41	1.449	4.583	**6.6**	43.56	2.569	8.124
2.2	4.84	1.483	4.690	**6.7**	44.89	2.588	8.185
2.3	5.29	1.517	4.796	**6.8**	46.24	2.608	8.246
2.4	5.76	1.549	4.899	**6.9**	47.61	2.627	8.307
2.5	6.25	1.581	5.000	**7.0**	49.00	2.646	8.367
2.6	6.76	1.612	5.099	**7.1**	50.41	2.665	8.426
2.7	7.29	1.643	5.196	**7.2**	51.84	2.683	8.485
2.8	7.84	1.673	5.292	**7.3**	53.29	2.702	8.544
2.9	8.41	1.703	5.385	**7.4**	54.76	2.720	8.602
3.0	9.00	1.732	5.477	**7.5**	56.25	2.739	8.660
3.1	9.61	1.761	5.568	**7.6**	57.76	2.757	8.718
3.2	10.24	1.789	5.657	**7.7**	59.29	2.775	8.775
3.3	10.89	1.817	5.745	**7.8**	60.84	2.793	8.832
3.4	11.56	1.844	5.831	**7.9**	62.41	2.811	8.888
3.5	12.25	1.871	5.916	**8.0**	64.00	2.828	8.944
3.6	12.96	1.897	6.000	**8.1**	65.61	2.846	9.000
3.7	13.69	1.924	6.083	**8.2**	67.24	2.864	9.055
3.8	14.44	1.949	6.164	**8.3**	68.89	2.881	9.110
3.9	15.21	1.975	6.245	**8.4**	70.56	2.898	9.165
4.0	16.00	2.000	6.325	**8.5**	72.25	2.915	9.220
4.1	16.81	2.025	6.403	**8.6**	73.96	2.933	9.274
4.2	17.64	2.049	6.481	**8.7**	75.69	2.950	9.327
4.3	18.49	2.074	6.557	**8.8**	77.44	2.966	9.381
4.4	19.36	2.098	6.633	**8.9**	79.21	2.983	9.434
4.5	20.25	2.121	6.708	**9.0**	81.00	3.000	9.487
4.6	21.16	2.145	6.782	**9.1**	82.81	3.017	9.539
4.7	22.09	2.168	6.856	**9.2**	84.64	3.033	9.592
4.8	23.04	2.191	6.928	**9.3**	86.49	3.050	9.644
4.9	24.01	2.214	7.000	**9.4**	88.36	3.066	9.695
5.0	25.00	2.236	7.071	**9.5**	90.25	3.082	9.747
5.1	26.01	2.258	7.141	**9.6**	92.16	3.098	9.798
5.2	27.04	2.280	7.211	**9.7**	94.09	3.114	9.849
5.3	28.09	2.302	7.280	**9.8**	96.04	3.130	9.899
5.4	29.16	2.324	7.348	**9.9**	98.01	3.146	9.950
5.5	30.25	2.345	7.416	**10**	100.00	3.162	10.000

Source: From *Modern Algebra and Trigonometry, Structure and Method,* Book 2, by Dolciani, Berman, and Wooton, copyright 1965 by Houghton Mifflin Company. Reprinted by permission.

A.34 CUBES AND CUBE ROOTS

N	N^3	$\sqrt[3]{N}$	$\sqrt[3]{10N}$	$\sqrt[3]{100N}$	N	N^3	$\sqrt[3]{N}$	$\sqrt[3]{10N}$	$\sqrt[3]{100N}$
1.0	1.000	1.000	2.154	4.642	**5.5**	166.375	1.765	3.803	8.193
1.1	1.331	1.032	2.224	4.791	**5.6**	175.616	1.776	3.826	8.243
1.2	1.728	1.063	2.289	4.932	**5.7**	185.193	1.786	3.849	8.291
1.3	2.197	1.091	2.351	5.066	**5.8**	195.112	1.797	3.871	8.340
1.4	2.744	1.119	2.410	5.192	**5.9**	205.379	1.807	3.893	8.387
1.5	3.375	1.145	2.466	5.313	**6.0**	216.000	1.817	3.915	8.434
1.6	4.096	1.170	2.520	5.429	**6.1**	226.981	1.827	3.936	8.481
1.7	4.913	1.193	2.571	5.540	**6.2**	238.328	1.837	3.958	8.527
1.8	5.832	1.216	2.621	5.646	**6.3**	250.047	1.847	3.979	8.573
1.9	6.859	1.239	2.668	5.749	**6.4**	262.144	1.857	4.000	8.618
2.0	8.000	1.260	2.714	5.848	**6.5**	274.625	1.866	4.021	8.662
2.1	9.261	1.281	2.759	5.944	**6.6**	287.496	1.876	4.041	8.707
2.2	10.648	1.301	2.802	6.037	**6.7**	300.763	1.885	4.062	8.750
2.3	12.167	1.320	2.844	6.127	**6.8**	314.432	1.895	4.082	8.794
2.4	13.824	1.339	2.884	6.214	**6.9**	328.509	1.904	4.102	8.837
2.5	15.625	1.357	2.924	6.300	**7.0**	343.000	1.913	4.121	8.879
2.6	17.576	1.375	2.962	6.383	**7.1**	357.911	1.922	4.141	8.921
2.7	19.683	1.392	3.000	6.463	**7.2**	373.248	1.931	4.160	8.963
2.8	21.952	1.409	3.037	6.542	**7.3**	389.017	1.940	4.179	9.004
2.9	24.389	1.426	3.072	6.619	**7.4**	405.224	1.949	4.198	9.045
3.0	27.000	1.442	3.107	6.694	**7.5**	421.875	1.957	4.217	9.086
3.1	29.791	1.458	3.141	6.768	**7.6**	438.976	1.966	4.236	9.126
3.2	32.768	1.474	3.175	6.840	**7.7**	456.533	1.975	4.254	9.166
3.3	35.937	1.489	3.208	6.910	**7.8**	474.552	1.983	4.273	9.205
3.4	39.304	1.504	3.240	6.980	**7.9**	493.039	1.992	4.291	9.244
3.5	42.875	1.518	3.271	7.047	**8.0**	512.000	2.000	4.309	9.283
3.6	46.656	1.533	3.302	7.114	**8.1**	531.441	2.008	4.327	9.322
3.7	50.653	1.547	3.332	7.179	**8.2**	551.368	2.017	4.344	9.360
3.8	54.872	1.560	3.362	7.243	**8.3**	571.787	2.025	4.362	9.398
3.9	59.319	1.574	3.391	7.306	**8.4**	592.704	2.033	4.380	9.435
4.0	64.000	1.587	3.420	7.368	**8.5**	614.125	2.041	4.397	9.473
4.1	68.921	1.601	3.448	7.429	**8.6**	636.056	2.049	4.414	9.510
4.2	74.088	1.613	3.476	7.489	**8.7**	658.503	2.057	4.431	9.546
4.3	79.507	1.626	3.503	7.548	**8.8**	681.472	2.065	4.448	9.583
4.4	85.184	1.639	3.530	7.606	**8.9**	704.969	2.072	4.465	9.619
4.5	91.125	1.651	3.557	7.663	**9.0**	729.000	2.080	4.481	9.655
4.6	97.336	1.663	3.583	7.719	**9.1**	753.571	2.088	4.498	9.691
4.7	103.823	1.675	3.609	7.775	**9.2**	778.688	2.095	4.514	9.726
4.8	110.592	1.687	3.634	7.830	**9.3**	804.357	2.103	4.531	9.761
4.9	117.649	1.698	3.659	7.884	**9.4**	830.584	2.110	4.547	9.796
5.0	125.000	1.710	3.684	7.937	**9.5**	857.375	2.118	4.563	9.830
5.1	132.651	1.721	3.708	7.990	**9.6**	884.736	2.125	4.579	9.865
5.2	140.608	1.732	3.733	8.041	**9.7**	912.673	2.133	4.595	9.899
5.3	148.877	1.744	3.756	8.093	**9.8**	941.192	2.140	4.610	9.933
5.4	157.464	1.754	3.780	8.143	**9.9**	970.299	2.147	4.626	9.967
5.5	166.375	1.765	3.803	8.193	**10**	1000.000	2.154	4.642	10.000

Source: From *Modern Algebra and Trigonometry, Structure and Method,* Book 2, by Dolciani, Berman, and Wooton, copyright 1965 by Houghton Mifflin Company. Reprinted by permission.

A.35 FOUR-PLACE LOGARITHMS

No.	0	1	2	3	4	5	6	7	8	9	Proportional Parts								
											1	2	3	4	5	6	7	8	9
100	.0000	.0004	.0009	.0013	.0017	.0022	.0026	.0030	.0035	.0039	0	1	1	2	2	3	3	3	4
101	0043	0048	0052	0056	0060	0065	0069	0073	0077	0082	0	1	1	2	2	3	3	3	4
102	0086	0090	0095	0099	0103	0107	0111	0116	0120	0124	0	1	1	2	2	3	3	3	4
103	0128	0133	0137	0141	0145	0149	0154	0158	0162	0166	0	1	1	2	2	3	3	3	4
104	0170	0175	0179	0183	0187	0191	0195	0199	0204	0208	0	1	1	2	2	2	3	3	4
105	.0212	.0216	.0220	.0224	.0228	.0233	.0237	.0241	.0245	.0249	0	1	1	2	2	2	3	3	4
106	0253	0257	0261	0265	0269	0273	0278	0282	0286	0290	0	1	1	2	2	2	3	3	4
107	0294	0298	0302	0306	0310	0314	0318	0322	0326	0330	0	1	1	2	2	2	3	3	4
108	0334	0338	0342	0346	0350	0354	0358	0362	0366	0370	0	1	1	2	2	2	3	3	4
109	0374	0378	0382	0386	0390	0394	0398	0402	0406	0410	0	1	1	2	2	2	3	3	4
10	.0000	.0043	.0086	.0128	.0170	.0212	.0253	.0294	.0334	.0374	4	8	12	17	21	25	29	33	37
11	0414	0453	0492	0531	0569	0607	0645	0682	0719	0755	4	8	11	15	19	23	27	30	34
12	0792	0828	0864	0899	0934	0969	1004	1038	1072	1106	3	7	10	14	17	21	24	28	31
13	1139	1173	1206	1239	1271	1303	1335	1367	1399	1430	3	6	10	13	16	19	23	26	29
14	1461	1492	1523	1553	1584	1614	1644	1673	1703	1732	3	6	9	12	15	18	21	24	27
15	.1761	.1790	.1818	.1847	.1875	.1903	.1931	.1959	1987	.2014	3	6	8	11	14	17	20	22	25
16	2041	2068	2095	2122	2148	2175	2201	2227	2253	2279	3	5	8	11	13	16	18	21	24
17	2304	2330	2355	2380	2405	2430	2455	2480	2504	2529	2	5	7	10	12	15	17	20	22
18	2553	2577	2601	2625	2648	2672	2695	2718	2742	2765	2	5	7	9	12	14	16	19	21
19	2788	2810	2833	2856	2878	2900	2923	2945	2967	2989	2	4	7	9	11	13	16	18	20
20	.3010	.3032	.3054	.3075	.3096	.3118	.3139	.3160	.3181	.3201	2	4	6	8	11	13	15	17	19
21	3222	3243	3263	3284	3304	3324	3345	3365	3385	3404	2	4	6	8	10	12	14	16	18
22	3424	3444	3464	3483	3502	3522	3541	3560	3579	3598	2	4	6	8	10	12	14	15	17
23	3617	3636	3655	3674	3692	3711	3729	3747	3766	3784	2	4	6	7	9	11	13	15	17
24	3802	3820	3838	3856	3874	3892	3909	3927	3945	3962	2	4	5	7	9	11	12	14	16
25	.3979	.3997	.4014	.4031	.4048	.4065	.4082	.4099	.4116	.4133	2	3	5	7	9	10	12	14	15
26	4150	4166	4183	4200	4216	4232	4249	4265	4281	4298	2	3	5	7	8	10	11	13	15
27	4314	4330	4346	4362	4378	4393	4409	4425	4440	4456	2	3	5	6	8	9	11	13	14
28	4472	4487	4502	4518	4533	4548	4564	4579	4594	4609	2	3	5	6	8	9	11	12	14
29	4624	4639	4654	4669	4683	4698	4713	4728	4742	4757	1	3	4	6	7	9	10	12	13
30	.4771	.4786	.4800	.4814	.4829	.4843	.4857	.4871	.4886	.4900	1	3	4	6	7	9	10	11	13
31	4914	4928	4942	4955	4969	4983	4997	5011	5024	5038	1	3	4	6	7	8	10	11	12
32	5051	5065	5079	5092	5105	5119	5132	5145	5159	5172	1	3	4	5	7	8	9	11	12
33	5185	5198	5211	5224	5237	5250	5263	5276	5289	5302	1	3	4	5	6	8	9	10	12
34	5315	5328	5340	5353	5366	5378	5391	5403	5416	5428	1	3	4	5	6	8	9	10	11
35	.5441	.5453	.5465	.5478	.5490	.5502	.5514	.5527	.5539	.5551	1	2	4	5	6	7	9	10	11
36	5563	5575	5587	5599	5611	5623	5635	5647	5658	5670	1	2	4	5	6	7	8	10	11
37	5682	5694	5705	5717	5729	5740	5752	5763	5775	5786	1	2	3	5	6	7	8	9	10
38	5798	5809	5821	5832	5843	5855	5866	5877	5888	5899	1	2	3	5	6	7	8	9	10
39	5911	5922	5933	5944	5955	5966	5977	5988	5999	6010	1	2	3	4	5	7	8	9	10
40	.6021	.6031	.6042	.6053	.6064	.6075	.6085	.6096	.6107	.6117	1	2	3	4	5	6	8	9	10
41	6128	6138	6149	6160	6170	6180	6191	6201	6212	6222	1	2	3	4	5	6	7	8	9
42	6232	6243	6253	6263	6274	6284	6294	6304	6314	6325	1	2	3	4	5	6	7	8	9
43	6335	6345	6355	6365	6375	6385	6395	6405	6415	6425	1	2	3	4	5	6	7	8	9
44	6435	6444	6454	6464	6474	6484	6493	6503	6513	6522	1	2	3	4	5	6	7	8	9
45	.6532	.6542	.6551	.6561	.6571	.6580	.6590	.6599	.6609	.6618	1	2	3	4	5	6	7	8	9
46	6628	6637	6646	6656	6665	6675	6684	6693	6702	6712	1	2	3	4	5	6	7	7	8
47	6721	6730	6739	6749	6758	6767	6776	6785	6794	6803	1	2	3	4	5	5	6	7	8
48	6812	6821	6830	6839	6848	6857	6866	6875	6884	6893	1	2	3	4	4	5	6	7	8
49	6902	6911	6920	6928	6937	6946	6955	6964	6972	6981	1	2	3	4	4	5	6	7	8
											1	2	3	4	5	6	7	8	9
No.	0	1	2	3	4	5	6	7	8	9	Proportional Parts								

Source: The Merck Index, Eighth Edition, copyright 1968 by Merck & Co., Inc., Rahway, New Jersey, U.S.A., by permission of the publishers.

A.35 (*continued*)

No.	0	1	2	3	4	5	6	7	8	9	Proportional Parts 1	2	3	4	5	6	7	8	9
50	.6990	.6998	.7007	.7016	.7024	.7033	.7042	.7050	.7059	.7067	1	2	3	3	4	5	6	7	8
51	7076	7084	7093	7101	7110	7118	7126	7135	7143	7152	1	2	3	3	4	5	6	7	8
52	7160	7168	7177	7185	7193	7202	7210	7218	7226	7235	1	2	2	3	4	5	6	7	7
53	7243	7251	7259	7267	7275	7284	7292	7300	7308	7316	1	2	2	3	4	5	6	6	7
54	7324	7332	7340	7348	7356	7364	7372	7380	7388	7396	1	2	2	3	4	5	6	6	7
55	.7404	.7412	.7419	.7427	.7435	.7443	.7451	.7459	.7466	.7474	1	2	2	3	4	5	5	6	7
56	7482	7490	7497	7505	7513	7520	7528	7536	7543	7551	1	2	2	3	4	5	5	6	7
57	7559	7566	7574	7582	7589	7597	7604	7612	7619	7627	1	2	2	3	4	5	5	6	7
58	7634	7642	7649	7657	7664	7672	7679	7686	7694	7701	1	1	2	3	4	4	5	6	7
59	7709	7716	7723	7731	7738	7745	7752	7760	7767	7774	1	1	2	3	4	4	5	6	7
60	.7782	.7789	.7796	.7803	.7810	.7818	.7825	.7832	.7839	.7846	1	1	2	3	4	4	5	6	6
61	7853	7860	7868	7875	7882	7889	7896	7903	7910	7917	1	1	2	3	4	4	5	6	6
62	7924	7931	7938	7945	7952	7959	7966	7973	7980	7987	1	1	2	3	3	4	5	6	6
63	7993	8000	8007	8014	8021	8028	8035	8041	8048	8055	1	1	2	3	3	4	5	5	6
64	8062	8069	8075	8082	8089	8096	8102	8109	8116	8122	1	1	2	3	3	4	5	5	6
65	.8129	.8136	.8142	.8149	.8156	.8162	.8169	.8176	.8182	.8189	1	1	2	3	3	4	5	5	6
66	8195	8202	8209	8215	8222	8228	8235	8241	8248	8254	1	1	2	3	3	4	5	5	6
67	8261	8267	8274	8280	8287	8293	8299	8306	8312	8319	1	1	2	3	3	4	5	5	6
68	8325	8331	8338	8344	8351	8357	8363	8370	8376	8382	1	1	2	3	3	4	4	5	6
69	8388	8395	8401	8407	8414	8420	8426	8432	8439	8445	1	1	2	2	3	4	4	5	6
70	.8451	.8457	.8463	.8470	.8476	.8482	.8488	.8494	.8500	.8506	1	1	2	2	3	4	4	5	6
71	8513	8519	8525	8531	8537	8543	8549	8555	8561	8567	1	1	2	2	3	4	4	5	5
72	8573	8579	8585	8591	8597	8603	8609	8615	8621	8627	1	1	2	2	3	4	4	5	5
73	8633	8639	8645	8651	8657	8663	8669	8675	8681	8686	1	1	2	2	3	4	4	5	5
74	8692	8698	8704	8710	8716	8722	8727	8733	8739	8745	1	1	2	2	3	4	4	5	5
75	.8751	.8756	.8762	.8768	.8774	.8779	.8785	.8791	.8797	.8802	1	1	2	2	3	3	4	5	5
76	8808	8814	8820	8825	8831	8837	8842	8848	8854	8859	1	1	2	2	3	3	4	5	5
77	8865	8871	8876	8882	8887	8893	8899	8904	8910	8915	1	1	2	2	3	3	4	4	5
78	8921	8927	8932	8938	8943	8949	8954	8960	8965	8971	1	1	2	2	3	3	4	4	5
79	8976	8982	8987	8993	8998	9004	9009	9015	9020	9025	1	1	2	2	3	3	4	4	5
80	.9031	.9036	.9042	.9047	.9053	.9058	.9063	.9069	.9074	.9079	1	1	2	2	3	3	4	4	5
81	9085	9090	9096	9101	9106	9112	9117	9122	9128	9133	1	1	2	2	3	3	4	4	5
82	9138	9143	9149	9154	9159	9165	9170	9175	9180	9186	1	1	2	2	3	3	4	4	5
83	9191	9196	9201	9206	9212	9217	9222	9227	9232	9238	1	1	2	2	3	3	4	4	5
84	9243	9248	9253	9258	9263	9269	9274	9279	9284	9289	1	1	2	2	3	3	4	4	5
85	.9294	.9299	.9304	.9309	.9315	.9320	.9325	.9330	.9335	.9340	1	1	2	2	3	3	4	4	5
86	9345	9350	9355	9360	9365	9370	9375	9380	9385	9390	1	1	2	2	3	3	4	4	5
87	9395	9400	9405	9410	9415	9420	9425	9430	9435	9440	0	1	1	2	2	3	3	4	4
88	9445	9450	9455	9460	9465	9469	9474	9479	9484	9489	0	1	1	2	2	3	3	4	4
89	9494	9499	9504	9509	9513	9518	9523	9528	9533	9538	0	1	1	2	2	3	3	4	4
90	.9542	.9547	.9552	.9557	.9562	.9566	.9571	.9576	.9581	.9586	0	1	1	2	2	3	3	4	4
91	9590	9595	9600	9605	9609	9614	9619	9624	9628	9633	0	1	1	2	2	3	3	4	4
92	9638	9643	9647	9652	9657	9661	9666	9671	9675	9680	0	1	1	2	2	3	3	4	4
93	9685	9689	9694	9699	9703	9708	9713	9717	9722	9727	0	1	1	2	2	3	3	4	4
94	9731	9736	9741	9745	9750	9754	9759	9763	9768	9773	0	1	1	2	2	3	3	4	4
95	.9777	.9782	.9786	.9791	.9795	.9800	.9805	.9809	.9814	.9818	0	1	1	2	2	3	3	4	4
96	9823	9827	9832	9836	9841	9845	9850	9854	9859	9863	0	1	1	2	2	3	3	4	4
97	9868	9872	9877	9881	9886	9890	9894	9899	9903	9908	0	1	1	2	2	3	3	4	4
98	9912	9917	9921	9926	9930	9934	9939	9943	9948	9952	0	1	1	2	2	3	3	4	4
99	9956	9961	9965	9969	9974	9978	9983	9987	9991	9996	0	1	1	2	2	3	3	3	4
No.	0	1	2	3	4	5	6	7	8	9	1	2	3	4	5	6	7	8	9
											Proportional Parts								

A.36 FOUR-PLACE ANTILOGARITHMS

Log₁₀	0	1	2	3	4	5	6	7	8	9	Proportional Parts 1	2	3	4	5	6	7	8	9
.00	1000	1002	1005	1007	1009	1012	1014	1016	1019	1021	0	0	1	1	1	1	2	2	2
.01	1023	1026	1028	1030	1033	1035	1038	1040	1042	1045	0	0	1	1	1	1	2	2	2
.02	1047	1050	1052	1054	1057	1059	1062	1064	1067	1069	0	0	1	1	1	1	2	2	2
.03	1072	1074	1076	1079	1081	1084	1086	1089	1091	1094	0	0	1	1	1	1	2	2	2
.04	1096	1099	1102	1104	1107	1109	1112	1114	1117	1119	0	1	1	1	1	2	2	2	2
.05	1122	1125	1127	1130	1132	1135	1138	1140	1143	1146	0	1	1	1	1	2	2	2	2
.06	1148	1151	1153	1156	1159	1161	1164	1167	1169	1172	0	1	1	1	1	2	2	2	2
.07	1175	1178	1180	1183	1186	1189	1191	1194	1197	1199	0	1	1	1	1	2	2	2	2
.08	1202	1205	1208	1211	1213	1216	1219	1222	1225	1227	0	1	1	1	1	2	2	2	3
.09	1230	1233	1236	1239	1242	1245	1247	1250	1253	1256	0	1	1	1	1	2	2	2	3
.10	1259	1262	1265	1268	1271	1274	1276	1279	1282	1285	0	1	1	1	1	2	2	2	3
.11	1288	1291	1294	1297	1300	1303	1306	1309	1312	1315	0	1	1	1	2	2	2	2	3
.12	1318	1321	1324	1327	1330	1334	1337	1340	1343	1346	0	1	1	1	2	2	2	2	3
.13	1349	1352	1355	1358	1361	1365	1368	1371	1374	1377	0	1	1	1	2	2	2	3	3
.14	1380	1384	1387	1390	1393	1396	1400	1403	1406	1409	0	1	1	1	2	2	2	3	3
.15	1413	1416	1419	1422	1426	1429	1432	1435	1439	1442	0	1	1	1	2	2	2	3	3
.16	1445	1449	1452	1455	1459	1462	1466	1469	1472	1476	0	1	1	1	2	2	2	3	3
.17	1479	1483	1486	1489	1493	1496	1500	1503	1507	1510	0	1	1	1	2	2	2	3	3
.18	1514	1517	1521	1524	1528	1531	1535	1538	1542	1545	0	1	1	1	2	2	2	3	3
.19	1549	1552	1556	1560	1563	1567	1570	1574	1578	1581	0	1	1	1	2	2	3	3	3
.20	1585	1589	1592	1596	1600	1603	1607	1611	1614	1618	0	1	1	1	2	2	3	3	3
.21	1622	1626	1629	1633	1637	1641	1644	1648	1652	1656	0	1	1	2	2	2	3	3	3
.22	1660	1663	1667	1671	1675	1679	1683	1687	1690	1694	0	1	1	2	2	2	3	3	3
.23	1698	1702	1706	1710	1714	1718	1722	1726	1730	1734	0	1	1	2	2	2	3	3	4
.24	1738	1742	1746	1750	1754	1758	1762	1766	1770	1774	0	1	1	2	2	2	3	3	4
.25	1778	1782	1786	1791	1795	1799	1803	1807	1811	1816	0	1	1	2	2	2	3	3	4
.26	1820	1824	1828	1832	1837	1841	1845	1849	1854	1858	0	1	1	2	2	3	3	3	4
.27	1862	1866	1871	1875	1879	1884	1888	1892	1897	1901	0	1	1	2	2	3	3	3	4
.28	1905	1910	1914	1919	1923	1928	1932	1936	1941	1945	0	1	1	2	2	3	3	4	4
.29	1950	1954	1959	1963	1968	1972	1977	1982	1986	1991	0	1	1	2	2	3	3	4	4
.30	1995	2000	2004	2009	2014	2018	2023	2028	2032	2037	0	1	1	2	2	3	3	4	4
.31	2042	2046	2051	2056	2061	2065	2070	2075	2080	2084	0	1	1	2	2	3	3	4	4
.32	2089	2094	2099	2104	2109	2113	2118	2123	2128	2133	0	1	1	2	2	3	3	4	4
.33	2138	2143	2148	2153	2158	2163	2168	2173	2178	2183	0	1	1	2	2	3	3	4	4
.34	2188	2193	2198	2203	2208	2213	2218	2223	2228	2234	1	1	2	2	3	3	4	4	5
.35	2239	2244	2249	2254	2259	2265	2270	2275	2280	2286	1	1	2	2	3	3	4	4	5
.36	2291	2296	2301	2307	2312	2317	2323	2328	2333	2339	1	1	2	2	3	3	4	4	5
.37	2344	2350	2355	2360	2366	2371	2377	2382	2388	2393	1	1	2	2	3	3	4	4	5
.38	2399	2404	2410	2415	2421	2427	2432	2438	2443	2449	1	1	2	2	3	3	4	5	5
.39	2455	2460	2466	2472	2477	2483	2489	2495	2500	2506	1	1	2	2	3	3	4	5	5
.40	2512	2518	2523	2529	2535	2541	2547	2553	2559	2564	1	1	2	2	3	4	4	5	5
.41	2570	2576	2582	2588	2594	2600	2606	2612	2618	2624	1	1	2	2	3	4	4	5	5
.42	2630	2636	2642	2649	2655	2661	2667	2673	2679	2685	1	1	2	2	3	4	4	5	6
.43	2692	2698	2704	2710	2716	2723	2729	2735	2742	2748	1	1	2	3	3	4	4	5	6
.44	2754	2761	2767	2773	2780	2786	2793	2799	2805	2812	1	1	2	3	3	4	4	5	6
.45	2818	2825	2831	2838	2844	2851	2858	2864	2871	2877	1	1	2	3	3	4	5	5	6
.46	2884	2891	2897	2904	2911	2917	2924	2931	2938	2944	1	1	2	3	3	4	5	5	6
.47	2951	2958	2965	2972	2979	2985	2992	2999	3006	3013	1	1	2	3	3	4	5	5	6
.48	3020	3027	3034	3041	3048	3055	3062	3069	3076	3083	1	1	2	3	4	4	5	6	6
.49	3090	3097	3105	3112	3119	3126	3133	3141	3148	3155	1	1	2	3	4	4	5	6	6
Log₁₀	0	1	2	3	4	5	6	7	8	9	1	2	3	4	5	6	7	8	9
											Proportional Parts								

Source: The Merck Index, Eighth Edition, copyright 1968 by Merck & Co., Inc., Rahway, New Jersey, U.S.A., by permission of the publishers.

A.36 (*continued*)

Log_{10}	0	1	2	3	4	5	6	7	8	9	PP 1	PP 2	PP 3	PP 4	PP 5	PP 6	PP 7	PP 8	PP 9
.50	3162	3170	3177	3184	3192	3199	3206	3214	3221	3228	1	1	2	3	4	4	5	6	7
.51	3236	3243	3251	3258	3266	3273	3281	3289	3296	3304	1	2	2	3	4	5	5	6	7
.52	3311	3319	3327	3334	3342	3350	3357	3365	3373	3381	1	2	2	3	4	5	5	6	7
.53	3388	3396	3404	3412	3420	3428	3436	3443	3451	3459	1	2	2	3	4	5	6	6	7
.54	3467	3475	3483	3491	3499	3508	3516	3524	3532	3540	1	2	2	3	4	5	6	6	7
.55	3548	3556	3565	3573	3581	3589	3597	3606	3614	3622	1	2	2	3	4	5	6	7	7
.56	3631	3639	3648	3656	3664	3673	3681	3690	3698	3707	1	2	3	3	4	5	6	7	8
.57	3715	3724	3733	3741	3750	3758	3767	3776	3784	3793	1	2	3	3	4	5	6	7	8
.58	3802	3811	3819	3828	3837	3846	3855	3864	3873	3882	1	2	3	4	4	5	6	7	8
.59	3890	3899	3908	3917	3926	3936	3945	3954	3963	3972	1	2	3	4	5	5	6	7	8
.60	3981	3990	3999	4009	4018	4027	4036	4046	4055	4064	1	2	3	4	5	6	6	7	8
.61	4074	4083	4093	4102	4111	4121	4130	4140	4150	4159	1	2	3	4	5	6	7	8	9
.62	4169	4178	4188	4198	4207	4217	4227	4236	4246	4256	1	2	3	4	5	6	7	8	9
.63	4266	4276	4285	4295	4305	4315	4325	4335	4345	4355	1	2	3	4	5	6	7	8	9
.64	4365	4375	4385	4395	4406	4416	4426	4436	4446	4457	1	2	3	4	5	6	7	8	9
.65	4467	4477	4487	4498	4508	4519	4529	4539	4550	4560	1	2	3	4	5	6	7	8	9
.66	4571	4581	4592	4603	4613	4624	4634	4645	4656	4667	1	2	3	4	5	6	7	9	10
.67	4677	4688	4699	4710	4721	4732	4742	4753	4764	4775	1	2	3	4	5	7	8	9	10
.68	4786	4797	4808	4819	4831	4842	4853	4864	4875	4887	1	2	3	4	6	7	8	9	10
.69	4898	4909	4920	4932	4943	4955	4966	4977	4989	5000	1	2	3	5	6	7	8	9	10
.70	5012	5023	5035	5047	5058	5070	5082	5093	5105	5117	1	2	4	5	6	7	8	9	11
.71	5129	5140	5152	5164	5176	5188	5200	5212	5224	5236	1	2	4	5	6	7	8	10	11
.72	5248	5260	5272	5284	5297	5309	5321	5333	5346	5358	1	2	4	5	6	7	9	10	11
.73	5370	5383	5395	5408	5420	5433	5445	5458	5470	5483	1	3	4	5	6	8	9	10	11
.74	5495	5508	5521	5534	5546	5559	5572	5585	5598	5610	1	3	4	5	6	8	9	10	12
.75	5623	5636	5649	5662	5675	5689	5702	5715	5728	5741	1	3	4	5	7	8	9	10	12
.76	5754	5768	5781	5794	5808	5821	5834	5848	5861	5875	1	3	4	5	7	8	9	11	12
.77	5888	5902	5916	5929	5943	5957	5970	5984	5998	6012	1	3	4	5	7	8	10	11	12
.78	6026	6039	6053	6067	6081	6095	6109	6124	6138	6152	1	3	4	6	7	8	10	11	13
.79	6166	6180	6194	6209	6223	6237	6252	6266	6281	6295	1	3	4	6	7	9	10	11	13
.80	6310	6324	6339	6353	6368	6383	6397	6412	6427	6442	1	3	4	6	7	9	10	12	13
.81	6457	6471	6486	6501	6516	6531	6546	6561	6577	6592	2	3	5	6	8	9	11	12	14
.82	6607	6622	6637	6653	6668	6683	6699	6714	6730	6745	2	3	5	6	8	9	11	12	14
.83	6761	6776	6792	6808	6823	6839	6855	6871	6887	6902	2	3	5	6	8	9	11	13	14
.84	6918	6934	6950	6966	6982	6998	7015	7031	7047	7063	2	3	5	6	8	10	11	13	15
.85	7079	7096	7112	7129	7145	7161	7178	7194	7211	7228	2	3	5	7	8	10	12	13	15
.86	7244	7261	7278	7295	7311	7328	7345	7362	7379	7396	2	3	5	7	8	10	12	13	15
.87	7413	7430	7447	7464	7482	7499	7516	7534	7551	7568	2	3	5	7	9	10	12	14	16
.88	7586	7603	7621	7638	7656	7674	7691	7709	7727	7745	2	4	5	7	9	11	12	14	16
.89	7762	7780	7798	7816	7834	7852	7870	7889	7907	7925	2	4	5	7	9	11	13	14	16
.90	7943	7962	7980	7998	8017	8035	8054	8072	8091	8110	2	4	6	7	9	11	13	15	17
.91	8128	8147	8166	8185	8204	8222	8241	8260	8279	8299	2	4	6	8	9	11	13	15	17
.92	8318	8337	8356	8375	8395	8414	8433	8453	8472	8492	2	4	6	8	10	12	14	15	17
.93	8511	8531	8551	8570	8590	8610	8630	8650	8670	8690	2	4	6	8	10	12	14	16	18
.94	8710	8730	8750	8770	8790	8810	8831	8851	8872	8892	2	4	6	8	10	12	14	16	18
.95	8913	8933	8954	8974	8995	9016	9036	9057	9078	9099	2	4	6	8	10	12	15	17	19
.96	9120	9141	9162	9183	9204	9226	9247	9268	9290	9311	2	4	6	8	11	13	15	17	19
.97	9333	9354	9376	9397	9419	9441	9462	9484	9506	9528	2	4	7	9	11	13	15	17	20
.98	9550	9572	9594	9616	9638	9661	9683	9705	9727	9750	2	4	7	9	11	13	16	18	20
.99	9772	9795	9817	9840	9863	9886	9908	9931	9954	9977	2	5	7	9	11	14	16	18	20

Log_{10} 0 1 2 3 4 5 6 7 8 9 — Proportional Parts 1 2 3 4 5 6 7 8 9

A.37 SUMS AND POWERS OF INTEGERS

x	Σx	Σx^2
1	1	1
2	3	5
3	6	14
4	10	30
5	15	55
6	21	91
7	28	140
8	36	204
9	45	285
10	55	385
11	66	506
12	78	650
13	91	819
14	105	1015
15	120	1240
16	136	1496
17	153	1785
18	171	2109
19	190	2470
20	210	2870
21	231	3311
22	253	3795
23	276	4324
24	300	4900
25	325	5525
26	351	6201
27	378	6930
28	406	7714
29	435	8555
30	465	9455
31	496	10416
32	528	11440
33	561	12529
34	595	13685
35	630	14910
36	666	16206
37	703	17575
38	741	19019
39	780	20540
40	820	22140

A.38 AREAS AND ORDINATES OF THE NORMAL CURVE IN TERMS OF x/σ

(1) z Standard Score $\left(\frac{x}{\sigma}\right)$	(2) A Area from Mean to $\frac{x}{\sigma}$	(3) B Area in Larger Portion	(4) C Area in Smaller Portion	(5) y Ordinate at $\frac{x}{\sigma}$
0.00	.0000	.5000	.5000	.3989
0.01	.0040	.5040	.4960	.3989
0.02	.0080	.5080	.4920	.3989
0.03	.0120	.5120	.4880	.3988
0.04	.0160	.5160	.4840	.3986
0.05	.0199	.5199	.4801	.3984
0.06	.0239	.5239	.4761	.3982
0.07	.0279	.5279	.4721	.3980
0.08	.0319	.5319	.4681	.3977
0.09	.0359	.5359	.4641	.3973
0.10	.0398	.5398	.4602	.3970
0.11	.0438	.5438	.4562	.3965
0.12	.0478	.5478	.4522	.3961
0.13	.0517	.5517	.4483	.3956
0.14	.0557	.5557	.4443	.3951
0.15	.0596	.5596	.4404	.3945
0.16	.0636	.5636	.4364	.3939
0.17	.0675	.5675	.4325	.3932
0.18	.0714	.5714	.4286	.3925
0.19	.0753	.5753	.4247	.3918
0.20	.0793	.5793	.4207	.3910
0.21	.0832	.5832	.4168	.3902
0.22	.0871	.5871	.4129	.3894
0.23	.0910	.5910	.4090	.3885
0.24	.0948	.5948	.4052	.3876
0.25	.0987	.5987	.4013	.3867
0.26	.1026	.6026	.3974	.3857
0.27	.1064	.6064	.3936	.3847
0.28	.1103	.6103	.3897	.3836
0.29	.1141	.6141	.3859	.3825
0.30	.1179	.6179	.3821	.3814
0.31	.1217	.6217	.3783	.3802
0.32	.1255	.6255	.3745	.3790
0.33	.1293	.6293	.3707	.3778
0.34	.1331	.6331	.3669	.3765

Source: Allen L. Edwards, *Experimental Design in Psychological Research*, Holt, Rinehart and Winston, revised edition, 1965, by permission of the author.

A.38 (*continued*)

(1) z STANDARD SCORE $\left(\frac{x}{\sigma}\right)$	(2) A AREA FROM MEAN TO $\frac{x}{\sigma}$	(3) B AREA IN LARGER PORTION	(4) C AREA IN SMALLER PORTION	(5) y ORDINATE AT $\frac{x}{\sigma}$
0.35	.1368	.6368	.3632	.3752
0.36	.1406	.6406	.3594	.3739
0.37	.1443	.6443	.3557	.3725
0.38	.1480	.6480	.3520	.3712
0.39	.1517	.6517	.3483	.3697
0.40	.1554	.6554	.3446	.3683
0.41	.1591	.6591	.3409	.3668
0.42	.1628	.6628	.3372	.3653
0.43	.1664	.6664	.3336	.3637
0.44	.1700	.6700	.3300	.3621
0.45	.1736	.6736	.3264	.3605
0.46	.1772	.6772	.3228	.3589
0.47	.1808	.6808	.3192	.3572
0.48	.1844	.6844	.3156	.3555
0.49	.1879	.6879	.3121	.3538
0.50	.1915	.6915	.3085	.3521
0.51	.1950	.6950	.3050	.3503
0.52	.1985	.6985	.3015	.3485
0.53	.2019	.7019	.2981	.3467
0.54	.2054	.7054	.2946	.3448
0.55	.2088	.7088	.2912	.3429
0.56	.2123	.7123	.2877	.3410
0.57	.2157	.7157	.2843	.3391
0.58	.2190	.7190	.2810	.3372
0.59	.2224	.7224	.2776	.3352
0.60	.2257	.7257	.2743	.3332
0.61	.2291	.7291	.2709	.3312
0.62	.2324	.7324	.2676	.3292
0.63	.2357	.7357	.2643	.3271
0.64	.2389	.7389	.2611	.3251
0.65	.2422	.7422	.2578	.3230
0.66	.2454	.7454	.2546	.3209
0.67	.2486	.7486	.2514	.3187
0.68	.2517	.7517	.2483	.3166
0.69	.2549	.7549	.2451	.3144

A.38 (*continued*)

(1) z Standard Score $\left(\frac{x}{\sigma}\right)$	(2) A Area from Mean to $\frac{x}{\sigma}$	(3) B Area in Larger Portion	(4) C Area in Smaller Portion	(5) y Ordinate at $\frac{x}{\sigma}$
0.70	.2580	.7580	.2420	.3123
0.71	.2611	.7611	.2389	.3101
0.72	.2642	.7642	.2358	.3079
0.73	.2673	.7673	.2327	.3056
0.74	.2704	.7704	.2296	.3034
0.75	.2734	.7734	.2266	.3011
0.76	.2764	.7764	.2236	.2989
0.77	.2794	.7794	.2206	.2966
0.78	.2823	.7823	.2177	.2943
0.79	.2852	.7852	.2148	.2920
0.80	.2881	.7881	.2119	.2897
0.81	.2910	.7910	.2090	.2874
0.82	.2939	.7939	.2061	.2850
0.83	.2967	.7967	.2033	.2827
0.84	.2995	.7995	.2005	.2803
0.85	.3023	.8023	.1977	.2780
0.86	.3051	.8051	.1949	.2756
0.87	.3078	.8078	.1922	.2732
0.88	.3106	.8106	.1894	.2709
0.89	.3133	.8133	.1867	.2685
0.90	.3159	.8159	.1841	.2661
0.91	.3186	.8186	.1814	.2637
0.92	.3212	.8212	.1788	.2613
0.93	.3238	.8238	.1762	.2589
0.94	.3264	.8264	.1736	.2565
0.95	.3289	.8289	.1711	.2541
0.96	.3315	.8315	.1685	.2516
0.97	.3340	.8340	.1660	.2492
0.98	.3365	.8365	.1635	.2468
0.99	.3389	.8389	.1611	.2444
1.00	.3413	.8413	.1587	.2420
1.01	.3438	.8438	.1562	.2396
1.02	.3461	.8461	.1539	.2371
1.03	.3485	.8485	.1515	.2347
1.04	.3508	.8508	.1492	.2323

A.38 (*continued*)

(1) z Standard Score $\left(\frac{x}{\sigma}\right)$	(2) A Area from Mean to $\frac{x}{\sigma}$	(3) B Area in Larger Portion	(4) C Area in Smaller Portion	(5) y Ordinate at $\frac{x}{\sigma}$
1.05	.3531	.8531	.1469	.2299
1.06	.3554	.8554	.1446	.2275
1.07	.3577	.8577	.1423	.2251
1.08	.3599	.8599	.1401	.2227
1.09	.3621	.8621	.1379	.2203
1.10	.3643	.8643	.1357	.2179
1.11	.3665	.8665	.1335	.2155
1.12	.3686	.8686	.1314	.2131
1.13	.3708	.8708	.1292	.2107
1.14	.3729	.8729	.1271	.2083
1.15	.3749	.8749	.1251	.2059
1.16	.3770	.8770	.1230	.2036
1.17	.3790	.8790	.1210	.2012
1.18	.3810	.8810	.1190	.1989
1.19	.3830	.8830	.1170	.1965
1.20	.3849	.8849	.1151	.1942
1.21	.3869	.8869	.1131	.1919
1.22	.3888	.8888	.1112	.1895
1.23	.3907	.8907	.1093	.1872
1.24	.3925	.8925	.1075	.1849
1.25	.3944	.8944	.1056	.1826
1.26	.3962	.8962	.1038	.1804
1.27	.3980	.8980	.1020	.1781
1.28	.3997	.8997	.1003	.1758
1.29	.4015	.9015	.0985	.1736
1.30	.4032	.9032	.0968	.1714
1.31	.4049	.9049	.0951	.1691
1.32	.4066	.9066	.0934	.1669
1.33	.4082	.9082	.0918	.1647
1.34	.4099	.9099	.0901	.1626
1.35	.4115	.9115	.0885	.1604
1.36	.4131	.9131	.0869	.1582
1.37	.4147	.9147	.0853	.1561
1.38	.4162	.9162	.0838	.1539
1.39	.4177	.9177	.0823	.1518

A.38 (*continued*)

(1) z Standard Score $\left(\frac{x}{\sigma}\right)$	(2) A Area from Mean to $\frac{x}{\sigma}$	(3) B Area in Larger Portion	(4) C Area in Smaller Portion	(5) y Ordinate at $\frac{x}{\sigma}$
1.40	.4192	.9192	.0808	.1497
1.41	.4207	.9207	.0793	.1476
1.42	.4222	.9222	.0778	.1456
1.43	.4236	.9236	.0764	.1435
1.44	.4251	.9251	.0749	.1415
1.45	.4265	.9265	.0735	.1394
1.46	.4279	.9279	.0721	.1374
1.47	.4292	.9292	.0708	.1354
1.48	.4306	.9306	.0694	.1334
1.49	.4319	.9319	.0681	.1315
1.50	.4332	.9332	.0668	.1295
1.51	.4345	.9345	.0655	.1276
1.52	.4357	.9357	.0643	.1257
1.53	.4370	.9370	.0630	.1238
1.54	.4382	.9382	.0618	.1219
1.55	.4394	.9394	.0606	.1200
1.56	.4406	.9406	.0594	.1182
1.57	.4418	.9418	.0582	.1163
1.58	.4429	.9429	.0571	.1145
1.59	.4441	.9441	.0559	.1127
1.60	.4452	.9452	.0548	.1109
1.61	.4463	.9463	.0537	.1092
1.62	.4474	.9474	.0526	.1074
1.63	.4484	.9484	.0516	.1057
1.64	.4495	.9495	.0505	.1040
1.65	.4505	.9505	.0495	.1023
1.66	.4515	.9515	.0485	.1006
1.67	.4525	.9525	.0475	.0989
1.68	.4535	.9535	.0465	.0973
1.69	.4545	.9545	.0455	.0957
1.70	.4554	.9554	.0446	.0940
1.71	.4564	.9564	.0436	.0925
1.72	.4573	.9573	.0427	.0909
1.73	.4582	.9582	.0418	.0893
1.74	.4591	.9591	.0409	.0878

A.38 (*continued*)

(1) z Standard Score $\left(\frac{x}{\sigma}\right)$	(2) A Area from Mean to $\frac{x}{\sigma}$	(3) B Area in Larger Portion	(4) C Area in Smaller Portion	(5) y Ordinate at $\frac{x}{\sigma}$
1.75	.4599	.9599	.0401	.0863
1.76	.4608	.9608	.0392	.0848
1.77	.4616	.9616	.0384	.0833
1.78	.4625	.9625	.0375	.0818
1.79	.4633	.9633	.0367	.0804
1.80	.4641	.9641	.0359	.0790
1.81	.4649	.9649	.0351	.0775
1.82	.4656	.9656	.0344	.0761
1.83	.4664	.9664	.0336	.0748
1.84	.4671	.9671	.0329	.0734
1.85	.4678	.9678	.0322	.0721
1.86	.4686	.9686	.0314	.0707
1.87	.4693	.9693	.0307	.0694
1.88	.4699	.9699	.0301	.0681
1.89	.4706	.9706	.0294	.0669
1.90	.4713	.9713	.0287	.0656
1.91	.4719	.9719	.0281	.0644
1.92	.4726	.9726	.0274	.0632
1.93	.4732	.9732	.0268	.0620
1.94	.4738	.9738	.0262	.0608
1.95	.4744	.9744	.0256	.0596
1.96	.4750	.9750	.0250	.0584
1.97	.4756	.9756	.0244	.0573
1.98	.4761	.9761	.0239	.0562
1.99	.4767	.9767	.0233	.0551
2.00	.4772	.9772	.0228	.0540
2.01	.4778	.9778	.0222	.0529
2.02	.4783	.9783	.0217	.0519
2.03	.4788	.9788	.0212	.0508
2.04	.4793	.9793	.0207	.0498
2.05	.4798	.9798	.0202	.0488
2.06	.4803	.9803	.0197	.0478
2.07	.4808	.9808	.0192	.0468
2.08	.4812	.9812	.0188	.0459
2.09	.4817	.9817	.0183	.0449

A.38 (*continued*)

(1) z Standard Score $\left(\frac{x}{\sigma}\right)$	(2) A Area from Mean to $\frac{x}{\sigma}$	(3) B Area in Larger Portion	(4) C Area in Smaller Portion	(5) y Ordinate at $\frac{x}{\sigma}$
2.10	.4821	.9821	.0179	.0440
2.11	.4826	.9826	.0174	.0431
2.12	.4830	.9830	.0170	.0422
2.13	.4834	.9834	.0166	.0413
2.14	.4838	.9838	.0162	.0404
2.15	.4842	.9842	.0158	.0396
2.16	.4846	.9846	.0154	.0387
2.17	.4850	.9850	.0150	.0379
2.18	.4854	.9854	.0146	.0371
2.19	.4857	.9857	.0143	.0363
2.20	.4861	.9861	.0139	.0355
2.21	.4864	.9864	.0136	.0347
2.22	.4868	.9868	.0132	.0339
2.23	.4871	.9871	.0129	.0332
2.24	.4875	.9875	.0125	.0325
2.25	.4878	.9878	.0122	.0317
2.26	.4881	.9881	.0119	.0310
2.27	.4884	.9884	.0116	.0303
2.28	.4887	.9887	.0113	.0297
2.29	.4890	.9890	.0110	.0290
2.30	.4893	.9893	.0107	.0283
2.31	.4896	.9896	.0104	.0277
2.32	.4898	.9898	.0102	.0270
2.33	.4901	.9901	.0099	.0264
2.34	.4904	.9904	.0096	.0258
2.35	.4906	.9906	.0094	.0252
2.36	.4909	.9909	.0091	.0246
2.37	.4911	.9911	.0089	.0241
2.38	.4913	.9913	.0087	.0235
2.39	.4916	.9916	.0084	.0229
2.40	.4918	.9918	.0082	.0224
2.41	.4920	.9920	.0080	.0219
2.42	.4922	.9922	.0078	.0213
2.43	.4925	.9925	.0075	.0208
2.44	.4927	.9927	.0073	.0203

A.38 (*continued*)

(1) z Standard Score $\left(\frac{x}{\sigma}\right)$	(2) A Area from Mean to $\frac{x}{\sigma}$	(3) B Area in Larger Portion	(4) C Area in Smaller Portion	(5) y Ordinate at $\frac{x}{\sigma}$
2.45	.4929	.9929	.0071	.0198
2.46	.4931	.9931	.0069	.0194
2.47	.4932	.9932	.0068	.0189
2.48	.4934	.9934	.0066	.0184
2.49	.4936	.9936	.0064	.0180
2.50	.4938	.9938	.0062	.0175
2.51	.4940	.9940	.0060	.0171
2.52	.4941	.9941	.0059	.0167
2.53	.4943	.9943	.0057	.0163
2.54	.4945	.9945	.0055	.0158
2.55	.4946	.9946	.0054	.0154
2.56	.4948	.9948	.0052	.0151
2.57	.4949	.9949	.0051	.0147
2.58	.4951	.9951	.0049	.0143
2.59	.4952	.9952	.0048	.0139
2.60	.4953	.9953	.0047	.0136
2.61	.4955	.9955	.0045	.0132
2.62	.4956	.9956	.0044	.0129
2.63	.4957	.9957	.0043	.0126
2.64	.4959	.9959	.0041	.0122
2.65	.4960	.9960	.0040	.0119
2.66	.4961	.9961	.0039	.0116
2.67	.4962	.9962	.0038	.0113
2.68	.4963	.9963	.0037	.0110
2.69	.4964	.9964	.0036	.0107
2.70	.4965	.9965	.0035	.0104
2.71	.4966	.9966	.0034	.0101
2.72	.4967	.9967	.0033	.0099
2.73	.4968	.9968	.0032	.0096
2.74	.4969	.9969	.0031	.0093
2.75	.4970	.9970	.0030	.0091
2.76	.4971	.9971	.0029	.0088
2.77	.4972	.9972	.0028	.0086
2.78	.4973	.9973	.0027	.0084
2.79	.4974	.9974	.0026	.0081

A.38 (*continued*)

(1) z Standard Score $\left(\frac{x}{\sigma}\right)$	(2) A Area from Mean to $\frac{x}{\sigma}$	(3) B Area in Larger Portion	(4) C Area in Smaller Portion	(5) y Ordinate at $\frac{x}{\sigma}$
2.80	.4974	.9974	.0026	.0079
2.81	.4975	.9975	.0025	.0077
2.82	.4976	.9976	.0024	.0075
2.83	.4977	.9977	.0023	.0073
2.84	.4977	.9977	.0023	.0071
2.85	.4978	.9978	.0022	.0069
2.86	.4979	.9979	.0021	.0067
2.87	.4979	.9979	.0021	.0065
2.88	.4980	.9980	.0020	.0063
2.89	.4981	.9981	.0019	.0061
2.90	.4981	.9981	.0019	.0060
2.91	.4982	.9982	.0018	.0058
2.92	.4982	.9982	.0018	.0056
2.93	.4983	.9983	.0017	.0055
2.94	.4984	.9984	.0016	.0053
2.95	.4984	.9984	.0016	.0051
2.96	.4985	.9985	.0015	.0050
2.97	.4985	.9985	.0015	.0048
2.98	.4986	.9986	.0014	.0047
2.99	.4986	.9986	.0014	.0046
3.00	.4987	.9987	.0013	.0044
3.01	.4987	.9987	.0013	.0043
3.02	.4987	.9987	.0013	.0042
3.03	.4988	.9988	.0012	.0040
3.04	.4988	.9988	.0012	.0039
3.05	.4989	.9989	.0011	.0038
3.06	.4989	.9989	.0011	.0037
3.07	.4989	.9989	.0011	.0036
3.08	.4990	.9990	.0010	.0035
3.09	.4990	.9990	.0010	.0034
3.10	.4990	.9990	.0010	.0033
3.11	.4991	.9991	.0009	.0032
3.12	.4991	.9991	.0009	.0031
3.13	.4991	.9991	.0009	.0030
3.14	.4992	.9992	.0008	.0029

A.38 *(continued)*

(1) z Standard Score $\left(\frac{x}{\sigma}\right)$	(2) A Area from Mean to $\frac{x}{\sigma}$	(3) B Area in Larger Portion	(4) C Area in Smaller Portion	(5) y Ordinate at $\frac{x}{\sigma}$
3.15	.4992	.9992	.0008	.0028
3.16	.4992	.9992	.0008	.0027
3.17	.4992	.9992	.0008	.0026
3.18	.4993	.9993	.0007	.0025
3.19	.4993	.9993	.0007	.0025
3.20	.4993	.9993	.0007	.0024
3.21	.4993	.9993	.0007	.0023
3.22	.4994	.9994	.0006	.0022
3.23	.4994	.9994	.0006	.0022
3.24	.4994	.9994	.0006	.0021
3.30	.4995	.9995	.0005	.0017
3.40	.4997	.9997	.0003	.0012
3.50	.4998	.9998	.0002	.0009
3.60	.4998	.9998	.0002	.0006
3.70	.4999	.9999	.0001	.0004

A.39 CHI SQUARE TABLE

n	·99	·98	·95	·90	·80	·70	·50	·30	·20	·10	·05	·02	·01	·001
1	$·0^3157$	$·0^3628$	·00393	·0158	·0642	·148	·455	1·074	1·642	2·706	3·841	5·412	6·635	10·827
2	·0201	·0404	·103	·211	·446	·713	1·386	2·408	3·219	4·605	5·991	7·824	9·210	13·815
3	·115	·185	·352	·584	1·005	1·424	2·366	3·665	4·642	6·251	7·815	9·837	11·345	16·266
4	·297	·429	·711	1·064	1·649	2·195	3·357	4·878	5·989	7·779	9·488	11·668	13·277	18·467
5	·554	·752	1·145	1·610	2·343	3·000	4·351	6·064	7·289	9·236	11·070	13·388	15·086	20·515
6	·872	1·134	1·635	2·204	3·070	3·828	5·348	7·231	8·558	10·645	12·592	15·033	16·812	22·457
7	1·239	1·564	2·167	2·833	3·822	4·671	6·346	8·383	9·803	12·017	14·067	16·622	18·475	24·322
8	1·646	2·032	2·733	3·490	4·594	5·527	7·344	9·524	11·030	13·362	15·507	18·168	20·090	26·125
9	2·088	2·532	3·325	4·168	5·380	6·393	8·343	10·656	12·242	14·684	16·919	19·679	21·666	27·877
10	2·558	3·059	3·940	4·865	6·179	7·267	9·342	11·781	13·442	15·987	18·307	21·161	23·209	29·588
11	3·053	3·609	4·575	5·578	6·989	8·148	10·341	12·899	14·631	17·275	19·675	22·618	24·725	31·264
12	3·571	4·178	5·226	6·304	7·807	9·034	11·340	14·011	15·812	18·549	21·026	24·054	26·217	32·909
13	4·107	4·765	5·892	7·042	8·634	9·926	12·340	15·119	16·985	19·812	22·362	25·472	27·688	34·528
14	4·660	5·368	6·571	7·790	9·467	10·821	13·339	16·222	18·151	21·064	23·685	26·873	29·141	36·123
15	5·229	5·985	7·261	8·547	10·307	11·721	14·339	17·322	19·311	22·307	24·996	28·259	30·578	37·697
16	5·812	6·614	7·962	9·312	11·152	12·624	15·338	18·418	20·465	23·542	26·296	29·633	32·000	39·252
17	6·408	7·255	8·672	10·085	12·002	13·531	16·338	19·511	21·615	24·769	27·587	30·995	33·409	40·790
18	7·015	7·906	9·390	10·865	12·857	14·440	17·338	20·601	22·760	25·989	28·869	32·346	34·805	42·312
19	7·633	8·567	10·117	11·651	13·716	15·352	18·338	21·689	23·900	27·204	30·144	33·687	36·191	43·820
20	8·260	9·237	10·851	12·443	14·578	16·266	19·337	22·775	25·038	28·412	31·410	35·020	37·566	45·315
21	8·897	9·915	11·591	13·240	15·445	17·182	20·337	23·858	26·171	29·615	32·671	36·343	38·932	46·797
22	9·542	10·600	12·338	14·041	16·314	18·101	21·337	24·939	27·301	30·813	33·924	37·659	40·289	48·268
23	10·196	11·293	13·091	14·848	17·187	19·021	22·337	26·018	28·429	32·007	35·172	38·968	41·638	49·728
24	10·856	11·992	13·848	15·659	18·062	19·943	23·337	27·096	29·553	33·196	36·415	40·270	42·980	51·179
25	11·524	12·697	14·611	16·473	18·940	20·867	24·337	28·172	30·675	34·382	37·652	41·566	44·314	52·620
26	12·198	13·409	15·379	17·292	19·820	21·792	25·336	29·246	31·795	35·563	38·885	42·856	45·642	54·052
27	12·879	14·125	16·151	18·114	20·703	22·719	26·336	30·319	32·912	36·741	40·113	44·140	46·963	55·476
28	13·565	14·847	16·928	18·939	21·588	23·647	27·336	31·391	34·027	37·916	41·337	45·419	48·278	56·893
29	14·256	15·574	17·708	19·768	22·475	24·577	28·336	32·461	35·139	39·087	42·557	46·693	49·588	58·302
30	14·953	16·306	18·493	20·599	23·364	25·508	29·336	33·530	36·250	40·256	43·773	47·962	50·892	59·703

For larger values of n, the expression $\sqrt{2\chi^2} - \sqrt{2n-1}$ may be used as a normal deviate with unit variance, remembering that the probability for χ^2 corresponds with that of a single tail of the normal curve.

Source: Table taken from Table IV of Fisher and Yates, *Statistical Tables for Biological, Agricultural and Medical Research,* published by Oliver and Boyd, Edinburgh, and by permission of the authors and publishers.

A.40 VALUES OF THE CORRELATION COEFFICIENT FOR DIFFERENT LEVELS OF SIGNIFICANCE

n	·1	·05	·02	·01	·001	*n*	·1	·05	·02	·01	·001
1	·98769	·99692	·999507	·999877	·9999988	16	·4000	·4683	·5425	·5897	·7084
2	·90000	·95000	·98000	·990000	·99900	17	·3887	·4555	·5285	·5751	·6932
3	·8054	·8783	·93433	·95873	·99116	18	·3783	·4438	·5155	·5614	·6787
4	·7293	·8114	·8822	·91720	·97406	19	·3687	·4329	·5034	·5487	·6652
5	·6694	·7545	·8329	·8745	·95074	20	·3598	·4227	·4921	·5368	·6524
6	·6215	·7067	·7887	·8343	·92493	25	·3233	·3809	·4451	·4869	·5974
7	·5822	·6664	·7498	·7977	·8982	30	·2960	·3494	·4093	·4487	·5541
8	·5494	·6319	·7155	·7646	·8721	35	·2746	·3246	·3810	·4182	·5189
9	·5214	·6021	·6851	·7348	·8471	40	·2573	·3044	·3578	·3932	·4896
10	·4973	·5760	·6581	·7079	·8233	45	·2428	·2875	·3384	·3721	·4648
11	·4762	·5529	·6339	·6835	·8010	50	·2306	·2732	·3218	·3541	·4433
12	·4575	·5324	·6120	·6614	·7800	60	·2108	·2500	·2948	·3248	·4078
13	·4409	·5139	·5923	·6411	·7603	70	·1954	·2319	·2737	·3017	·3799
14	·4259	·4973	·5742	·6226	·7420	80	·1829	·2172	·2565	·2830	·3568
15	·4124	·4821	·5577	·6055	·7246	90	·1726	·2050	·2422	·2673	·3375
						100	·1638	·1946	·2301	·2540	·3211

Source: Table taken from Table VI of Fisher and Yates, *Statistical Tables for Biological, Agricultural and Medical Research*, published by Oliver and Boyd, Edinburgh, and by permission of the authors and publishers.

A.41 THE 5 (ROMAN TYPE) AND 1 (BOLDFACE TYPE) PER CENT POINTS FOR THE DISTRIBUTION OF F*

n_2	n_1 degrees of freedom (for greater mean square)																							
	1	2	3	4	5	6	7	8	9	10	11	12	14	16	20	24	30	40	50	75	100	200	500	∞
1	161	200	216	225	230	234	237	239	241	242	243	244	245	246	248	249	250	251	252	253	253	254	254	254
	4,052	**4,999**	**5,403**	**5,625**	**5,764**	**5,859**	**5,928**	**5,981**	**6,022**	**6,056**	**6,082**	**6,106**	**6,142**	**6,169**	**6,208**	**6,234**	**6,258**	**6,286**	**6,302**	**6,323**	**6,334**	**6,352**	**6,361**	**6,366**
2	18.51	19.00	19.16	19.25	19.30	19.33	19.36	19.37	19.38	19.39	19.40	19.41	19.42	19.43	19.44	19.45	19.46	19.47	19.47	19.48	19.49	19.49	19.50	19.50
	98.49	**99.00**	**99.17**	**99.25**	**99.30**	**99.33**	**99.34**	**99.36**	**99.38**	**99.40**	**99.41**	**99.42**	**99.43**	**99.44**	**99.45**	**99.46**	**99.47**	**99.48**	**99.48**	**99.49**	**99.49**	**99.49**	**99.50**	**99.50**
3	10.13	9.55	9.28	9.12	9.01	8.94	8.88	8.84	8.81	8.78	8.76	8.74	8.71	8.69	8.66	8.64	8.62	8.60	8.58	8.57	8.56	8.54	8.54	8.53
	34.12	**30.82**	**29.46**	**28.71**	**28.24**	**27.91**	**27.67**	**27.49**	**27.34**	**27.23**	**27.13**	**27.05**	**26.92**	**26.83**	**26.69**	**26.60**	**26.50**	**26.41**	**26.35**	**26.27**	**26.23**	**26.18**	**26.14**	**26.12**
4	7.71	6.94	6.59	6.39	6.26	6.16	6.09	6.04	6.00	5.96	5.93	5.91	5.87	5.84	5.80	5.77	5.74	5.71	5.70	5.68	5.66	5.65	5.64	5.63
	21.20	**18.00**	**16.69**	**15.98**	**15.52**	**15.21**	**14.98**	**14.80**	**14.66**	**14.54**	**14.45**	**14.37**	**14.24**	**14.15**	**14.02**	**13.93**	**13.83**	**13.74**	**13.69**	**13.61**	**13.57**	**13.52**	**13.48**	**13.46**
5	6.61	5.79	5.41	5.19	5.05	4.95	4.88	4.82	4.78	4.74	4.70	4.68	4.64	4.60	4.56	4.53	4.50	4.46	4.44	4.42	4.40	4.38	4.37	4.36
	16.26	**13.27**	**12.06**	**11.39**	**10.97**	**10.67**	**10.45**	**10.27**	**10.15**	**10.05**	**9.96**	**9.89**	**9.77**	**9.68**	**9.55**	**9.47**	**9.38**	**9.29**	**9.24**	**9.17**	**9.13**	**9.07**	**9.04**	**9.02**
6	5.99	5.14	4.76	4.53	4.39	4.28	4.21	4.15	4.10	4.06	4.03	4.00	3.96	3.92	3.87	3.84	3.81	3.77	3.75	3.72	3.71	3.69	3.68	3.67
	13.74	**10.92**	**9.78**	**9.15**	**8.75**	**8.47**	**8.26**	**8.10**	**7.98**	**7.87**	**7.79**	**7.72**	**7.60**	**7.52**	**7.39**	**7.31**	**7.23**	**7.14**	**7.09**	**7.02**	**6.99**	**6.94**	**6.90**	**6.88**
7	5.59	4.74	4.35	4.12	3.97	3.87	3.79	3.73	3.68	3.63	3.60	3.57	3.52	3.49	3.44	3.41	3.38	3.34	3.32	3.29	3.28	3.25	3.24	3.23
	12.25	**9.55**	**8.45**	**7.85**	**7.46**	**7.19**	**7.00**	**6.84**	**6.71**	**6.62**	**6.54**	**6.47**	**6.35**	**6.27**	**6.15**	**6.07**	**5.98**	**5.90**	**5.85**	**5.78**	**5.75**	**5.70**	**5.67**	**5.65**
8	5.32	4.46	4.07	3.84	3.69	3.58	3.50	3.44	3.39	3.34	3.31	3.28	3.23	3.20	3.15	3.12	3.08	3.05	3.03	3.00	2.98	2.96	2.94	2.93
	11.26	**8.65**	**7.59**	**7.01**	**6.63**	**6.37**	**6.19**	**6.03**	**5.91**	**5.82**	**5.74**	**5.67**	**5.56**	**5.48**	**5.36**	**5.28**	**5.20**	**5.11**	**5.06**	**5.00**	**4.96**	**4.91**	**4.88**	**4.86**
9	5.12	4.26	3.86	3.63	3.48	3.37	3.29	3.23	3.18	3.13	3.10	3.07	3.02	2.98	2.93	2.90	2.86	2.82	2.80	2.77	2.76	2.73	2.72	2.71
	10.56	**8.02**	**6.99**	**6.42**	**6.06**	**5.80**	**5.62**	**5.47**	**5.35**	**5.26**	**5.18**	**5.11**	**5.00**	**4.92**	**4.80**	**4.73**	**4.64**	**4.56**	**4.51**	**4.45**	**4.41**	**4.36**	**4.33**	**4.31**
10	4.96	4.10	3.71	3.48	3.33	3.22	3.14	3.07	3.02	2.97	2.94	2.91	2.86	2.82	2.77	2.74	2.70	2.67	2.64	2.61	2.59	2.56	2.55	2.54
	10.04	**7.56**	**6.55**	**5.99**	**5.64**	**5.39**	**5.21**	**5.06**	**4.95**	**4.85**	**4.78**	**4.71**	**4.60**	**4.52**	**4.41**	**4.33**	**4.25**	**4.17**	**4.12**	**4.05**	**4.01**	**3.96**	**3.93**	**3.91**
11	4.84	3.98	3.59	3.36	3.20	3.09	3.01	2.95	2.90	2.86	2.82	2.79	2.74	2.70	2.65	2.61	2.57	2.53	2.50	2.47	2.45	2.42	2.41	2.40
	9.65	**7.20**	**6.22**	**5.67**	**5.32**	**5.07**	**4.88**	**4.74**	**4.63**	**4.54**	**4.46**	**4.40**	**4.29**	**4.21**	**4.10**	**4.02**	**3.94**	**3.86**	**3.80**	**3.74**	**3.70**	**3.66**	**3.62**	**3.60**
12	4.75	3.88	3.49	3.26	3.11	3.00	2.92	2.85	2.80	2.76	2.72	2.69	2.64	2.60	2.54	2.50	2.46	2.42	2.40	2.36	2.35	2.32	2.31	2.30
	9.33	**6.93**	**5.95**	**5.41**	**5.06**	**4.82**	**4.65**	**4.50**	**4.39**	**4.30**	**4.22**	**4.16**	**4.05**	**3.98**	**3.86**	**3.78**	**3.70**	**3.61**	**3.56**	**3.49**	**3.46**	**3.41**	**3.38**	**3.36**
13	4.67	3.80	3.41	3.18	3.02	2.92	2.84	2.77	2.72	2.67	2.63	2.60	2.55	2.51	2.46	2.42	2.38	2.34	2.32	2.28	2.26	2.24	2.22	2.21
	9.07	**6.70**	**5.74**	**5.20**	**4.86**	**4.62**	**4.44**	**4.30**	**4.19**	**4.10**	**4.02**	**3.96**	**3.85**	**3.78**	**3.67**	**3.59**	**3.51**	**3.42**	**3.37**	**3.30**	**3.27**	**3.21**	**3.18**	**3.16**

*Source: Reprinted by permission from *Statistical Methods*, 6th edition, by George W. Snedecor and William G. Cochran, © 1967 by The Iowa State University Press, Ames, Iowa.

A.41 (*continued*)

n_2	1	2	3	4	5	6	7	8	9	10	11	12	14	16	20	24	30	40	50	75	100	200	500	∞
	n_1 degrees of freedom (for greater mean square)																							
14	4.60	3.74	3.34	3.11	2.96	2.85	2.77	2.70	2.65	2.60	2.56	2.53	2.48	2.44	2.39	2.35	2.31	2.27	2.24	2.21	2.19	2.16	2.14	2.13
	8.86	**6.51**	**5.56**	**5.03**	**4.69**	**4.46**	**4.28**	**4.14**	**4.03**	**3.94**	**3.86**	**3.80**	**3.70**	**3.62**	**3.51**	**3.43**	**3.34**	**3.26**	**3.21**	**3.14**	**3.11**	**3.06**	**3.02**	**3.00**
15	4.54	3.68	3.29	3.06	2.90	2.79	2.70	2.64	2.59	2.55	2.51	2.48	2.43	2.39	2.33	2.29	2.25	2.21	2.18	2.15	2.12	2.10	2.08	2.07
	8.68	**6.36**	**5.42**	**4.89**	**4.56**	**4.32**	**4.14**	**4.00**	**3.89**	**3.80**	**3.73**	**3.67**	**3.56**	**3.48**	**3.36**	**3.29**	**3.20**	**3.12**	**3.07**	**3.00**	**2.97**	**2.92**	**2.89**	**2.87**
16	4.49	3.63	3.24	3.01	2.85	2.74	2.66	2.59	2.54	2.49	2.45	2.42	2.37	2.33	2.28	2.24	2.20	2.16	2.13	2.09	2.07	2.04	2.02	2.01
	8.53	**6.23**	**5.29**	**4.77**	**4.44**	**4.20**	**4.03**	**3.89**	**3.78**	**3.69**	**3.61**	**3.55**	**3.45**	**3.37**	**3.25**	**3.18**	**3.10**	**3.01**	**2.96**	**2.89**	**2.86**	**2.80**	**2.77**	**2.75**
17	4.45	3.59	3.20	2.96	2.81	2.70	2.62	2.55	2.50	2.45	2.41	2.38	2.33	2.29	2.23	2.19	2.15	2.11	2.08	2.04	2.02	1.99	1.97	1.96
	8.40	**6.11**	**5.18**	**4.67**	**4.34**	**4.10**	**3.93**	**3.79**	**3.68**	**3.59**	**3.52**	**3.45**	**3.35**	**3.27**	**3.16**	**3.08**	**3.00**	**2.92**	**2.86**	**2.79**	**2.76**	**2.70**	**2.67**	**2.65**
18	4.41	3.55	3.16	2.93	2.77	2.66	2.58	2.51	2.46	2.41	2.37	2.34	2.29	2.25	2.19	2.15	2.11	2.07	2.04	2.00	1.98	1.95	1.93	1.92
	8.28	**6.01**	**5.09**	**4.58**	**4.25**	**4.01**	**3.85**	**3.71**	**3.60**	**3.51**	**3.44**	**3.37**	**3.27**	**3.19**	**3.07**	**3.00**	**2.91**	**2.83**	**2.78**	**2.71**	**2.68**	**2.62**	**2.59**	**2.57**
19	4.38	3.52	3.13	2.90	2.74	2.63	2.55	2.48	2.43	2.38	2.34	2.31	2.26	2.21	2.15	2.11	2.07	2.02	2.00	1.96	1.94	1.91	1.90	1.88
	8.18	**5.93**	**5.01**	**4.50**	**4.17**	**3.94**	**3.77**	**3.63**	**3.52**	**3.43**	**3.36**	**3.30**	**3.19**	**3.12**	**3.00**	**2.92**	**2.84**	**2.76**	**2.70**	**2.63**	**2.60**	**2.54**	**2.51**	**2.49**
20	4.35	3.49	3.10	2.87	2.71	2.60	2.52	2.45	2.40	2.35	2.31	2.28	2.23	2.18	2.12	2.08	2.04	1.99	1.96	1.92	1.90	1.87	1.85	1.84
	8.10	**5.85**	**4.94**	**4.43**	**4.10**	**3.87**	**3.71**	**3.56**	**3.45**	**3.37**	**3.30**	**3.23**	**3.13**	**3.05**	**2.94**	**2.86**	**2.77**	**2.69**	**2.63**	**2.56**	**2.53**	**2.47**	**2.44**	**2.42**
21	4.32	3.47	3.07	2.84	2.68	2.57	2.49	2.42	2.37	2.32	2.28	2.25	2.20	2.15	2.09	2.05	2.00	1.96	1.93	1.89	1.87	1.84	1.82	1.81
	8.02	**5.78**	**4.87**	**4.37**	**4.04**	**3.81**	**3.65**	**3.51**	**3.40**	**3.31**	**3.24**	**3.17**	**3.07**	**2.99**	**2.88**	**2.80**	**2.72**	**2.63**	**2.58**	**2.51**	**2.47**	**2.42**	**2.38**	**2.36**
22	4.30	3.44	3.05	2.82	2.66	2.55	2.47	2.40	2.35	2.30	2.26	2.23	2.18	2.13	2.07	2.03	1.98	1.93	1.91	1.87	1.84	1.81	1.80	1.78
	7.94	**5.72**	**4.82**	**4.31**	**3.99**	**3.76**	**3.59**	**3.45**	**3.35**	**3.26**	**3.18**	**3.12**	**3.02**	**2.94**	**2.83**	**2.75**	**2.67**	**2.58**	**2.53**	**2.46**	**2.42**	**2.37**	**2.33**	**2.31**
23	4.28	3.42	3.03	2.80	2.64	2.53	2.45	2.38	2.32	2.28	2.24	2.20	2.14	2.10	2.04	2.00	1.96	1.91	1.88	1.84	1.82	1.79	1.77	1.76
	7.88	**5.66**	**4.76**	**4.26**	**3.94**	**3.71**	**3.54**	**3.41**	**3.30**	**3.21**	**3.14**	**3.07**	**2.97**	**2.89**	**2.78**	**2.70**	**2.62**	**2.53**	**2.48**	**2.41**	**2.37**	**2.32**	**2.28**	**2.26**
24	4.26	3.40	3.01	2.78	2.62	2.51	2.43	2.36	2.30	2.26	2.22	2.18	2.13	2.09	2.02	1.98	1.94	1.89	1.86	1.82	1.80	1.76	1.74	1.73
	7.82	**5.61**	**4.72**	**4.22**	**3.90**	**3.67**	**3.50**	**3.36**	**3.25**	**3.17**	**3.09**	**3.03**	**2.93**	**2.85**	**2.74**	**2.66**	**2.58**	**2.49**	**2.44**	**2.36**	**2.33**	**2.27**	**2.23**	**2.21**
25	4.24	3.38	2.99	2.76	2.60	2.49	2.41	2.34	2.28	2.24	2.20	2.16	2.11	2.06	2.00	1.96	1.92	1.87	1.84	1.80	1.77	1.74	1.72	1.71
	7.77	**5.57**	**4.68**	**4.18**	**3.86**	**3.63**	**3.46**	**3.32**	**3.21**	**3.13**	**3.05**	**2.99**	**2.89**	**2.81**	**2.70**	**2.62**	**2.54**	**2.45**	**2.40**	**2.32**	**2.29**	**2.23**	**2.19**	**2.17**
26	4.22	3.37	2.98	2.74	2.59	2.47	2.39	2.32	2.27	2.22	2.18	2.15	2.10	2.05	1.99	1.95	1.90	1.85	1.82	1.78	1.76	1.72	1.70	1.69
	7.72	**5.53**	**4.64**	**4.14**	**3.82**	**3.59**	**3.42**	**3.29**	**3.17**	**3.09**	**3.02**	**2.96**	**2.86**	**2.77**	**2.66**	**2.58**	**2.50**	**2.41**	**2.36**	**2.28**	**2.25**	**2.19**	**2.15**	**2.13**

A.41 (*continued*)

n_2	n_1 degrees of freedom (for greater mean square)																							
	1	2	3	4	5	6	7	8	9	10	11	12	14	16	20	24	30	40	50	75	100	200	500	∞
27	4.21	3.35	2.96	2.73	2.57	2.46	2.37	2.30	2.25	2.20	2.16	2.13	2.08	2.03	1.97	1.93	1.88	1.84	1.80	1.76	1.74	1.71	1.68	1.67
	7.68	**5.49**	**4.60**	**4.11**	**3.79**	**3.56**	**3.39**	**3.26**	**3.14**	**3.06**	**2.98**	**2.93**	**2.83**	**2.74**	**2.63**	**2.55**	**2.47**	**2.38**	**2.33**	**2.25**	**2.21**	**2.16**	**2.12**	**2.10**
28	4.20	3.34	2.95	2.71	2.56	2.44	2.36	2.29	2.24	2.19	2.15	2.12	2.06	2.02	1.96	1.91	1.87	1.81	1.78	1.75	1.72	1.69	1.67	1.65
	7.64	**5.45**	**4.57**	**4.07**	**3.76**	**3.53**	**3.36**	**3.23**	**3.11**	**3.03**	**2.95**	**2.90**	**2.80**	**2.71**	**2.60**	**2.52**	**2.44**	**2.35**	**2.30**	**2.22**	**2.18**	**2.13**	**2.09**	**2.06**
29	4.18	3.33	2.93	2.70	2.54	2.43	2.35	2.28	2.22	2.18	2.14	2.10	2.05	2.00	1.94	1.90	1.85	1.80	1.77	1.73	1.71	1.68	1.65	1.64
	7.60	**5.42**	**4.54**	**4.04**	**3.73**	**3.50**	**3.33**	**3.20**	**3.08**	**3.00**	**2.92**	**2.87**	**2.77**	**2.68**	**2.57**	**2.49**	**2.41**	**2.32**	**2.27**	**2.19**	**2.15**	**2.10**	**2.06**	**2.03**
30	4.17	3.32	2.92	2.69	2.53	2.42	2.34	2.27	2.21	2.16	2.12	2.09	2.04	1.99	1.93	1.89	1.84	1.79	1.76	1.72	1.69	1.66	1.64	1.62
	7.56	**5.39**	**4.51**	**4.02**	**3.70**	**3.47**	**3.30**	**3.17**	**3.06**	**2.98**	**2.90**	**2.84**	**2.74**	**2.66**	**2.55**	**2.47**	**2.38**	**2.29**	**2.24**	**2.16**	**2.13**	**2.07**	**2.03**	**2.01**
32	4.15	3.30	2.90	2.67	2.51	2.40	2.32	2.25	2.19	2.14	2.10	2.07	2.02	1.97	1.91	1.86	1.82	1.76	1.74	1.69	1.67	1.64	1.61	1.59
	7.50	**5.34**	**4.46**	**3.97**	**3.66**	**3.42**	**3.25**	**3.12**	**3.01**	**2.94**	**2.86**	**2.80**	**2.70**	**2.62**	**2.51**	**2.42**	**2.34**	**2.25**	**2.20**	**2.12**	**2.08**	**2.02**	**1.98**	**1.96**
34	4.13	3.28	2.88	2.65	2.49	2.38	2.30	2.23	2.17	2.12	2.08	2.05	2.00	1.95	1.89	1.84	1.80	1.74	1.71	1.67	1.64	1.61	1.59	1.57
	7.44	**5.29**	**4.42**	**3.93**	**3.61**	**3.38**	**3.21**	**3.08**	**2.97**	**2.89**	**2.82**	**2.76**	**2.66**	**2.58**	**2.47**	**2.38**	**2.30**	**2.21**	**2.15**	**2.08**	**2.04**	**1.98**	**1.94**	**1.91**
36	4.11	3.26	2.86	2.63	2.48	2.36	2.28	2.21	2.15	2.10	2.06	2.03	1.98	1.93	1.87	1.82	1.78	1.72	1.69	1.65	1.62	1.59	1.56	1.55
	7.39	**5.25**	**4.38**	**3.89**	**3.58**	**3.35**	**3.18**	**3.04**	**2.94**	**2.86**	**2.78**	**2.72**	**2.62**	**2.54**	**2.43**	**2.35**	**2.26**	**2.17**	**2.12**	**2.04**	**2.00**	**1.94**	**1.90**	**1.87**
38	4.10	3.25	2.85	2.62	2.46	2.35	2.26	2.19	2.14	2.09	2.05	2.02	1.96	1.92	1.85	1.80	1.76	1.71	1.67	1.63	1.60	1.57	1.54	1.53
	7.35	**5.21**	**4.34**	**3.86**	**3.54**	**3.32**	**3.15**	**3.02**	**2.91**	**2.82**	**2.75**	**2.69**	**2.59**	**2.51**	**2.40**	**2.32**	**2.22**	**2.14**	**2.08**	**2.00**	**1.97**	**1.90**	**1.86**	**1.84**
40	4.08	3.23	2.84	2.61	2.45	2.34	2.25	2.18	2.12	2.07	2.04	2.00	1.95	1.90	1.84	1.79	1.74	1.69	1.66	1.61	1.59	1.55	1.53	1.51
	7.31	**5.18**	**4.31**	**3.83**	**3.51**	**3.29**	**3.12**	**2.99**	**2.88**	**2.80**	**2.73**	**2.66**	**2.56**	**2.49**	**2.37**	**2.29**	**2.20**	**2.11**	**2.05**	**1.97**	**1.94**	**1.88**	**1.84**	**1.81**
42	4.07	3.22	2.83	2.59	2.44	2.32	2.24	2.17	2.11	2.06	2.02	1.99	1.94	1.89	1.82	1.78	1.73	1.68	1.64	1.60	1.57	1.54	1.51	1.49
	7.27	**5.15**	**4.29**	**3.80**	**3.49**	**3.26**	**3.10**	**2.96**	**2.86**	**2.77**	**2.70**	**2.64**	**2.54**	**2.46**	**2.35**	**2.26**	**2.17**	**2.08**	**2.02**	**1.94**	**1.91**	**1.85**	**1.80**	**1.78**
44	4.06	3.21	2.82	2.58	2.43	2.31	2.23	2.16	2.10	2.05	2.01	1.98	1.92	1.88	1.81	1.76	1.72	1.66	1.63	1.58	1.56	1.52	1.50	1.48
	7.24	**5.12**	**4.26**	**3.78**	**3.46**	**3.24**	**3.07**	**2.94**	**2.84**	**2.75**	**2.68**	**2.62**	**2.52**	**2.44**	**2.32**	**2.24**	**2.15**	**2.06**	**2.00**	**1.92**	**1.88**	**1.82**	**1.78**	**1.75**
46	4.05	3.20	2.81	2.57	2.42	2.30	2.22	2.14	2.09	2.04	2.00	1.97	1.91	1.87	1.80	1.75	1.71	1.65	1.62	1.57	1.54	1.51	1.48	1.46
	7.21	**5.10**	**4.24**	**3.76**	**3.44**	**3.22**	**3.05**	**2.92**	**2.82**	**2.73**	**2.66**	**2.60**	**2.50**	**2.42**	**2.30**	**2.22**	**2.13**	**2.04**	**1.98**	**1.90**	**1.86**	**1.80**	**1.76**	**1.72**
48	4.04	3.19	2.80	2.56	2.41	2.30	2.21	2.14	2.08	2.03	1.99	1.96	1.90	1.86	1.79	1.74	1.70	1.64	1.61	1.56	1.53	1.50	1.47	1.45
	7.19	**5.08**	**4.22**	**3.74**	**3.42**	**3.20**	**3.04**	**2.90**	**2.80**	**2.71**	**2.64**	**2.58**	**2.48**	**2.40**	**2.28**	**2.20**	**2.11**	**2.02**	**1.96**	**1.88**	**1.84**	**1.78**	**1.73**	**1.70**

A.41 (*continued*)

n_2	n_1 degrees of freedom (for greater mean square) 1	2	3	4	5	6	7	8	9	10	11	12	14	16	20	24	30	40	50	75	100	200	500	∞
50	4.03	3.18	2.79	2.56	2.40	2.29	2.20	2.13	2.07	2.02	1.98	1.95	1.90	1.85	1.78	1.74	1.69	1.63	1.60	1.55	1.52	1.48	1.46	1.44
	7.17	**5.06**	**4.20**	**3.72**	**3.41**	**3.18**	**3.02**	**2.88**	**2.78**	**2.70**	**2.62**	**2.56**	**2.46**	**2.39**	**2.26**	**2.18**	**2.10**	**2.00**	**1.94**	**1.86**	**1.82**	**1.76**	**1.71**	**1.68**
55	4.02	3.17	2.78	2.54	2.38	2.27	2.18	2.11	2.05	2.00	1.97	1.93	1.88	1.83	1.76	1.72	1.67	1.61	1.58	1.52	1.50	1.46	1.43	1.41
	7.12	**5.01**	**4.16**	**3.68**	**3.37**	**3.15**	**2.98**	**2.85**	**2.75**	**2.66**	**2.59**	**2.53**	**2.43**	**2.35**	**2.23**	**2.15**	**2.06**	**1.96**	**1.90**	**1.82**	**1.78**	**1.71**	**1.66**	**1.64**
60	4.00	3.15	2.76	2.52	2.37	2.25	2.17	2.10	2.04	1.99	1.95	1.92	1.86	1.81	1.75	1.70	1.65	1.59	1.56	1.50	1.48	1.44	1.41	1.39
	7.08	**4.98**	**4.13**	**3.65**	**3.34**	**3.12**	**2.95**	**2.82**	**2.72**	**2.63**	**2.56**	**2.50**	**2.40**	**2.32**	**2.20**	**2.12**	**2.03**	**1.93**	**1.87**	**1.79**	**1.74**	**1.68**	**1.63**	**1.60**
65	3.99	3.14	2.75	2.51	2.36	2.24	2.15	2.08	2.02	1.98	1.94	1.90	1.85	1.80	1.73	1.68	1.63	1.57	1.54	1.49	1.46	1.42	1.39	1.37
	7.04	**4.95**	**4.10**	**3.62**	**3.31**	**3.09**	**2.93**	**2.79**	**2.70**	**2.61**	**2.54**	**2.47**	**2.37**	**2.30**	**2.18**	**2.09**	**2.00**	**1.90**	**1.84**	**1.76**	**1.71**	**1.64**	**1.60**	**1.56**
70	3.98	3.13	2.74	2.50	2.35	2.23	2.14	2.07	2.01	1.97	1.93	1.89	1.84	1.79	1.72	1.67	1.62	1.56	1.53	1.47	1.45	1.40	1.37	1.35
	7.01	**4.92**	**4.08**	**3.60**	**3.29**	**3.07**	**2.91**	**2.77**	**2.67**	**2.59**	**2.51**	**2.45**	**2.35**	**2.28**	**2.15**	**2.07**	**1.98**	**1.88**	**1.82**	**1.74**	**1.69**	**1.62**	**1.56**	**1.53**
80	3.96	3.11	2.72	2.48	2.33	2.21	2.12	2.05	1.99	1.95	1.91	1.88	1.82	1.77	1.70	1.65	1.60	1.54	1.51	1.45	1.42	1.38	1.35	1.32
	6.96	**4.88**	**4.04**	**3.56**	**3.25**	**3.04**	**2.87**	**2.74**	**2.64**	**2.55**	**2.48**	**2.41**	**2.32**	**2.24**	**2.11**	**2.03**	**1.94**	**1.84**	**1.78**	**1.70**	**1.65**	**1.57**	**1.52**	**1.49**
100	3.94	3.09	2.70	2.46	2.30	2.19	2.10	2.03	1.97	1.92	1.88	1.85	1.79	1.75	1.68	1.63	1.57	1.51	1.48	1.42	1.39	1.34	1.30	1.28
	6.90	**4.82**	**3.98**	**3.51**	**3.20**	**2.99**	**2.82**	**2.69**	**2.59**	**2.51**	**2.43**	**2.36**	**2.26**	**2.19**	**2.06**	**1.98**	**1.89**	**1.79**	**1.73**	**1.64**	**1.59**	**1.51**	**1.46**	**1.43**
125	3.92	3.07	2.68	2.44	2.29	2.17	2.08	2.01	1.95	1.90	1.86	1.83	1.77	1.72	1.65	1.60	1.55	1.49	1.45	1.39	1.36	1.31	1.27	1.25
	6.84	**4.78**	**3.94**	**3.47**	**3.17**	**2.95**	**2.79**	**2.65**	**2.56**	**2.47**	**2.40**	**2.33**	**2.23**	**2.15**	**2.03**	**1.94**	**1.85**	**1.75**	**1.68**	**1.59**	**1.54**	**1.46**	**1.40**	**1.37**
150	3.91	3.06	2.67	2.43	2.27	2.16	2.07	2.00	1.94	1.89	1.85	1.82	1.76	1.71	1.64	1.59	1.54	1.47	1.44	1.37	1.34	1.29	1.25	1.22
	6.81	**4.75**	**3.91**	**3.44**	**3.14**	**2.92**	**2.76**	**2.62**	**2.53**	**2.44**	**2.37**	**2.30**	**2.20**	**2.12**	**2.00**	**1.91**	**1.83**	**1.72**	**1.66**	**1.56**	**1.51**	**1.43**	**1.37**	**1.33**
200	3.89	3.04	2.65	2.41	2.26	2.14	2.05	1.98	1.92	1.87	1.83	1.80	1.74	1.69	1.62	1.57	1.52	1.45	1.42	1.35	1.32	1.26	1.22	1.19
	6.76	**4.71**	**3.88**	**3.41**	**3.11**	**2.90**	**2.73**	**2.60**	**2.50**	**2.41**	**2.34**	**2.28**	**2.17**	**2.09**	**1.97**	**1.88**	**1.79**	**1.69**	**1.62**	**1.53**	**1.48**	**1.39**	**1.33**	**1.28**
400	3.86	3.02	2.62	2.39	2.23	2.12	2.03	1.96	1.90	1.85	1.81	1.78	1.72	1.67	1.60	1.54	1.49	1.42	1.38	1.32	1.28	1.22	1.16	1.13
	6.70	**4.66**	**3.83**	**3.36**	**3.06**	**2.85**	**2.69**	**2.55**	**2.46**	**2.37**	**2.29**	**2.23**	**2.12**	**2.04**	**1.92**	**1.84**	**1.74**	**1.64**	**1.57**	**1.47**	**1.42**	**1.32**	**1.24**	**1.19**
1000	3.85	3.00	2.61	2.38	2.22	2.10	2.02	1.95	1.89	1.84	1.80	1.76	1.70	1.65	1.58	1.53	1.47	1.41	1.36	1.30	1.26	1.19	1.13	1.08
	6.66	**4.62**	**3.80**	**3.34**	**3.04**	**2.82**	**2.66**	**2.53**	**2.43**	**2.34**	**2.26**	**2.20**	**2.09**	**2.01**	**1.89**	**1.81**	**1.71**	**1.61**	**1.54**	**1.44**	**1.38**	**1.28**	**1.19**	**1.11**
∞	3.84	2.99	2.60	2.37	2.21	2.09	2.01	1.94	1.88	1.83	1.79	1.75	1.69	1.64	1.57	1.52	1.46	1.40	1.35	1.28	1.24	1.17	1.11	1.00
	6.64	**4.60**	**3.78**	**3.32**	**3.02**	**2.80**	**2.64**	**2.51**	**2.41**	**2.32**	**2.24**	**2.18**	**2.07**	**1.99**	**1.87**	**1.79**	**1.69**	**1.59**	**1.52**	**1.41**	**1.36**	**1.25**	**1.15**	**1.00**

A.42 DISTRIBUTION OF t

Probability.

n	·9	·8	·7	·6	·5	·4	·3	·2	·1	·05	·02	·01	·001
1	·158	·325	·510	·727	1·000	1·376	1·963	3·078	6·314	12·706	31·821	63·657	636·619
2	·142	·289	·445	·617	·816	1·061	1·386	1·886	2·920	4·303	6·965	9·925	31·598
3	·137	·277	·424	·584	·765	·978	1·250	1·638	2·353	3·182	4·541	5·841	12·924
4	·134	·271	·414	·569	·741	·941	1·190	1·533	2·132	2·776	3·747	4·604	8·610
5	·132	·267	·408	·559	·727	·920	1·156	1·476	2·015	2·571	3·365	4·032	6·869
6	·131	·265	·404	·553	·718	·906	1·134	1·440	1·943	2·447	3·143	3·707	5·959
7	·130	·263	·402	·549	·711	·896	1·119	1·415	1·895	2·365	2·998	3·499	5·408
8	·130	·262	·399	·546	·706	·889	1·108	1·397	1·860	2·306	2·896	3·355	5·041
9	·129	·261	·398	·543	·703	·883	1·100	1·383	1·833	2·262	2·821	3·250	4·781
10	·129	·260	·397	·542	·700	·879	1·093	1·372	1·812	2·228	2·764	3·169	4·58[illegible]
11	·129	·260	·396	·540	·697	·876	1·088	1·363	1·796	2·201	2·718	3·106	4·43[illegible]
12	·128	·259	·395	·539	·695	·873	1·083	1·356	1·782	2·179	2·681	3·055	4·318
13	·128	·259	·394	·538	·694	·870	1·079	1·350	1·771	2·160	2·650	3·012	4·221
14	·128	·258	·393	·537	·692	·868	1·076	1·345	1·761	2·145	2·624	2·977	4·140
15	·128	·258	·393	·536	·691	·866	1·074	1·341	1·753	2·131	2·602	2·947	4·073
16	·128	·258	·392	·535	·690	·865	1·071	1·337	1·746	2·120	2·583	2·921	4·015
17	·128	·257	·392	·534	·689	·863	1·069	1·333	1·740	2·110	2·567	2·898	3·965
18	·127	·257	·392	·534	·688	·862	1·067	1·330	1·734	2·101	2·552	2·878	3·922
19	·127	·257	·391	·533	·688	·861	1·066	1·328	1·729	2·093	2·539	2·861	3·883
20	·127	·257	·391	·533	·687	·860	1·064	1·325	1·725	2·086	2·528	2·845	3·850
21	·127	·257	·391	·532	·686	·859	1·063	1·323	1·721	2·080	2·518	2·831	3·819
22	·127	·256	·390	·532	·686	·858	1·061	1·321	1·717	2·074	2·508	2·819	3·792
23	·127	·256	·390	·532	·685	·858	1·060	1·319	1·714	2·069	2·500	2·807	3·767
24	·127	·256	·390	·531	·685	·857	1·059	1·318	1·711	2·064	2·492	2·797	3·745
25	·127	·256	·390	·531	·684	·856	1·058	1·316	1·708	2·060	2·485	2·787	3·725
26	·127	·256	·390	·531	·684	·856	1·058	1·315	1·706	2·056	2·479	2·779	3·707
27	·127	·256	·389	·531	·684	·855	1·057	1·314	1·703	2·052	2·473	2·771	3·690
28	·127	·256	·389	·530	·683	·855	1·056	1·313	1·701	2·048	2·467	2·763	3·674
29	·127	·256	·389	·530	·683	·854	1·055	1·311	1·699	2·045	2·462	2·756	3·659
30	·127	·256	·389	·530	·683	·854	1·055	1·310	1·697	2·042	2·457	2·750	3·646
40	·126	·255	·388	·529	·681	·851	1·050	1·303	1·684	2·021	2·423	2·704	3·551
60	·126	·254	·387	·527	·679	·848	1·046	1·296	1·671	2·000	2·390	2·660	3·460
120	·126	·254	·386	·526	·677	·845	1·041	1·289	1·658	1·980	2·358	2·617	3·373
∞	·126	·253	·385	·524	·674	·842	1·036	1·282	1·645	1·960	2·326	2·576	3·291

Source: Table taken from Table III of Fisher and Yates, *Statistical Tables for Biological, Agricultural and Medical Research,* published by Oliver and Boyd, Edinburgh, and by permission of the authors and publishers.

A.43 FACTORS USED IN THE CALCULATION OF SEDIMENTATION COEFFICIENT AND MOLECULAR WEIGHT IN SECTION 2.26 ON CENTRIFUGAL FORCE

rpm	$\omega^2 \times 10^{-6}$	$F \times 10^{10}$
8,000	0.701837	546.80
9,000	0.888267	432.04
10,000	1.09663	349.95
11,000	1.32692	289.21
12,000	1.57914	243.02
12,590	1.73818	220.82
13,000	1.85330	207.07
13,410	1.97203	194.64
14,000	2.14939	178.55
14,290	2.23939	171.40
15,000	2.46741	155.53
15,220	2.54036	151.09
16,000	2.80737	136.70
16,200	2.87798	133.37
17,000	3.16925	121.10
17,250	3.26312	117.63
17,980	3.54524	108.27
18,000	3.55307	108.01
19,160	4.02568	94.346
20,000	4.38647	87.488
20,410	4.56827	84.021
21,740	5.18282	74.059
22,000	5.30768	72.303
23,150	5.87694	65.312
24,000	6.31658	60.755
24,630	6.65253	57.697
25,980	7.40177	51.857
26,000	7.41315	51.768
27,690	8.40820	45.650
28,000	8.59756	44.636
29,500	9.54347	40.219
30,000	9.86965	38.883
31,410	10.8192	35.477
32,000	11.2294	34.175
33,450	12.2710	31.280
34,000	12.6770	30.272
35,600	13.8980	27.618
36,000	14.2122	27.002

A.43 (*continued*)

rpm	$\omega^2 \times 10^{-6}$	$F \times 10^{10}$
37,020	15.0290	25.540
39,460	17.0756	22.479
40,000	17.5460	21.872
42,040	19.3815	19.804
44,000	21.2306	18.076
44,770	21.9803	17.463
47,660	24.9093	15.409
48,000	25.2662	15.189
50,740	28.2333	13.595
52,000	29.6527	12.942
52,640	30.3868	12.632
56,000	34.3901	11.159
56,100	34.5129	11.121
59,780	39.1891	9.7944
60,000	39.4785	9.7208

Bibliography

ALGEBRA AND TRIGONOMETRY

Bryant, S. J.; Karush, J.; Nower, L.; and Saltz, D. *College Algebra and Trigonometry*, Goodyear Pub. Co., Pacific Palisades, Calif., 1970.

Burnside, W. S., and Panton, A. W. *The Theory of Equations with an Introduction to the Theory of Binary Algebraic Forms*, vols. 1 and 2, Dover Publ., New York, 1960.

Dadourin, H. M. *Plane Trigonometry*, Addison-Wesley Press, Reading, Mass., 1951.

Drooyan, Irving; Hadel, Walter; and Fleming, Frank. *Elementary Algebra Structure and Skills*, John Wiley & Sons, New York, 1966.

Fehr, Howard F.; Carnahan, Walter H.; and Beberman, Max. *Algebra, Course One*, D. C. Heath and Co., Lexington, Mass., 1955.

Weisner, Louis. *Introduction to the Theory of Equations*, Macmillan Co., New York, 1938.

ASTRONOMY

Rudaux, Lucien, and De Vaucouleurs, G. *Larousse Encyclopedia of Astronomy*, 2nd ed., Prometheus Press, New York, 1962.

Spitz, Armand N., and Gaynor, Frank. *Dictionary of Astronomy and Astronautics*, Philosophical Library, New York, 1959.

Weigert, A., and Zimmermann, H. *A Concise Encyclopedia of Astronomy*, American Elsevier Pub. Co., New York, 1968.

BIOLOGY

Baily, N. T. *The Mathematical Approach to Biology and Medicine*, John Wiley & Sons, New York, 1967.

Rashevsky, N. *Mathematical Biology*, Charles C. Thomas, Springfield, Ill., 1964.

Smith, J. M. *Mathematical Ideas in Biology*, Cambridge Univ. Press, New York, 1968.

Stibitz, G. R. *Mathematics in Medicine and the Life Sciences*, Year Book Medical Pub., Chicago, 1966.

CALCULUS

Agnew, R. P. *Calculus, Analytic Geometry and Calculus with Vectors*, McGraw-Hill Book Co., New York, 1962.

Bass, J. *Exercises in Mathematics*, Translated by Scripta Technica, Academic Press, New York, 1966.

Courant, Richard, and Fritz, John. *Introduction to Calculus and Analysis*, vol. 1, John Wiley & Sons, New York, 1965.

Fort, Tomlinson. *Calculus*, D. C. Heath and Co., Lexington, Mass., 1951.

Johnson, R. E., and Kiokemeister, F. L. *Calculus with Analytic Geometry*, Allyn & Bacon, Boston, 1964.

Munroe, M. Evans. *Calculus*, W. B. Saunders Co., Philadelphia, 1970.

Toralballa, Leopoldo V. *Calculus with Analytical Geometry and Linear Algebra*, Academic Press, New York, 1967.

CHEMISTRY

Andres, Donald Hitch, and Kokes, Richard J. *Fundamental Chemistry*, 2nd ed., John Wiley & Sons, New York, 1965.

Compton, Charles. *An Introduction to Chemistry*, D. Van Nostrand Co., Princeton, N. J., 1958.

Conway, B. E. *Electrochemical Data*, Greenwood Press, Westport, Conn., 1969.

Daniels, Farrington. *Outlines of Physical Chemistry*, John Wiley & Sons, New York, 1948.

Handbook of Chemistry and Physics, 48th ed., Chemical Rubber Co., Cleveland, 1967–1968.

Hawes, B. W. V., and Davies, N. H. *Calculations in Physical Chemistry*, 2nd ed., English Univ. Press, London, 1965.

Martin, R. B. *Introduction to Biophysical Chemistry*, McGraw-Hill Book Co., New York, 1964.

Shoemaker, D. P., and Garland, C. W. *Experiments in Physical Chemistry*, 2nd ed., McGraw-Hill Book Co., New York, 1967.

Standard Methods for the Examination of Water, Sewage and Industrial Wastes, 10th ed., American Public Health Assoc., New York, 1955.

Tanford, C. *Physical Chemistry of Macromolecules*, John Wiley & Sons, New York, 1961.

Wallace, S. B. *Principles of Physical Chemistry*, Appleton-Century-Crofts, Inc., New York, 1958.

Yphantis, D. A. "Advances in Ultracentrifugal Analysis" in *Annals of the New York Academy of Sciences.* 164 (1964):1.

DIFFERENTIAL EQUATIONS

Agnew, Ralph Palmer. *Differential Equations*, 2nd ed., McGraw-Hill Book Co., New York, 1960.

Carrier, George F., and Pearson, Carl E. *Ordinary Differential Equations*, Blaisdell Pub. Co., Waltham, Mass., 1968.

Coddington, Earl A. *An Introduction to Ordinary Differential Equations*, Prentice-Hall, Englewood Cliffs, N. J., 1961.

Halanay, A. *Differential Equations: Stability, Oscillations, Time Lags*, Academic Press, New York, 1966.

Kells, Lyman M. *Elementary Differential Equations*, 4th ed., McGraw-Hill Book Co., New York, 1954.

Tenenbaum, Morris, and Pollard, Harry. *Ordinary Differential Equations*, Harper & Row, New York, 1963.

Weinberger, H. F. *A First Course in Partial Differential Equations with Complex Variables and Transform Methods*, Blaisdell Pub. Co., Waltham, Mass., 1965.

ELEMENTARY MATHEMATICS

Armstrong, James. *Mathematics for Elementary School Teachers*, Harper & Row, New York, 1968.

Betz, William. *Everyday General Mathematics*, Book 2, Ginn & Co., Boston, 1960.

Bouwsma, Ward D.; Corle, Clyde G.; and Clemson, Davis F. *Basic Mathematics for Elementary Teachers*, Ronald Press Co., New York, 1967.

Brumfiel, Charles F.; Eicholz, Robert E.; and Shanks, Merrill E. *Fundamental Concepts of Elementary Mathematics*, Addison-Wesley Pub. Co., Reading, Mass., 1962.

Byrne, J. Richard. *Modern Elementary Mathematics*, McGraw-Hill Book Co., New York, 1966.

Fehr, Howard F., and Schult, Veryl. *Mathematics at Work*, Book 1, 2nd ed., D. C. Heath and Co., Lexington, Mass., 1962.

——. *Mathematics in Life*, Book 2, 2nd ed., D. C. Heath and Co., Lexington, Mass., 1962.

Garstens, Helen L., and Jackson, Stanley B. *Mathematics for Elementary School Teachers*, Macmillan Co., New York, 1967.

Hart, Walter W.; Schult, Veryl; and Irvin, Lee. *Mathematics in Daily Use*, 3rd ed., rev., D. C. Heath and Co., Lexington, Mass., 1961.

Henderson, Kenneth B., and Pingry, Robert E. *Using Mathematics 7*, Teachers Edition, McGraw-Hill Book Co., New York, 1961.

——. *Using Mathematics 8*, McGraw-Hill Book Co., New York, 1961.

Schaaf, William L. *Basic Concepts of Elementary Mathematics*, 2nd ed., John Wiley & Sons, New York, 1965.

Wheeler, Ruric E. *Modern Mathematics: An Elementary Approach*, Brooks/Cole Pub. Co., Belmont, Calif., 1967.

GENERAL MATHEMATICS

Bergamini, David, et al. *Mathematics*, Life Science Library, New York, Time, Inc., 1963.

Daus, Paul H.; Gleason, John M.; and Whyburn, William M. *Basic Mathematics for War and Industry*, Macmillan Co., New York, 1944.

Denbow, Carl H., and Goedicke, Victor. *Foundations of Mathematics*, Harper Brothers, New York, 1959.

Dubisch, Roy. *The Teaching of Mathematics from Intermediate Algebra through First Year Calculus*, John Wiley & Sons, New York, 1963.

Gelfond, A. O. *The Solution of Equations in Integers*, translated from Russian and edited by J. B. Roberts, W. H. Freeman & Co., San Francisco, 1961.

Geographical Conversion Tables, compiled and edited by D. H. K. Amiran and A. P. Schick, Aschmann and Scheller AG, Zurich, Switzerland, 1961.

Logsdon, Mayme I. *A Mathematician Explains*, Univ. of Chicago Press, Chicago, 1936.

Mathematics Dictionary, edited by Glenn James and Robert C. James, Multilingual Edition, 3rd ed., D. Van Nostrand Co., Princeton, N. J., 1968.

Merck Index, 8th ed., Merck & Co., Rahway, N. J., 1968.

Meserve, Bruce E., and Sobel, Max A. *Elements of Mathematics*, Prentice-Hall, Englewood Cliffs, N. J., 1968.

A Metric America: A Decision Whose Time Has Come. (Report to Congress) U.S. Dept. of Commerce, National Bureau of Standards, Washington, D.C., U.S. Government Printing Office No. C13.10:345, 1971.

Meyers, Lester. *High-Speed Math*, D. Van Nostrand Co., Princeton, N. J., 1957.

Niven, Ivan. *Numbers: Rational and Irrational*, L. W. Singer Co., New York, 1961.

Palmer, Claude Irwin, and Bibb, Samuel Fletcher. *Practical Mathematics*, 5th ed., McGraw-Hill Book Co., New York, 1970.

Person, Russell V. *Essentials of Mathematics*, John Wiley & Sons, New York, 1961.

Rider, P. R. *First Year Mathematics for Colleges*, 2nd ed., Macmillan Co., New York, 1962.

Youse, Bevan K. *Arithmetic: A Modern Approach*, Prentice-Hall, Englewood Cliffs, N. J., 1963.

GEOMETRY

Fadell, Albert G. *Calculus with Analytic Geometry*, D. Van Nostrand Co., Princeton, N. J., 1964.

Fehr, Howard F., and Carnahan, Walter H. *Geometry*, D. C. Heath and Co., Lexington, Mass., 1961.

Hart, Walter W.; Schult, Veryl; and Swain, Henry. *Plane Geometry and Supplements*, D. C. Heath and Co., Lexington, Mass., 1959.

Lavis, David R. *Modern College Geometry*, Addison-Wesley, Reading, Mass., 1957.

Miller, Leslie H. *College Geometry*, Appleton-Century-Crofts, New York, 1957.

Prenowitz, Walter, and Jordan, Meyer. *Basic Concepts of Geometry*, Blaisdell Pub. Co., Waltham, Mass., 1965.

PHYSICS

American Institute of Physics Handbook, 2nd ed., McGraw-Hill Book Co., New York, 1963.

Asimov, Isaac. *Understanding Physics: Motion, Sound and Heat*, vol. 1; *Light, Magnetism and Electricity*, vol. 2; *Electron, Proton, Neutron*, vol. 3; George Allen & Unwin, London, 1967.

Elliot, L. P., and Wilcox, W. F. *Physics: A Modern Approach*, Macmillan Co., New York, 1958.

Shortley, G., and Williams, D. *Elements of Physics for Students of Science and Engineering*, Prentice-Hall, New York, 1953.

Weber, R. L.; White, N. W.; and Manning, K. V. *College Physics*, 2nd ed., McGraw-Hill Book Co., New York, 1952.

SLIDE RULE

Asimov, Isaac. *An Easy Introduction to the Slide Rule*, Houghton Mifflin Co., Boston, 1965.

Clark, J. J. *The Slide Rule and Logarithmic Tables*, Frederick J. Drake & Co., Chicago, 1957.

Harris, Charles O. *Slide Rule Simplified*, American Technical Society, Chicago, 1943.

Hartung, Maurice L. *How to Use Log Log Slide Rules*, Pickett & Eckel, Chicago, 1953.

STATISTICS

Alexander, Howard W. *Elements of Mathematical Statistics*, John Wiley & Sons, New York, 1961.

Amos, Jimmy; Brown, Foster Lloyd; and Mink, Oscar G. *Statistical Concepts: A Basic Program*, Harper & Row, New York, 1965.

Arley, Niels, and Buch, Rander, K. *Introduction to the Theory of Probability and Statistics*, John Wiley & Sons, New York, 1956.

Bancroft, Huldah. *Introduction to Biostatistics*, Hoeber-Harper Pub., New York, 1957.

Birbaum, Z. W. *Introduction to Probability and Mathematical Statistics*, Harper Brothers, New York, 1962.

Bliss, C. I., and Calhoun, D. W. *An Outline of Biometry*, Yale Co-Operative Corporation, 1954.

Bourke, Geoffrey J., and McGilvray, James. *Interpretation and Uses of Medical Statistics*, Blackwell Scientific Publ., Oxford, 1969.

Croxton, Frederick. *Elementary Statistics with Applications in Medicine and the Biological Sciences*, Dover Publ., New York, 1953.

Croxton, Frederick; Cowden, D. J.; and Bokh, B. W. *Practical Business Statistics*, 4th ed., Prentice-Hall, Englewood Cliffs, N.J., 1969.

Dixon, Wilfrid J., and Massey, Frank J. *Introduction to Statistical Analysis*, 2nd ed., McGraw-Hill Book Co., New York, 1957.

Edwards, Allen L. *Experimental Design in Psychological Research*, Holt, Rinehart & Winston, New York, 1965.

Finney, D. J. *Statistical Method in Biological Assay*, Hafner Pub. Co., New York, 1952.

Fisher, R. A. *Statistical Methods for Research Workers*, 6th ed., Oliver and Boyd, Edinburgh, 1936.

Fisher, R. A., and Yates, F. *Statistical Tables for Biological, Agricultural and Medical Research*, 6th ed., Oliver and Boyd, Edinburgh, 1963.

Goldstein, Avram. *Biostatistics: An Introductory Text*, Macmillan Co., New York, 1964.

Hald, A. *Statistical Theory with Engineering Applications*, John Wiley & Sons, New York, 1952.

Handbook of Probability and Statistics, 2nd ed., Chemical Rubber Co., Cleveland, 1968.

Hill, A. Bradford. *Statistical Methods in Clinical and Preventive Medicine*, E & S Livingstone, Edinburgh, 1962.

———. *Principles of Medical Statistics*, 7th ed., Oxford Univ. Press, 1961.

Hoel, Paul G. *Introduction to Mathematical Statistics*, 3rd ed., John Wiley & Sons, New York, 1962.

Hoffman, Robert G. *Statistics for Medical Students*, Charles C. Thomas, Springfield, Ill., 1963.

Holman, H. H. *Biological Research Methods*, 2nd ed., Oliver and Boyd, Edinburgh, 1969.

Introductory Probability and Statistical Inference: An Experimental Course, prepared for the Commission on Mathematics College Entrance Examination Board, New York, 1959.

Kendall, M. G., and Buckland, W. R. *A Dictionary of Statistical Terms*, prepared for the International Statistical Institute with the assistance of UNESCO, Hafner Pub. Co., New York, 1957.

Kiem, Iris Mabel. "Biostatistics with Applications in Medicine," *Univ. of Miami Medical School Manual*, 1961.

Mack, Cornelius. *Essentials of Statistics for Scientists and Technologists*, Plenum Press, New York, 1967.

Mainland, Donald. *Elementary Medical Statistics*, W. B. Saunders Co., Philadelphia, 1952.

———. *The Treatment of Clinical and Laboratory Data*, Oliver and Boyd, Edinburgh, 1938.

Moore, F. J.; Cramer, F. B.; and Knowles, R. G. *Statistics for Medical Students and Investigators in the Clinical and Biological Sciences*, Blakiston, Philadelphia, 1951.

Moroney, M. J. *Facts from Figures*, 3rd ed., rev., Penguin Books, Baltimore, 1956.

Mosteller, Frederick; Rourke, Robert E. K.; and Thomas, George B., Jr. *Probability with Statistical Applications*, Addison-Wesley Pub. Co., Reading, Mass., 1961.

Oldham, P. D. *Measurement in Medicine*, English Universities Press, London, 1968.

Ostle, Bernard. *Statistics in Research: Basic Concepts and Techniques for Research Workers*, 2nd ed., Iowa State Univ. Press, Ames, 1963.

Pearl, Raymond. *Introduction to Medical Biometry and Statistics*, 3rd ed., W. B. Saunders Co., Philadelphia, 1940.

Remington, Richard D., and Schork, M. Anthony. *Statistics with Applications to the Biological and Health Sciences*, Prentice-Hall, Englewood Cliffs, N.J., 1970.

Schor, Stanley. *Fundamentals of Biostatistics*, G. P. Putnam's Sons, New York, 1968.

Smart, J. V. *Elements of Medical Statistics*, Charles C. Thomas, Springfield, Ill., 1963.

Snedecor, George W., and Cochran, William G. *Statistical Methods*, 6th ed., Iowa State Univ. Press, Ames, 1967.

Wallis, W. Allen, and Roberts, Harry V. *The Nature of Statistics*, Free Press, New York, 1962.

——. *Statistics: A New Approach*, Free Press, New York, 1956.

Whipple, George Chandler. *Vital Statistics*, John Wiley & Sons, New York, 1919.

Woods, Hilda M., and Russell, William T. *An Introduction to Medical Statistics*, P. S. King & Son, London, 1931.

Youden, W. J. *Statistical Methods for Chemists*, John Wiley & Sons, New York, 1951.

JOURNAL ARTICLES–GENERAL MATHEMATICS

Anderson, Johnston A. "A Geometrical Approach to Natural Logarithms," *Mathematics Teaching*, no. 56, Autumn 1971, p. 19.

Assad, Saleh. "From Graph to Formula," *Mathematics Teacher* 64 (3, Mar. 1971): 231.

Brady, W. G. "Complex Roots of a Quadratic Equation Graphically," *Mathematics Teacher* 63 (3, Mar. 1970): 229.

Coltharp, Forrest L. "Properties of Polygonal Regions," *Arithmetic Teacher* 19 (2, Feb. 1972): 117.

Damaskos, Nickander J. "A Case Study in Mathematics: The Cone Problem," *Mathematics Teacher* 62 (8, Dec. 1962): 642.

Duncan, Hilda F. "Division by Zero," *Arithmetic Teacher* 18 (6, Oct. 1971): 381.

Flanders, Harley. "Analysis of Calculus Problems," *Mathematics Teacher* 65 (1, Jan. 1972): 9.

Henning, Harley B. "Geometric Solutions to Quadratic and Cubic Equations," *Mathematics Teacher* 65 (2, Feb. 1972): 113.

Henry, Boyd. "Do We Need Separate Rules to Compute in Decimal Notation?" *Arithmetic Teacher* 18 (1, Jan. 1971): 40.

Iman, Ronald L. "Use of Summation Operators for the Derivation of Common Formulas," *Mathematics Teacher* 63 (4, Apr. 1970): 296.

Knight, Carlton W., and Schweitzer, James P. "Using Stream Flow to Develop Measuring Skills, *Arithmetic Teacher* 19 (2, Feb. 1972): 88.

Lay, L. Clark. "An Elementary Theory of Equations," *Arithmetic Teacher* 18 (7, Nov. 1971): 457.

Lazerick, Beth Ellen. "The Conversion Game," *Arithmetic Teacher* 18 (1, Jan. 1971): 54.

Niman, John. "A Game Introduction to the Binary Numeration System," *Arithmetic Teacher* 18 (8, Dec. 1971): 600.

Phillip, S. Jones. "Binary System," *Mathematics Teacher* 46 (1953): 575.

Ropes, George H. "Cubic Equations for High School," *Mathematics Teacher* 63 (4, Apr. 1970): 356.

Ruchlis, Hy. "Putting Reality into Mathematics," *Mathematics Teacher* 64 (4, Apr. 1971): 369.

Sullivan, John H. "Problem Solving Using the Sphere," *Arithmetic Teacher* 16 (1, Jan. 1969): 29.

Warner, Elizabeth V. "An Approximation Method of Finding Square Roots, *Arithmetic Teacher* 18 (3, Mar. 1971): 155.

Wilkinson, Jack. "Teaching General Mathematics: A Semi-Laboratory Approach," *Mathematics Teacher* 63 (7, Nov. 1970): 571.

Wiscamb, Margaret. "'B-ary' Fractions," *Mathematics Teacher* 63 (3, Mar. 1970): 244.

Index